MW00845093

Mechanical Engineering Series

Frederick F. Ling
Series Editor

Springer
New York
Berlin
Heidelberg
Barcelona
Hong Kong
London
Milan
Paris
Singapore
Tokyo

Mechanical Engineering Series

Introductory Attitude Dynamics
F.P. Rimrott

Balancing of High-Speed Machinery
M.S. Darlow

Theory of Wire Rope, 2nd ed.
G.A. Costello

Theory of Vibration: An Introduction, 2nd ed.
A.A. Shabana

Theory of Vibration: Discrete and Continuous Systems, 2nd ed.
A.A. Shabana

Laser Machining: Theory and Practice
G. Chryssolouris

Underconstrained Structural Systems
E.N. Kuznetsov

Principles of Heat Transfer in Porous Media, 2nd ed.
M. Kaviany

Mechatronics: Electromechanics and Contromechanics
D.K. Miu

Structural Analysis of Printed Circuit Board Systems
P.A. Engel

Kinematic and Dynamic Simulation of Multibody Systems: The Real-Time Challenge
J. García de Jalón and E. Bayo

High Sensitivity Moiré: Experimental Analysis for Mechanics and Materials
D. Post, B. Han, and P. Ifju

Principles of Convective Heat Transfer
M. Kaviany

Laminar Viscous Flow
V.N. Constantinescu

(continued after index)

Ilene J. Busch-Vishniac

Electromechanical Sensors and Actuators

With 378 Figures

Ilene J. Busch-Vishniac
Dean, Whiting School of Engineering
Johns Hopkins University
3400 N. Charles Street
Baltimore, MD 21218-2681, USA

Series Editor
Frederick F. Ling
Ernest F. Gloyna Regents Chair in Engineering
Department of Mechanical Engineering
The University of Texas at Austin
Austin, TX 78712-1063, USA
and
William Howard Hart Professor Emeritus
Department of Mechanical Engineering,
Aeronautical Engineering and Mechanics
Rensselaer Polytechnic Institute
Troy, NY 12180-3590, USA

Library of Congress Cataloging-in-Publication Data
Busch-Vishniac, Ilene J.
Electromechanical sensors and actuators / Ilene J. Busch-Vishniac.
p. cm. — (Mechanical engineering series)
Includes bibliographical references and index.
ISBN 0-387-98495-X (alk. paper)
1. Transducers. 2. Detectors. 3. Actuators. I. Title.
II. Series: Mechanical engineering series (Berlin, Germany)
TK7872.T6B87 1998
681′.2—dc21 98-11966

Printed on acid-free paper.

Production managed by Timothy Taylor; manufacturing supervised by Jacqui Ashri.
Photocomposed copy prepared using the compositor's TEX files.
Printed and bound by Edwards Brothers, Inc., Ann Arbor, MI.
Printed in the United States of America.

9 8 7 6 5 4 3 2 1

ISBN 0-387-98495-X Springer-Verlag New York Berlin Heidelberg SPIN 10658156

For Ethan, Cady, and Miriam, the lights on my journey.

Mechanical Engineering Series

Frederick F. Ling
Series Editor

Series Preface

Mechanical engineering, an engineering discipline borne of the needs of the industrial revolution, is once again asked to do its substantial share in the call for industrial renewal. The general call is urgent as we face profound issues of productivity and competitiveness that require engineering solutions, among others. The Mechanical Engineering Series features graduate texts and research monographs intended to address the need for information in contemporary areas of mechanical engineering.

The series is conceived as a comprehensive one that covers a broad range of concentrations important to mechanical engineering graduate education and research. We are fortunate to have a distinguished roster of consulting editors on the advisory board, each an expert in one of the areas of concentration. The names of the consulting editors are listed on the facing page of this volume. The areas of concentration are: applied mechanics; biomechanics; computational mechanics; dynamic systems and control; energetics; mechanics of materials; processing; thermal science; and tribology.

I am pleased to present this volume in the Series: *Electromechanical Sensors and Actuators*, by Ilene Busch-Vishniac. The selection of this volume underscores again the interest of the Mechanical Engineering series to provide our readers with topical monographs as well as graduate texts in a wide variety of fields.

Austin, Texas — Frederick F. Ling

Preface

In the past few decades improvements in electronics have outpaced improvements in sensors and actuators so significantly that in virtually all measurement and control systems it is the sensors and actuators which account for the bulk of the expense, the majority of the size, and the lion's share of the system failures. For improvements to be forthcoming, it clearly is necessary to understand how sensors and actuators work. The aim of this book is to provide a detailed description of the various mechanisms that can be exploited in the design of electromechanical sensors and actuators.

There are a number of books available on sensors and measurements, but this book is a significant departure from the norm. The typical book on sensors or on measurements is either a compendium of commercially available devices, or a discussion of techniques used in sensing various parameters (e.g., acceleration, force, displacement, etc.). What these approaches obscure is that different sorts of devices might well use the same fundamental coupling mechanism to link electrical and mechanical behavior. For instance, an ultrasonic cleaner and an accelerometer might be nearly identical, despite their seemingly disparate applications. In this book, the focus is on the fundamental coupling mechanisms. As each mechanism is presented, commercially available sensors and actuators from a wide a range of applications are given.

The standard treatment of measurement also draws a strong distinction between sensing and actuation, although the devices used might be nearly identical in means of operation. For instance, a solenoid can be used either to sense a position (sensor) or to impose a position (actuator). While the optimal geometric design and choice of materials might well depend on which of these two operations the transducer is aimed to perform, the fundamental means by which the electrical and mechanical domains are coupled are identical. In this book, both sensors and actuators are treated, and the analysis developed applies uniformly to both unless otherwise stated.

By focusing on the fundamental coupling between electrical and mechanical energy domains, this book arms the reader with an arsenal that is appropriate for transducer design. Because the book grew out of notes used in a graduate course on electromechanical sensors and actuators at The University of Texas, the analytical developments are the sort we would find in a textbook rather than a handbook. Additionally, a large number of examples of commercially available transducers are presented. These examples serve to define the state-of-the-art electromechanical transduction, and provide a vehicle for discussion of some of the more practical issues associated with transducer design and use. A significant amount of the material contained within has never been previously published, so this book is a combination of a textbook and a research monograph, appropriate for both students and practitioners with an interest in the measurement and control of mechanical systems.

This book is divided into three parts. Chapters 1 and 2 form the introductory part of the book, entitled Basic Tools for Transducer Modeling. Chapters 3 to 8 constitute the Transduction Mechanisms part of the book. Each chapter discusses a particular class of transduction mechanisms. For example, Chapter 3 deals with transduction based on changes in the energy stored in an electric field. Each chapter in this part of the book provides an analytical description of the transduction mechanisms and examples of devices using that method. These examples incorporate a wide variety of applications, sizes, scales, and sophistication. Chapters 9 to 11 form the Transducer Theory and Description part of the book. This section deals with 2-port theory, which applies generally to transducers, and with the meaning of various specifications associated with transducers.

Baltimore, Maryland — Ilene J. Busch-Vishniac

Acknowledgments

This manuscript is the culmination of years of conversations with leading authorities in the field of transduction. Without the help of my colleagues this text would never have appeared. I am particularly indebted to Elmer Hixson for introducing me to his way of thinking about sensors and actuators, and to David Blackstock and Mark Hamilton for their encouragement while I labored over this book. I am also grateful to my students, who inspired the writing and helped me find errors.

Most of this book was written while I was on the faculty of The University of Texas at Austin. The final stages were completed at Boston University, where I spent a year on sabbatical thanks to the generosity of Allan Pierce.

A great deal of the painstaking work for this book required the careful attention of Cindy Pflughoft, the world's greatest secretary. Without her willingness to track down sources and organize my files of catalogs and papers, this book would not have been possible.

Baltimore, MD Ilene J. Busch-Vishniac

Contents

Series Preface vii

Preface ix

Acknowledgments xi

I Basic Tools for Transducer Modeling 1

1 Introduction 3

1.1 What Is a Transducer? 3
1.2 Why Study Transducers? 7
1.3 Division of Transducers into Sensors and Actuators 8
1.4 Transducers as Part of a Measurement or Control System . . . 11
1.5 Historical Perspective 13
1.6 Organization of this Book 14

2 System Models 17

2.1 System Analogies 18
2.2 Ideal 1-Port Elements 22
2.2.1 Comments on Modeling and 1-Port Elements 27
2.3 Circuit Models 29
2.4 Bond Graph Models 34
2.5 Nonenergic 2-Port Elements 43
2.6 Multiport Energic Elements 45

2.7 Model Analysis 50
2.8 References 52

II Transduction Mechanisms **55**

3 Transduction Based on Changes in the Energy Stored in an Electric Field **59**
3.1 Electric Fields and Forces 59
3.2 Transducers Made with a Variable Gap Parallel Plate Capacitor . 63
3.2.1 Relating Displacement and Voltage 66
3.2.2 Relating Force and Voltage 74
3.2.3 Relating Force and Charge 78
3.3 Transducers Using Other Means of Varying the Capacitance in a Parallel Plate Capacitor 79
3.3.1 Variable Area 79
3.3.2 Varying Permittivity 85
3.4 Transducers with Cylindrical Geometry 89
3.5 Gradient Transduction Using Two Dielectrics 92
3.6 Electrostrictive Transduction 96
3.7 Summary 97
3.8 References 98

4 Transduction Based on Changes in the Energy Stored in a Magnetic Field **100**
4.1 Magnetic Systems 100
4.1.1 Magnetic Materials 105
4.1.2 Magnetic Circuits 106
4.2 Variable Reluctance Transducers with Varying Gap 107
4.2.1 Relating Displacement and MAGNETOMOTANCE 110
4.2.2 Relating Force and Magnetomotance 116
4.2.3 Relating Force and Flux 117
4.3 Variable Reluctance Transducers with Varying Permeability and Area 121
4.3.1 Variable Area 122
4.3.2 Variable Permeability 123
4.4 Gradient Transduction with Two Ferromagnetic Materials . . . 129
4.5 Magnetostrictive Transduction 131
4.6 Eddy Current Transducers 134
4.7 Summary 137
4.8 References 138

5 Piezoelectricity and Pyroelectricity **140**
5.1 Piezoelectric Relations 142
5.2 Piezoelectric Materials 147

5.3 Piezoelectric Structures in Transducers 154
5.4 Models of Piezoelectricity 159
5.5 Examples of Piezoelectric Transducers 163
5.5.1 Examples of Broadband Transducers Using One Piezoelectric Element or a Stack of Parallel Piezoelectric Elements 164
5.5.2 Examples of Resonant Transducers Using a Single Piezoelectric Element or Element Stack 170
5.5.3 Example of a Transducer Using Multiple Nonparallel Piezoelectric Elements 172
5.5.4 Examples of Transducers which Are Unconventional . 173
5.6 Pyroelectricity 177
5.7 Summary 179
5.8 References 181

6 Linear Inductive Transduction Mechanisms **184**
6.1 Piezomagnetism 185
6.2 Pyromagnetism 187
6.3 Charged Particle Interactions 188
6.3.1 Examples of Transducers Based on the Motion of Charged Particles in a Magnetic Field 190
6.4 Hall Effect Transducers 195
6.4.1 Example Hall Effect Sensors 200
6.5 Summary 203
6.6 References 206

7 Transduction Based on Changes in the Energy Dissipated **207**
7.1 Conductive Switches 208
7.2 Continuously Variable Conductivity Transducers 210
7.3 Potentiometric Devices 213
7.4 Piezoresistivity 216
7.4.1 Material Description 217
7.4.2 Strain Gauge Structures 219
7.4.3 Electrical Operation 225
7.5 Thermoresistivity 228
7.6 Thermoelectricity 234
7.6.1 Seebeck Effect 234
7.6.2 Peltier and Thomson Effects 240
7.7 Magnetoresistivity 242
7.8 Shape Memory Alloys in Transduction 243
7.9 Summary 247
7.10 References 247

8 Optomechanical Sensors **250**
8.1 Quantum Detectors 252

8.1.1 Photoconductive Sensors 253
8.2 Fiber Optic Waveguide Fundamentals 256
8.3 Intensity-Modulated Fiber Optic Sensors 262
8.4 Phase-Modulated Sensors 271
8.5 Photostriction 273
8.6 Summary 274
8.7 References 275

III Analysis of Transducers **277**

9 2-port Theory **279**
9.1 Basic 2-Port Equations 279
9.2 Reciprocity 285
9.3 Connected 2-Ports 288
9.4 Transfer Matrix and Sensitivity 291
9.5 Wave Matrix Representation and Efficiency 293
9.6 Summary 299
9.7 References 299

10 Response Characteristics **300**
10.1 Response Characteristics Defined 300
10.1.1 Example of Transducer Selection Based on Specifications 308
10.2 Calibration 310
10.3 Frequency and Time Scaling 316
10.4 Summary 318
10.5 References 318

11 Practical Considerations **320**
11.1 Digitization of Analog Signals 321
11.2 Signal Conditioning 325
11.2.1 Filtering 325
11.2.2 Amplifying 330
11.3 Novel Sensing/Actuation Techniques 331
11.3.1 Frequency Detection Schemes 331
11.3.2 Lapsed Time Detection 334
11.4 Spatially Distributed Transducers 335
11.5 Summary 337
11.6 References 337

Index **339**

Part I

Basic Tools for Transducer Modeling

1
Introduction

1.1 What Is a Transducer?

A typical technical field had evolved from infancy through the efforts of a number of researchers. As the evolution progresses, a consensus on the scope of the field and the fundamental definitions of associated objects or parameters emerge. When the technical area is dominated by a single major discipline, such as mechanical engineering or physics, the consensus on definitions and scope can be achieved quite rapidly. However, when the emerging technical area is truly multidisciplinary, as is the case with the technical field of transduction, then the consensus can be difficult to forge. Indeed, the field of transduction has involved virtually all of the major technical disciplines, each having its own perspective on the fundamental definition of a transducer. As a result there are many different definitions of transducers in use. The most common of these definitions are included in what follows:

1. A transducer is a device which transforms nonelectrical energy into electrical energy or vice versa. (See, e.g., Middlehoek and Hoogerwerf [1].)
2. A transducer is a device which transforms energy from one domain into another. (See for example Rosenberg and Karnopp [2].) Typical energy domains are mechanical, electrical, chemical, fluid, and thermal.
3. A transducer is a device which transforms energy from one type to another, even if both energy types are in the same domain. (See Busch-Vishniac [3] for instance.)

These three definitions are alike, in that each stipulates that a transducer is a *device*. For example, the irreversible conversion of mechanical energy into thermal

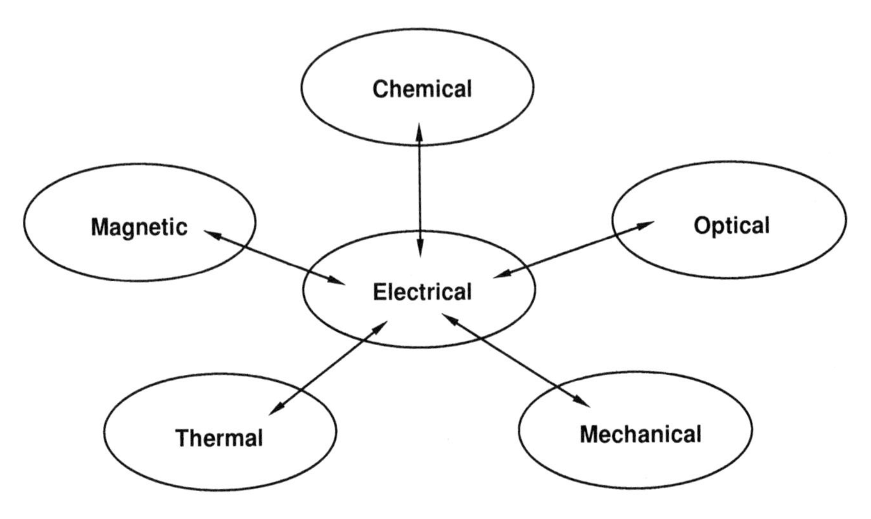

FIGURE 1.1. First definition of a transducer.

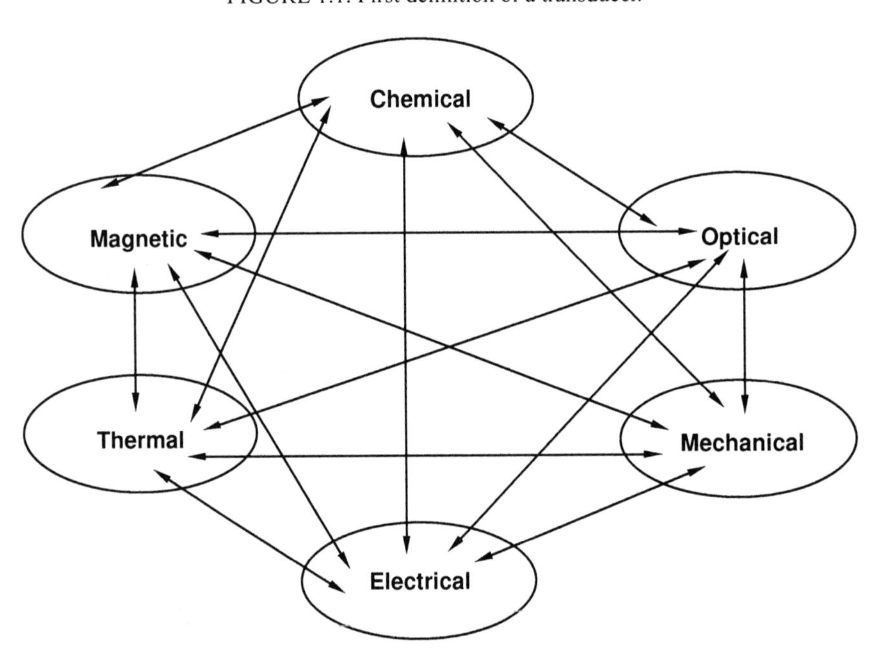

FIGURE 1.2. Second definition of a transducer.

energy due to friction is not a transducer because it does not involve a device. Friction can be, however, a key source of constraints on the functioning of a transducer.

The three definitions above differ primarily in the degree of restriction. As presented, they progress from very restrictive to very general. The first definition, which is shown graphically in Fig. 1.1, would not include a device such as a mercury thermometer or a mechanical pressure gauge, since all conversion must be to or from the electrical domain. The second definition is pictured in Fig. 1.2 and might include a mercury thermometer but not a Hall effect sensor, which couples magnetic and electric energy (both of which are normally considered to be in the electrical domain). Note that Fig. 1.2 differs from Fig. 1.1 in that the electrical domain no longer has a special position compared to other energy domains.

The problem with the definitions shown in Figs. 1.1 and 1.2 is that there is a certain amount of arbitrariness to the delineation of energy domains. For instance, while some modelers consider magnetic, optical, and electrical energy to be distinct domains, others argue that they are all manifestations of the same energy (as demonstrated by the fact that they all obey Maxwell's laws), and thus they are of the same domain. Further, some modelers would state that the thermal, fluid, and translational and rotational energy are all just different manifestations of mechanical energy. While these philosophical debates are interesting, they are moot from the perspective of understanding transduction, because the focus is on conversion of energy in a device by such a means as to make it useful in monitoring a system or imposing a state on a system. Indeed, the modeling technique that will be used in this book has been applied to many different energy domains, and to various manifestations of energy in the same broad domain with virtually no change required by shifting of the energy types. In order to avoid the arguments of what constitutes a conversion of energy from one domain to another, we will use an unusually broad definition of a transducer, namely that given in the third definition above and graphically illustrated in Fig. 1.3. This view of transducers enhances Fig. 1.2 by the addition of loops within a single energy domain, thus permitting us to include devices which couple energy types within a single broad domain. This choice has the advantage of permitting us to avoid semantic arguments about what constitutes an energy domain, but has the disadvantage of including devices that are not generally thought of as transducers. For instance, a wheel, which converts rotational energy to translational energy, would be a transducer according to the definition given in Fig. 1.3. While this disadvantage might seem significant, it might actually be an advantage in disguise, as the broadening of the definition of a transducer shows that some of our traditional distinctions between devices are rather arbitrary. The very same techniques that will be applied here to classical transducers, could be applied equally well to devices which are not usually considered to be transducers but which meet this very broad transducer definition.

The definition of a transducer upon which we have settled can now be made more formal. We define a transducer as a multiport device in which the input impedance(s) is (are) not equal to the output impedance(s). The gist of this definition is picture in Fig. 1.4 below, in which a 2-port device is considered. The transducer has at least two locations or ports at which it exchanges energy with the environment. It is convenient to think of at least one of these ports as providing input power and at least one as supplying output power. The power at each port is defined by two variables, such as a current and a voltage for an electrical port or a velocity and a force for a mechanical port. The ratio analogous to the electrical

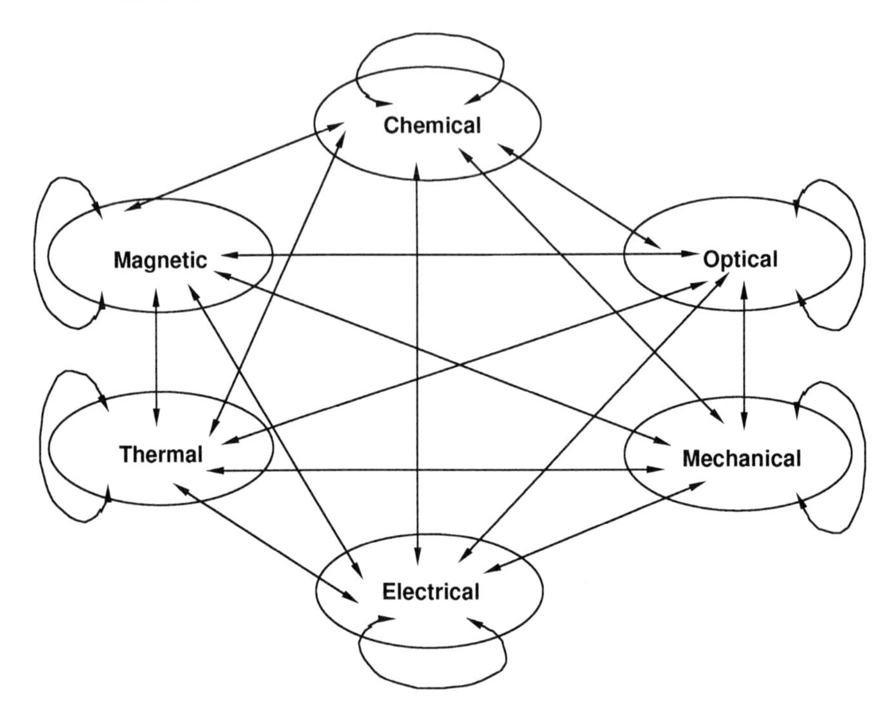

FIGURE 1.3. Third definition of a transducer.

FIGURE 1.4. Generic view of a transducer.

impedance (voltage to current) is defined as the impedance. In a transducer with more than two ports, there would be multiple input ports and/or multiple output ports, but the port impedances would be different for each port.

This text will concentrate on electromechanical transducers. In such devices one port is electrical and one is mechanical. This means that the distinction between the definitions of a transducer presented above evaporates. Electromechanical transducers are one of the most common of all transducer types because naturally occurring events are primarily mechanical or chemical in nature, but we prefer to use electrical signals so that we may take advantage of the speed and linearity of electrical systems. It should be noted that this focus is chosen solely for the sake of convenience. The techniques developed in this text apply quite broadly. By narrowing our focus to electromechanical transducers we simply limit the number of different transduction mechanisms that must be considered.

1.2 Why Study Transducers?

In general, there has been an increasing interest in transducers over the last decade. The causes for this enhanced attention may be used to explain why it is important to understand how transducers operate.

Perhaps the biggest source of interest in transduction is the exponentially growing number of transducers being found in common products. To some extent, this growth in the transducer market is due to an increase in the number of items available that inherently rely on transducers. For example, every telephone contains two transducers: a microphone which converts mechanical (fluid) energy to electrical energy, and an earphone which converts electrical energy to mechanical (fluid) energy. As another example, every thermostat has a mechanism for sensing temperature. In addition, there has been a rapid increase in the number of items that are being modified to incorporate transducers in order to make them more commercially viable. Typically, the assumption is that the addition of a transducer into a device which previously was transducer-less means an improvement in the quality of the product. For example, just a decade or two ago, the vast majority of clothing dryers for the home operated open-loop, i.e., with cycle lengths determined solely by time. Now a significant number of dryers incorporate a humidity sensor or a temperature sensor that permits the dryer to run until the clothes are dry. Another example of a common object with a rapidly growing number of transducers is the automobile. In the last decade transducers have been added to permit the development of antilock braking systems, airbag deployment systems, and even keyless door locks.

A second source of rising interest in transducers has been the success of the conversion of electronics to microelectronic circuits. Most transducers have at least one electrical port and are packaged with significant electronics attached. The last 20 years have seen considerable change in electronics. Thanks to silicon fabrication techniques, even complicated electrical circuits are now considered to be inexpensive, reliable, and relatively easy to design and construct. In the same 20 years there has been relatively little improvement in transducers. As a result, transducers are normally the weak link in a system and the gap between transducers and the processing electronics in terms of reliability, cost, and power is increasing at an alarming rate. As recognition of this problem has surfaced, there has been a call to improve transducer technology. One approach has led a number of researchers to seek to duplicate these successes, that microelectronics fabrication approaches brought about in electronics, by directly incorporating these techniques into transducer manufacturing.

A third source of interest in transducers is the escalating demand for the automatic control of processes. All controllers need actuators and, if the control is closed-loop, sensors are needed as well. This trend toward integration of more transducers is quite pervasive, including simple devices, such as outdoor lighting which turns on automatically, as well as complicated equipment, such as a new military airplane. In many cases the desire for automation has preceeded the availability of appropriate sensors and actuators. This situation has led to pressure for more research on and development of transducers.

Further complicating the desire for automation is the increasing demand for stringent performance characteristics in machines, particularly in terms of cost, reliability, and error tolerance. Over the last decade, we have come to appreciate that it is difficult to achieve stringent performance goals in a controlled system without including the effect of the sensors, actuators, and control electronics (hardware and software) on this performance. This has led to a mechatronics view of the world, and a refocusing on electromechanical interfaces. In other words, where it was once assumed that we could pick a sensor off the shelf and use it without significantly affecting system performance, we now appreciate that the sensor choice can play a major role in the overall performance of the system. This is leading to increased demands for high quality sensors and actuators, and also provides impetus to a trend toward custom transducers chosen specifically for the application at hand.

Another impetus for increased interest and work in the area of transduction has been the development of new materials and techniques. Materials capable of converting one form of energy to another are at the heart of many transducers. As new materials of this type are developed, they permit the design of new types of transducers. The last two decades have seen the rapid introduction of new materials, especially polymers and polymer composites. The development of these materials has outstripped the work needed to incorporate them into sensors and actuators. Examples of materials which offer potential advantages in transducers are new magnetostrictive materials and piezoelectric polymers. Similarly, techniques that make it possible to produce major changes in parameters of electromechanical transducers have the potential to create new classes of transducers. For instance, the use of microelectronic techniques in the manufacturing of transducers could revolutionize electromechanical transduction for applications requiring small-scale devices.

1.3 Division of Transducers into Sensors and Actuators

Transducers may generally be divided into two classes: sensors, which monitor a system; and actuators, which impose a condition on a system. Sensors and actuators are comprehensive classes of transducers, i.e., any transducer in operation is functioning at any given moment either as a sensor or as an actuator. Some transducers can operate as a sensor or as an actuator, but not as both simultaneously. Such transducers are said to be reversible. A typical example is a loudspeaker (actuator), which can be used to sense motions of the speaker diaphragm. Another example of a reversible transducer is an accelerometer which is designed to sense vibration, but can be used as a shaker.

Sensors are devices which monitor a parameter of a system, hopefully without disturbing that parameter. Examples of electromechanical sensors include hot-wire anemometers (which measure flow velocity), microphones (which measure fluid pressure), and accelerometers (which measure the acceleration of a structure). In real sensors it is not possible to achieve the ideal of a transducer which monitors a parameter while having absolutely no affect on it, because the act of making a measurement requires either adding energy to the system or subtracting energy from the system. We can minimize the effect of a sensor on the system by minimizing the energy exchange. For this reason, most sensors are low-power devices. Most sensors are also small, in order to minimize the load they impose on the system that they are monitoring. Exactly what constitutes *small* or *low power* depends on the particular situation. For example, typical airbag restraint systems in automobiles rely on acceleration switches which are about the size of a thumbnail. These sensors are clearly quite small relative to either the automobile overall, or to the airbag. Further, the power required for the sensor is a negligible fraction of that used to move the car.

Actuators are devices which impose a state on a system, hopefully independent of the load applied to them. Examples of actuators are motors (which impose a torque), force heads (which impose a force), and pumps (which impose either a pressure or a fluid velocity). The ideal of a load-independent actuator can never be achieved for a transducer with finite energy, because there will always be a load which exceeds its energy handling ability. This can be seen easily by consideration of Fig. 1.5. In Fig. 1.5 an example of an ideal velocity actuator is presented. Note that it is capable of imposing a velocity level, V_0, regardless of the mass of the object to which it is attached, i.e., independent of the load. Since the kinetic energy of the mass is given by the product of $mV^2/2$, and V_0 must be supplied even for an infinite mass, the ideal velocity actuator must be able to supply infinite kinetic energy to the mass. Clearly this statement is not an accurate reflection of reality. In any real actuator, shown by the dashed line in Fig. 1.5, the actuator might function like the ideal for low-mass attachments, but as the mass exceeds a certain threshold, in this case m_0, the velocity supplied deviates from that desired. As the mass of the attachment increases still further, the velocity actuator deviates further, eventually being unable to move the load at all. In general, we minimize the effect of the load seen by the actuator by having high-power actuators, so that typical loads cause only a small perturbation. This desire to minimize the effect of the load imposed by the system on the actuator performance usually leads to large actuators. Exactly what constitutes *large* and *high power* depends on the particular circumstances. For instance, a shaker used to shock test a circuit board can be substantially smaller than that used to test the main components of the space shuttle.

The definitions above make it clear that the two major practical distinctions between sensors and actuators are the power level involved and the size required. It is generally considered to be easier to build a low-power device than a high-power device. This puts sensors in an advantageous position relative to actuators and overall, the state-of-the-art of sensors is more advanced than that of actuators. Indeed, the need for high power for actuators has led historically to articulation of the *staging problem* in which the tradeoffs between using many lower power actuators versus using a single higher power actuator are considered. A classic

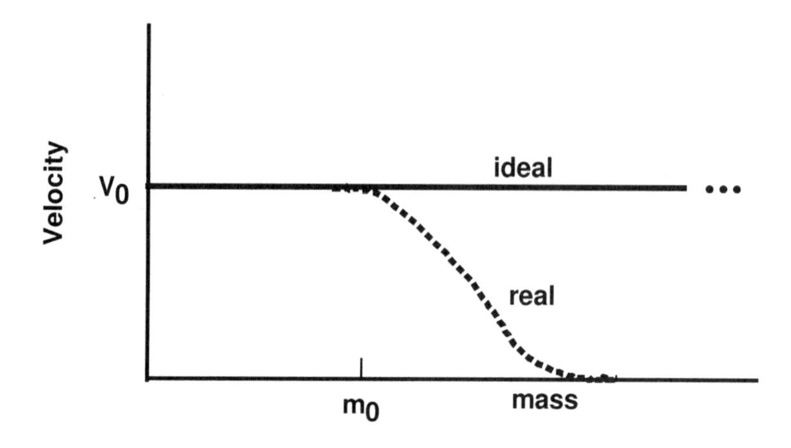

FIGURE 1.5. Ideal versus real actuator.

example of the application of this problem is the use of a motor to turn the hands of a clock. If the clock has second, minute, and hour hands, we could use either a single motor with gearing to turn all three simultaneously, or use three separate motors linked by the appropriate electronics for timing synchrony. The appropriate choice in each staging problem is determined by the specific application needs. In the clock example, we would consider the relative cost of the larger motor and gear train, as compared to three smaller motors and the controlling electronics, but the single motor and gear train almost always wins.

While the power level generally is the main issue distinguishing the state-of-the-art of sensors from actuators, physical size can also play an important role. At conventional scales for electromechanical sensors and actuators, there is little difference in the difficulty presented in manufacturing. However, for very large or very small transducers, fabrication can become a major issue. Indeed, one of the traits that separates solid-state transducers from their more conventional cousins is that the choice of using microelectronic fabrication techniques rather than conventional fabrication approaches means a transition to a regime in which fabrication considerations severely constrain acceptable designs.

The size and power distinctions between ideal sensors and actuators provide an interesting background from which to discuss the trend to create sensors and actuators using microelectronic fabrication techniques. For the most part, these solid-state sensors and actuators use the same transduction mechanisms that are used on macroscopic scales. A notable exception to this is the growing trend for thermoelectrical actuation. Thermal actuation generally is not used on macroscopic scales because the power needs are prohibitive and the system would operate too slowly. On microscopic scales, these problems are largely eliminated by the use of very small structures.

From the discussion above, it is clear that there are good reasons to require sensors that have a tiny footprint. This suggests that it is quite logical to miniaturize sensors, and microelectronic fabrication methods provide a reliable means of accomplishing this. However, from the discussion above it is far from clear that it is logical to seek to miniaturize actuators. For such a miniaturization to be justified there must be a compelling reason based on either geometry (a big actuator can not fit) or on consideration of the staging problem. While some applications which must use small actuators are emerging, there has been little attention paid to the staging problem, as research on microscopic actuators has been conducted. For instance, substantial funding has been directed to research on microactuators which would be used to coat submarines or airplane wings. Virtually no consideration has been given as to how one would power these devices, or interconnect them. Once these thorny issues are added to the problem, it might well be that more conventional approaches, using a few large actuators, possess substantial advantages over the approach of many very small actuators.

The distinctions between sensors and actuators also explain why it is so difficult to build a transducer which functions well, as both a sensor and an actuator. If the design is optimized for operation as a sensor, meaning low power and small size, then the device is not well suited to operating as an actuator. Similarly, if the transducer is designed to function primarily as an actuator, with high power and large size, then it is not likely to make an adequate sensor. Because it is extremely difficult to conceive of a transducer design that produces a device that can function well, both as a sensor and as an actuator, very few transducers are used in a manner which takes advantage of their reversible nature. Instead, it is usually preferable to use two transducers, each optimized for a particular performance as either a sensor or an actuator.

1.4 Transducers as Part of a Measurement or Control System

Transducers are most frequently used as part of a measurement or control system. Such a system is depicted in Fig. 1.6. This system has four main parts: the system under study, the sensing and actuation stage, the conditioning stage, and the display and storage stage.

The sensing portion of the system is designed to extract information about some parameter of the system under study. We assume that there is a one-to-one correspondence between the signal produced by the sensor and the parameter we wish to monitor. This correspondence need not be strictly linear, although we tend to seek a linear relationship in order to simplify the signal processing required later. A good signal-to-parameter correspondence requires that the sensor be stable, have good repeatability, have low distortion, and be operated in an environment with little noise.

The actuating portion of the measurement system is designed to impose a state on a selected parameter of the system of interest. Here we assume that there is a one-to-one correspondence between the signal applied to the actuator and the resulting value of the system parameter. Again, the correspondence need not be

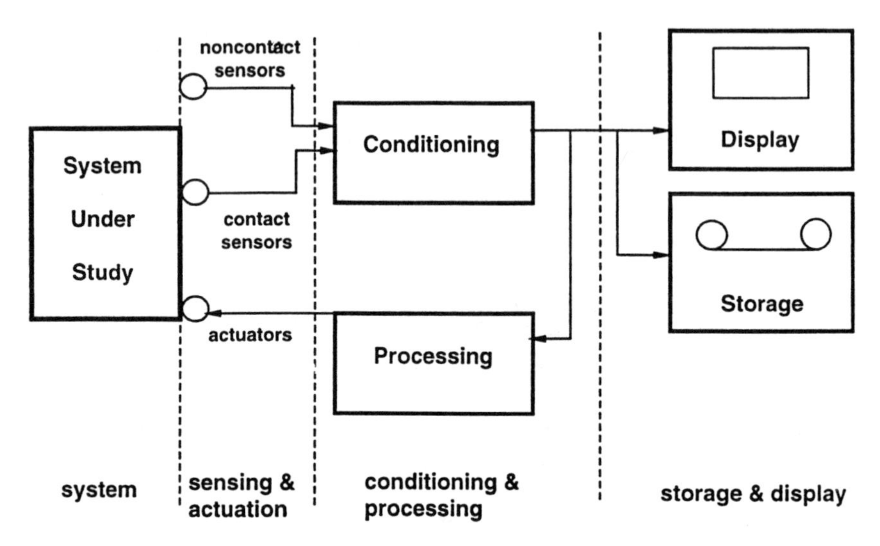

FIGURE 1.6. Measurement and control system.

linear, although we usually seek a linear relationship in order to simplify the system processing needs. Just as in the sensor, a good signal-to-parameter correspondence can be achieved only with an actuator which is stable, repeatable, of low distortion, and low in noise. These requirements impose tougher constraints on actuators than on sensors, as the usual need for high power makes it more difficult to achieve stability (temperature effects cause drift, for example).

The conditioning stage of the measurement system exists to change the sensing signal into a form ideally suited for the required task, and to produce the appropriate actuation signal. This portion of the measurement or control system typically involves amplification and intentional reshaping of the signal such as in filtering. In the last two decades, great strides have been made in conditioning algorithms and hardware. Measurement and control systems have improved substantially thanks almost entirely to these new developments, including advanced digital signal processing techniques such as those described in Bendat and Piersol [4].

The storage and display portion of the system is used to preserve information for some immediate or future use. In well-established control systems, this stage is often absent. However, in systems which are being monitored for compliance with error standards, or which are being debugged, this data can be an essential tool.

In most measurement and control systems, the sensors and actuators are the least reliable elements. Thanks to the advances in electronics, they are also often the most expensive items. Hence, there is strong motivation to improve our knowledge of transducer functioning, and to produce more reliable and less costly transducers.

1.5 Historical Perspective

Electromechanical sensors and actuators may be put into two broad categories that describe the means by which they couple electrical and mechanical energy: geometric and material. In the first category, electrical and mechanical fields are coupled, due primarily to a particular geometry which is exploited. For instance, we can construct a variety of sensors and actuators which exploit the geometry of a parallel plate capacitor. The second broad category of electromechanical transducers couples electrical and mechanical fields through some property that is displayed by a class of materials. For instance, certain metals and rare earth materials exhibit magnetostriction, a property that can be used to link mechanical and magnetic behavior.

Over the years that electromechanical transducers have been in existence, most of the critical discoveries and innovations have resulted from the introduction of new transduction materials. To name just a few of these discoveries: Seebeck found what is now called the Seebeck effect in bismuth and copper in 1821; Villari and Joule discovered magnetostriction in nickel in the 1840s; Siemens studied thermoresistivity in platinum in 1871; Hall discovered and described the mechanism we call the Hall effect in 1879; piezoelectricity was discovered in quartz by Jacques and Pierre Curie in 1880 and in barium titanate by Gray in 1947; and Smith characterized piezoresistivity in silicon in 1954. While there are notable exceptions, most often the materials which have been incorporated with great success into transducers have not originally been studied for their potential use in sensors and actuators. Thus, the changes that these materials have brought about in transduction have relied on two milestones: the development or discovery of the material; and the design of a transducer taking advantage of the new material. As a result, there has been traditionally a significant delay from the time a material is first characterized until transducers taking advantage of its unique properties emerge. While this pattern remains true today, improved communications and a pressing need for progress in electromechanical sensors and actuators has dramatically reduced the time from discovery of a new material to its introduction in a transducer.

In today's world, there are a number of trends which combine to make it an exciting time to be working in the area of transduction. For instance, it is clear that sensors, and particularly sensors with intelligence, have become quite pervasive in industrialized nations. Every car manufactured must have crash sensors. Light switches, more and more frequently, are outfitted with motion sensors which turn off lights to conserve power. Home computers routinely come with microphones and speakers. Most new homes have smoke detectors in place. Clearly, the market for transducers is growing very rapidly and consumers increasingly expect products to be smart and to monitor their performance. In part, this expectation is responsible for increasing demands being placed on transducers. In part, it is also responsible for the creation of electronic devices or intelligent devices where their need is questionable. (Does anybody really need an electric pencil sharpener?)

As the capabilities of transducers have improved, it is becoming possible to apply them to do things not previously possible. For instance, hearing aids, so small that they fit deep into the ear canal where they are not visible externally, are now common. Microscopes with sensors, able to detect individual atoms on a surface, are also available. Because of the demands imposed by an aging population, it is likely that we will see progressively more applications of electromechanical transducers in biomedical arenas.

Another trend in electromechanical transduction is an increased emphasis on actuators rather than sensors. This trend is understandable given that it is far easier to make nearly ideal sensors than it is to make nearly ideal actuators. Because of this distinction, the advances in actuation have been few and far between while the progress in sensors has been steady for many decades. In particular, we desperately need actuators which are more energy efficient, more reliable, and more precise.

1.6 Organization of this Book

Traditionally transducers have been classed in three ways. First we may categorize and discuss them by their use, such as microphones, accelerometers, displacement measurement, etc. This is the most common classification scheme found in books on measurement or on sensors. However, this approach is not helpful for a fundamental understanding of transduction processes because the physical principles may be quite similar in devices with very different applications. As an example, consider a parallel plate capacitor such as shown in Fig. 1.7. The energy stored in the capacitor depends on the capacitance which is given by

$$C = \frac{\epsilon A}{d}, \tag{1}$$

where A is the cross-sectional area of the plates, ϵ is the electrical permittivity of the material between the plates, and d is the separation between the plates. From a design perspective, this suggests that we could easily use the capacitor as a sensor, by allowing any one of the three parameters on the right-hand side of (1) to vary in a manner which provides a one-to-one correspondence between the parameter value and the desired quantity to be measured (the measurand). Regardless of how this variation is accomplished, the fundamental transduction mechanism is unchanged. Consider the commercial examples available using this approach. One transducer exploiting a variation in permittivity is a humidity sensor, which uses a material in which ϵ is a function of moisture content. A sensor in which the measurand relates to the effective area variation is a position sensor in which one capacitor plate is fixed while the other moves laterally, thus effectively changing the area of plate overlap. Still another manifestation of this transduction mechanism is a condenser microphone. In this transducer one of the plates moves in response to sound pressure, thus changing the size of the gap between the plates. While the microphone, humidity sensor, and displacement sensor are all based on the same transduction principle, this would be far from obvious were they to be classed by their application.

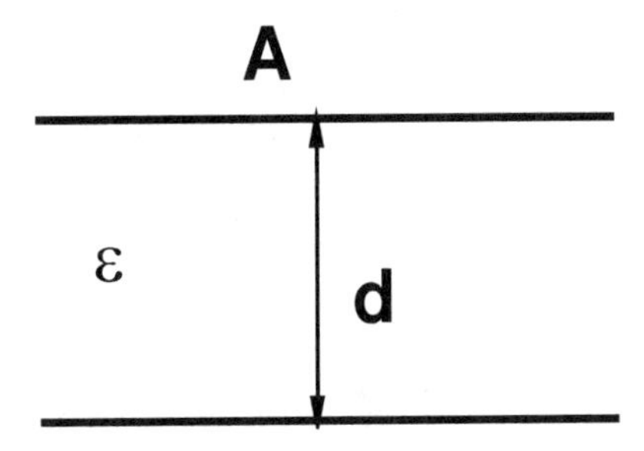

FIGURE 1.7. Sensors based on parallel plate capacitor.

Another problem with classing transducers by application, is that it necessarily limits discussions to a survey of commercially available types of transducers. This stifles creativity and can leave questions about why certain types of transducers are not useful for a particular application. For instance, many pressure sensors sold today exploit piezoresistivity by allowing the pressure to strain a membrane. This strain causes a measured change in electrical resistance. However, the choice by manufacturers of pressure sensors to use piezoresistivity is dictated by the fact that their specifications are best met by currently available piezoresistive approaches. Were a new magnetostrictive material to be discovered tomorrow which better met the needs of pressure sensing, it would not be long before many of these same manufacturers would choose to change their product line to include the new material and transduction approach.

A second method of classing transducers has been by basic energy type. Capacitive devices are those which rely on energy in an electric field in some manner. Inductive devices rely on magnetic fields. Resistive devices are those which clearly work through the irreversible conversion of energy. Although this classification scheme permits discussion of the mechanisms of transduction rather than applications, the categories are so broad as to make comparison of transducers difficult. In other words, two capacitive devices might have nothing more in common than the fact that they rely on changes in an electric field.

A third alternative, and the one we will use in this text, is to categorize transducers by the material or structural behavior which leads to transduction. This effectively introduces subclasses into the capacitive, inductive, and resistive categories defined above. As we will discuss, there are many such subclasses. By using this categorization we permit people to see similarities between various transducers, while understanding the details of the transduction mechanism associated with each class.

The following chapter discusses modeling with particular attention paid to electromechanical transducers. The modeling technique presented will be used routinely through this book. As will be shown by example, the development of a model can simplify analysis and make it possible to draw conclusions quickly. Following this introduction to a unified modeling approach, various electromechanical transduction mechanisms will be discussed in Part II. In each case, commercially available examples applying the principles will be given. The final part of the book deals with the performance characteristics of transducers and presents material which holds regardless of the particular transduction mechanism involved.

References

1. S. Middelhoek and A. C. Hoogerwerf, *Classifying Solid-State Sensors: The Sensor Effect Cube*, Sensors and Actuators, **10**, 1 (1986).

2. R. C. Rosenberg and D. C. Karnopp, *Introduction to Physical System Dynamics* (McGraw-Hill, New York, 1983).

3. I. J. Busch-Vishniac, *Bond Graph Models of Sound and Vibration Systems,* J. Acoust. Soc. Am., **85**, 1750 (1989).

4. J. S. Bendat and A. G. Piersol, *Random Data*, 2nd ed. (Wiley, New York, 1986).

2

System Models

The academic discipline of transduction necessarily covers a broad range of energy domains. In the course of this book we will discuss translational, rotational, fluid, thermal, magnetic, electrical, and optical systems. In order to have a common language and common tools, it is necessary to use a system modeling procedure which can easily span these domains.

Modeling of physical systems is a branch of the field of system dynamics, which concentrates on the description and analysis of complex physical entities. Modern approaches to physical system modeling use energy-based methods, much as modern physics relies on Lagrangian and Hamiltonian mechanics rather than Newtonian or classical mechanics.

In system modeling we typically use connected, ideal, lumped elements (such as masses, springs, and dashpots) to represent the system. The number of elements in the system model is not necessarily equal to the number of identifiable physical parts in the real system. For example, consider a piece of foam rubber which is being compressed. If the foam rubber is part of a larger system, we may choose simply to include the foam's springiness in the model. In a more accurate model of the foam rubber being compressed, we would note that some of the energy is dissipated as heat in the material. We could include the dissipated energy by developing a model with both a spring and a dashpot. In a still more sophisticated model of the foam rubber, we might recognize that the center of mass of the foam moves as it is compressed, and include this phenomenon by including an ideal mass in the spring and dashpot model.

There are three primary advantages of using discrete lumped elements to model complex physical systems. First, there are many well-developed types of models which rely on such elements. These include circuit models, linear graphs, and bond graph models. Second, discrete lumped element models produce a finite number of ordinary differential equations which describe the system. These equations are solved easily, particularly if the constitutive equations describing the individual elements are linear. Third, these models have been shown to produce good results in many cases, from automotive power trains (see Hrovat and Tobler [1]) to vacuum cleaners (see Remmerswaal and Pacejka [2]) and acoustic transducers (see Hanish [3]).

There are, however, two major disadvantages to discrete lumped element models of physical systems. First, since we use a finite number of discrete elements to model a physical continuum, the validity of the model is restricted. As a result, most lumped element models are only accurate for relatively low frequencies. Second, there are structures and processes in which the continuous nature of the medium plays a major role in determining the system response. The propagation of waves in structures is one example of this. In order to model accurately such wave propagation using discrete elements, it is necessary to use an infinite number of elements so that the ordinary differential equations are converted to partial differential equations in the limit of shrinking element size. For further information on the general advantages and disadvantages of lumped element models, see [4]–[8].

Throughout this book we will use discrete lumped element models to describe transducers. In this chapter we describe the modeling procedures that we will use universally. This presentation will be necessarily brief, and interested readers are referred to the list of texts, on system modeling, at the end of the chapter.

2.1 System Analogies

In developing a logical set of analogies between energy domains, we begin with the definition of fundamental variables. There are two sets of fundamental variables which are defined for each energy domain: power conjugate variables and Hamiltonian variables.

Power conjugate variables are two variables which when multiplied yield the power in a given energy domain. This definition permits a large number of choices of variable sets for any given energy domain, but it is traditional to choose those variables most easily measured. For example, in mechanical systems in translation, we could use force and velocity, or time rate of change of force and displacement. The former pair is the standard choice. The standard power conjugate variables are listed below:

- Translational: force, F, and velocity, V.
- Rotational: torque, τ, and angular velocity, Ω.
- Electrical: voltage, e, and current, i.
- Magnetic: magnetomotive force (also called magnetomotance), M, and time rate of change of magnetic flux, $d\phi/dt$.
- Fluid: pressure, P, and volume flow rate, Q.

- Thermal: temperature, T, and time rate of change of entropy, ds/dt.

Of these sets, the magnetic power conjugate variables are likely to be the least familiar. They will be discussed in further detail in Chapter 4, when they are needed for the first time. Additionally, it should be noted that the thermal variables violate the suggestion that the choice of power conjugate variables be restricted to those parameters which are easily measured. Entropy and its rate of change, are not system variables which can be directly measured. For this reason, some system dynamicists have chosen to represent thermal systems using parameters which are easily measured and interpreted, but are not power conjugate variables. However, the price paid for using thermal system variables which are not power conjugates is that many of the conclusions that can be drawn generally for the other types of systems cannot be applied to the thermal systems. It is the view of this author that the desire for a directly measurable parameter is less important than the ease of interpretation of the physical model. Thus this text will strictly adhere to power conjugate variables for thermal systems.

Having defined the power conjugate variables, we are now free to choose how we wish to relate them. Once this choice is made, the remaining aspects of the system of analogies are determined.

Historically there have been three popular choices for relating the power conjugate variables in the various energy domains. One analogy is attributed to Firestone [9] and [10] and is known as the Firestone or the mobility analogy. In the Firestone analogy, the linear velocity, angular velocity, volume velocity, and voltage are treated as analogous (thermal and magnetic variables were not included in Firestone's work). Similarly, force, torque, pressure, and current are analogous variables.

Firestone or Mobility Analogy

- V, Ω, Q, and e are analogous.
- F, τ, P, and i are analogous.

Using this analogy, the electrical impedance, the ratio of voltage to current, is analogous not to the mechanical and fluid impedances but to the mechanical and fluid mobilities, ratios of velocity to force, and volume flow rate to pressure.

A variation of the Firestone analogy that was introduced by Trent [11] is known as the through-and-across analogy. In this system of analogies, the linear velocity, angular velocity, pressure, and voltage are treated as analogous variables. They are collectively referred to as across-variables because they are relative measures typically referenced across an ideal element. The complementary variables, force, torque, volume velocity, and current, are called through-variables because they are not relative measures and are often transmitted through an ideal element. (For example, a spring transmits forces.)

Through-and-Across Analogy

- V, Ω, P, and e are across-variables.
- F, τ, Q, and i are through-variables.

In the Trent analogy the electrical and fluid impedances are analogous to the mechanical mobility.

The third common choice for system analogies is attributed to Maxwell [12] and referred to as the Maxwell or impedance analogy. In this choice, we relate the linear velocity, angular velocity, volume velocity, and current. These variables are referred to collectively as flow variables, because each can be conceived in terms of the motion of an object. Similarly, the force, torque, pressure, and voltage are analogous variables. They are collectively called effort variables.

Maxwell or Impedance Analogy

- V, Ω, Q, and i are flow variables.
- F, τ, P, and e are effort variables.

In the Maxwell analogy, the electrical, fluid, and mechanical impedances are analogous.

In this book we will use the Maxwell analogy exclusively. We extend it to include thermal and magnetic systems by declaring the magnetomotive force and temperature to be effort variables, and the time rates of change of magnetic flux and entropy to be flow variables.

Expanded Maxwell or Impedance Analogy

- V, Ω, Q, i, $d\phi/dt$, and ds/dt are flow variables.
- F, τ, P, e, M, and T are effort variables.

The Maxwell analogy has three advantages when compared to the Trent and Firestone analogies. First, it preserves the analogy between impedances in the mechanical, fluid, and electrical domains. Second, it preserves the analogy between force and pressure. In the Trent analogy, pressure and force are not like variables despite the fact that pressure is defined as a force per unit area. Third, there is no problem associated with the mass analogy in electrical systems. Using the Trent and Firestone analogies, we argue that since the velocity of a mass is always measured relative to the Earth, all electrical elements analogous to masses must be grounded. Nongrounded capacitors have no mechanical analog. Hence many electrical systems have no mechanical or fluid analog. This problem does not arise with the Maxwell analogy because it does not distinguish between relative and absolute measures. It should be noted that this lack of distinction between relative and absolute measures is a mixed blessing. It permits greater breadth of coverage in modeling, but it introduces a need for careful interpretation of the modeling elements.

The major disadvantage of the Maxwell analogy is that the topology established in electrical systems is not preserved in going to the analogous mechanical system and vice versa. For example, two resistors in series are shown in Fig. 2.1. These elements have the same current running through them, and since current is analogous to velocity, the circuit is analogous to a mechanical system with two dampers in parallel, as is shown in Fig. 2.2. This can cause confusion to the novice system modeller.

FIGURE 2.1. Two resistors electrically in series.

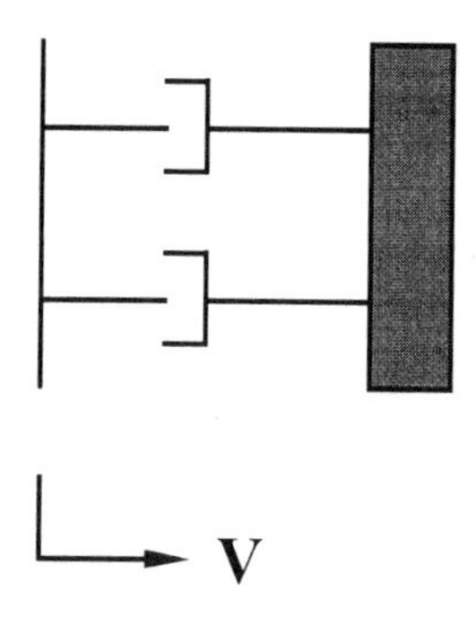

FIGURE 2.2. Two dampers mechanically in parallel.

Having determined a set of power conjugate variables, and a means of associating them with one another, we may now define consistent Hamiltonian or energy variables. These variables are the system descriptors which, when differentiated in time, yield the power conjugate variables. They are referred to as the Hamiltonian variables because they are the variables that commonly appear in a system Hamiltonian. **Hamiltonian variables** are shown below:

- Translational: displacement, x, and linear momentum, p.
- Rotational: angular displacement, θ, and angular momentum, h.
- Electrical: charge, q, and flux linkage, λ.
- Magnetic: magnetic flux, ϕ (only one defined).
- Fluid: volume, U, and pressure–momentum, Γ.
- Thermal: entropy, s (only one defined).

Note that in two of the energy domains only one Hamilton variable is defined. In these domains the variable analogous to a linear momentum is not defined because it is not useful. There is simply no phenomenon which is reasonably modeled as thermal momentum or magnetic momentum. Instead, in the thermal and magnetic domains it is sufficient to characterize the storage of potential energy and the dissipation of energy. Note also that the pressure–momentum as defined above generally is not given a name in fluids or dynamics texts. We introduce this definition, and the symbol Γ, for convenience. The pressure–momentum is simply the linear momentum of a fluid flow, divided by the cross-sectional area of the flow.

Using Maxwell's terminology, we divide the energy variables into two classes: generalized displacement and generalized momentum. The displacement variables are those energy variables which when differentiated in time yield flow variables: linear displacement, angular displacement, charge, magnetic flux, vol-

TABLE 2.1. Power conjugate and Hamiltonian variables associated with various energy domains.

	Trans.	Rot.	Elect.	Mag.	Fluid	Thermal
Effort	F	τ	e	M	P	T
Flow	V	Ω	i	$\dot{\phi}$	Q	$\dot{s}$
Displacement	x	θ	q	ϕ	U	s
Momentum	p	h	λ		Γ	

ume, and entropy. The generalized momentum variables are those variables which when differentiated in time yield the effort variables: linear momentum, angular momentum, flux linkage, and pressure–momentum.

Table 2.1 summarizes the analogy system that we will use. It lists the power conjugate variables and Hamilton variables for each energy domain.

2.2 Ideal 1-Port Elements

Having defined the variables associated with various energy domains, we can now consistently define the building blocks of discrete lumped element models, namely, the elements. We begin with 1-port elements, i.e., with elements which have only one port through which energy is exchanged with the environment.

One-port elements come in two general forms: passive elements and active elements. Active elements increase or decrease the total energy present in the system. In other words, active elements provide a means for energy exchange with the **environment**, i.e., with everything outside the system boundaries. Thus they are sources or sinks. These elements are further divided into two types—ideal sources (sinks) of flow, and ideal sources (sinks) of effort. The **ideal flow source** is an element which specifies the flow variable at a port independent of the effort at that port. The **ideal effort source** specifies the effort variable at a port independent of the flow at that port. Clearly real sources are not the same as the ideal 1-port sources, because the availability of finite energy ensures that as the effort increases at a flow source, or the flow becomes high at an effort source, the source level must be affected. Were real sources the same as ideal sources, we could power the space shuttle with the same battery as that used to run a portable radio. This is clearly nonsensical.

Passive elements respond to the energy they receive by either storing it in a form suitable for performing work, or by irreversibly converting it to a form in which it cannot be used for work (i.e., converting the energy to heat in a process we call dissipation). The storage of energy can take two distinctly different forms in many energy domains. For example, in the mechanical domain it is possible to store either potential or kinetic energy. Thus, there are three different passive

1-port elements, each associated with a particular response to the energy seen. We use a different ideal 1-port element to represent each of the three passive energetic responses, but we can define them independently of the energy domains.

First consider the storage of energy in the form which we call kinetic energy for mechanical systems. We define the element capable of this type of energy storage as an **ideal inertance**. It is described by a constitutive relation of the form

$$f = g(p), \tag{1}$$

where f is a flow variable, p is a generalized momentum, and g is any function. A mechanical example of an ideal 1-port inertance is a mass. An electrical example is an inductor. Note that these examples are not arbitrarily chosen. Since both a mass and an inductor are described by constitutive equations of the form of (1), they must be analogous for our system of analogies to be logical. The energy stored by the inertance is given by

$$\mathcal{E} = \int dt\, ef = \int f\, dp \quad , \tag{2}$$

where e is the effort. If the constitutive relation is linear, then (1) may be replaced by

$$f = \frac{p}{I}, \tag{3}$$

where I is defined as the inertance or inductance, and the energy stored is given by

$$\mathcal{E} = \frac{p^2}{2I}. \tag{4}$$

Next consider the storage of energy in the form we call potential energy in mechanical systems. We define the element capable of this as an **ideal capacitance**. It is described by the following constitutive equation:

$$e = g(q), \tag{5}$$

where q is the generalized displacement. Note that q is often a relative measure. A mechanical example of a 1-port capacitance is an ideal spring, and here q is the difference in displacement of the spring ends. An electrical example of a 1-port capacitance is a capacitor. The fact that the spring and capacitor are analogous is a consequence of choosing to use the Maxwell analogy. Since both can be described by (5), they must be analogous. The energy stored in the capacitance is given by

$$\mathcal{E} = \int dt\, ef = \int e\, dq. \tag{6}$$

If the constitutive relation is linear, then (5) is replaced by

$$e = \frac{q}{C}, \tag{7}$$

where C is the capacitance. The energy stored in the linear element then is given by

$$\mathcal{E} = \frac{q^2}{2C}. \tag{8}$$

Note that in this form the capacitance for a mechanical spring is the compliance, which is the inverse of the stiffness.

The third ideal 1-port energic element is an **ideal resistance**. This element irreversibly converts energy to a form unsuitable for work. It is described by the constitutive equation

$$e = g(f), \tag{9}$$

or its inverse relation, where we also insist that $ef > 0$. This latter requirement ensures that energy is being removed from the system rather than added to it. Examples of ideal resistances include mechanical dashpots and electrical resistors. In a linear element, (9) is rewritten as

$$e = Rf, \tag{10}$$

where R is the resistance.

Mechanical and electrical examples of the ideal 1-port elements introduced so far are almost certainly familiar to all readers. Fluid, magnetic, and thermal elements are less likely to be so. We will discuss the fluid and thermal elements here, but we delay introducing the magnetic elements until they are needed (in Chapter 4).

The fluid elements are best understood by comparison to the mechanical analogs. Just as a solid has mass, so does a fluid. Hence if a fluid moves, it has inertia and we should be able to express this as an ideal inertance. Figure 2.3 illustrates this concept. If we consider a pipe with uniform cross-section through which an incompressible fluid flows, then application of Newton's second law of mechanics yields that

$$F = (P_2 - P_1)A = m\frac{dV}{dt}, \tag{11}$$

where A is the pipe cross-sectional area, and m is the mass of the moving fluid. We rewrite this equation as

$$(P_2 - P_1)A = \rho l A\frac{d(Q/A)}{dt}, \tag{12}$$

where ρ is the mass density of the fluid, and l is the pipe length. Rewriting this equation yields

$$\Delta P = \frac{\rho l}{A}\frac{dQ}{dt}. \tag{13}$$

Integration in time finally produces the desired form:

$$\Gamma = \frac{\rho l}{A}Q, \tag{14}$$

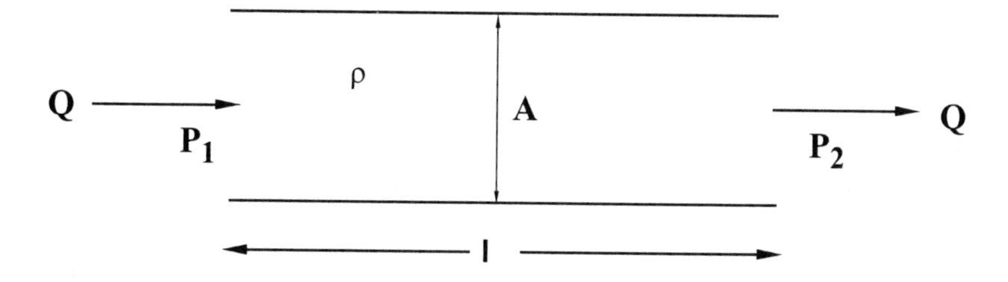

FIGURE 2.3. Fluid inertance element.

from which we conclude, by comparison with (3), that the pipe fluid inertance is given by $I = (\rho l)/A$.

A fluid capacitance is analogous to a mechanical capacitance. There are two common forms of fluid capacitance, both related to the storage of potential energy: spring-like compressibility, and pressure increases with depth due to the presence of a gravitational field. The former is analogous to the mechanical spring, and the latter to the storage of potential energy (and thus the capacitance) in a pendulum.

An example of a fluid spring-like capacitance is the adiabatic compression of an ideal gas. For such a gas we have

$$PU^{\gamma} = \text{constant}, \tag{15}$$

where γ is the ratio of specific heat at constant pressure to that at constant volume ($\gamma = 1.4$ for diatomic gases such as air). If we allow the gas pressure to change we can relate the pressure and volume changes by differentiating (15). This yields

$$U^{\gamma}\Delta P + \gamma P U^{\gamma-1}\Delta U = 0, \tag{16}$$

which we rewrite as

$$\Delta P = -\frac{\gamma P}{U}\,\Delta U. \tag{17}$$

We compare (17) to (7) and note that we have the correct form if we define the pressure change to be the pressure variable (effort) of interest and the volume change to be the volume variable (displacement) of interest. Note that this constitutive equation is not linear because of the appearance of P and U in the factor multiplying ΔU. We may produce an approximate linear constitutive relation by using equilibrium values for P and U in this factor. Note also the minus sign in Eq. (17). This sign indicates that the fluid behavior is indeed spring-like, supplying a restoring force that opposes the motion. The minus sign is not included as part of the fluid capacitance, just as a spring stiffness is taken to be positive even though the spring resists compression and extension.

Fluid resistance is akin to mechanical resistance. It describes processes that result in significant head loss, such as the flow of fluids around bends or through orifices and valves. Like its mechanical counterpart, fluid resistance is less likely to be described by a linear constitutive equation than the inertance and capacitance elements.

While the above discussion shows that it is relatively easy to develop 1-port mechanical, electrical, and fluid elements, the same statement cannot be made for thermal elements. The problem with 1-port ideal thermal elements is that thermodynamic systems generally are described using two (not one) independent variables. This suggests that ideal thermal elements should have two-ports rather than one. For example, if we choose pressure and temperature as the independent thermodynamic variables, then we would express the energy stored or lost to the system in terms of those two variables. This would lead to ideal elements with two ports: one with temperature and entropy flow rate as the power conjugate variables, and the other with pressure and volume flow rate as the power conjugate variables. There have been attempts to circumvent this problem and permit the description of thermal behavior using 1-port ideal elements (see for example Karnopp et al. [4]). These attempts generally have relied upon using thermal port variables other than temperature and entropy flow rate. The normal alternative choice is to use temperature and heat flow rate. With this choice, the port variables are no longer power conjugate variables, since the heat flow rate has units of power, and this has significant negative repercussions on system model development. However, with the use of this alternative choice of port variables, we can model the thermal behavior using 1-port elements. In this book we choose not to develop models using temperature and heat flow rate as the port variables. Thus, thermal behavior requires the introduction of 2-port elements, and will be presented in detail in the section of this chapter which discusses multiport elements (Section 2.6).

Despite the problems presented by consistent modeling of thermal system behavior, we can make qualitative statements at this time which relate thermal behavior to analogous mechanical, electrical, and fluid behavior. We begin with consideration of energy storage and dissipation in a thermal system. The storage of thermal energy in an object is typified by that object's ability to perform work through raising the temperature of other objects with which it makes contact. For example, the fact that a boiling pot of water can be used to raise the temperature of an egg which is dropped into it, means that the water stores energy. This energy storage is potential energy, and is analogous to mechanical and fluid springs and electrical capacitors. Since there is no other way for a thermal system to store energy, there is nothing analogous to a mass or inductor in a purely thermal system. It is for this reason that the thermal momentum variable is not defined.

In addition to the storage of energy, purely thermal systems are also capable of irreversibly, altering the amount of energy which is available for work. This is not only analogous to all of the dissipative elements previously defined, but, in fact, is a much more accurate definition of the dissipation process. In physical systems in which we choose to ignore thermal behavior, it is sufficient to think of dissipation as releasing heat to the environment. In models in which we choose to consider the thermal behavior, we must perform bookkeeping on the thermal energy as well as on the mechanical, electrical, and fluid energies. Under such circumstances we must divide the thermal energy into that which can be used to perform work and that which has been made irreversibly unavailable for the performance of work. The former represents capacitive energy storage, while the latter is resistive energy.

TABLE 2.2. Ideal 1-port energy storing and dissipating elements.

	Constitutive Equation	Mechanical	Electrical	Fluid
Capacitance	$e = g(q)$	spring	capacitor	fluid compliance
Inertance	$f = g(p)$	mass	inductor	mass flow
Resistance	$e = g(f)$	dashpot	resistor	head loss

A typical example of a resistive thermal element is a thermal insulator. As we shall discuss in Section 2.6, such an insulator is modeled using a 2-port thermal resistance.

Table 2.2 summarizes the ideal 1-port energy storing and energy dissipating elements. It also shows the types of elements specific to the mechanical, electrical, and fluid energy domains.

2.2.1 *Comments on Modeling and 1-Port Elements*

The process of developing the model of a physical system may be divided into two key tasks: determining what elements represent each of the significant physical phenomena in the system; and developing a model which connects these elements appropriately. This section has focused on the first of these tasks, namely, the representation of physical phenomena by ideal elements. However, the discussion here has considered only very simple system examples. A reasonable question to ask is how we goes about defining the ideal elements needed to model a complicated system. Alas there is no rule which guides this part of the modeling procedure, and two people modeling the same system often come up with very different models. The following procedure is recommended: first, divide the system into its physical components; second, determine the important phenomena associated with each physical component. Typically, we can list assumptions being made which permit some phenomena to be ignored. Often, the process of defining the physical phenomena is adequate to identify the type of element. Third, for phenomena not yet identified by a set of ideal elements, use mathematical expressions to describe the physical phenomena. These expressions could be very specific, closed form, linear or nonlinear expressions, or could simply be statements that some parameter is a function of some other parameter or parameters. Fourth, use these mathematical expressions to define the ideal elements of which the system model is comprised. For instance, using the first column of Table 2.2 allows for the determination of ideal 1-port elements in mechanical, electrical, and fluid systems. Finally, check the list of ideal elements which has been generated and modify as needed in order to strike the desired balance between inclusiveness and simplicity. In particular, we should consider all of the elements of a given type and ask whether the component-wise modeling approach has resulted in too many or too few elements. For example, it may not be sensible to include an element to represent a small mass in a system model which already includes several larger masses. Nor is it typical that the model of a complicated physical system results in no elements of particular type (e.g., ideal capacitance). When the initial list of ideal elements does not produce any

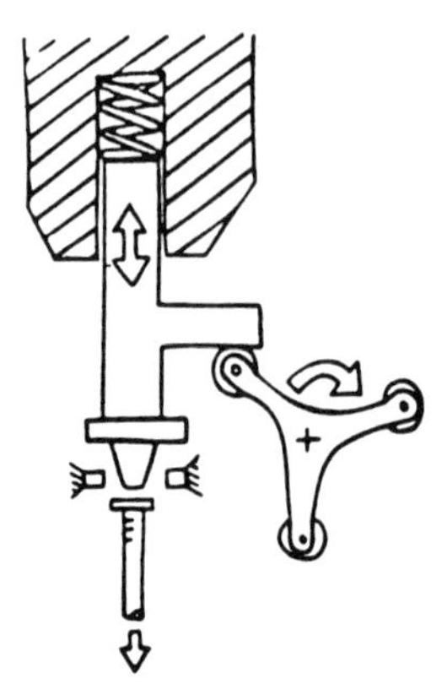

FIGURE 2.4. A view of a nail gun (taken from K. T. Ulrich and S. D. Eppinger, **Product Design and Development** (McGraw Hill, New York, 1995)) Reproduced with permission by The McGraw-Hill Companies.

element of some particular type, then we should re-examine the system either to identify the dominant source of behavior of that type, or to verify that that sort of behavior may be legitimately neglected.

Let us demonstrate the element identification process described above through consideration of a nail gun, as shown in Fig. 2.4. In this nail gun a motor turns a mechanism which moves a plunger up and then releases it. A spring at one end of the plunger drives the plunger down upon release, forcing it into contact with the nail. A washer at the opposite end of the plunger stops the plunger downward motion when it comes into contact with stops. We begin the process of determining what ideal elements would be used to model this nail gun by identifying the physical components: the motor, the metal part which revolves, the three nylon wheels attached to the rotating metal part, the plunger, the spring, the washer at the plunger tip, and the stops.

Next we consider the physical phenomena important in each component. The motor is assumed to be an ideal source of rotational velocity. The metal part which revolves has inertia. We will assume that flexing in the metal part is negligible. Hence, this part is modeled as an ideal inertance. The nylon wheels rub on the plunger tab (one wheel at most at any moment), so energy dissipation is probably important (i.e., ideal resistance in the model). In addition, each of the wheels have inertia which may be modeled with an ideal inertance. The plunger has mass (ideal inertance), and is assumed not to deform in operation. In other words, it is assumed to move as a rigid body. The spring is modeled as an ideal spring (ideal capacitance), as we assume that its mass is small relative to that of the plunger. The washer at the plunger tip deforms significantly on impact, but we assume that it does not deform permanently (as it would if it were a pure damper). This suggests that the washer includes both spring-like and damping behavior, modeled using an ideal capacitance and an ideal resistance. The stops cause deformation of the washer in braking, but don't particularly move themselves. Nor do they deform significantly. Thus we can neglect them in the system model. Note that in this system it is possible to determine the ideal elements which are to be used in the model without resorting to a mathematical description of the physical phenomena. This is because the elements can be determined by simple comparison with the ideal mechanical elements, namely the mass, spring, and dashpot.

Finally, we consider the list of ideal elements that we have: one ideal source (motor), five ideal inertance elements (one plunger, three wheels, one rotating piece), two ideal capacitance elements (spring and washer), and two ideal resistance elements (nylon wheel and washer). This list meets the criterion of having at least one element of each type. It is not a terribly sophisticated model, but there is room for simplification if desired. Specifically, the number of inertance elements is higher than necessary for a consistently simple model of nail gun. The nylon wheels are probably very light-weight compared to the plunger and metal rotating piece. Thus the inertia of the nylon wheels could be lumped with that of the metal rotating part, reducing the number of inertance elements to two.

Before leaving this example, we should note that one important element of the model has been neglected here, namely, the element which describes the conversion from rotational motion to translational motion. In this particular system, we assume that this conversion takes place without storage or loss of energy. Transformation mechanisms of this sort are described in Section 2.5. We can always determine that such a transformation element is needed by looking at the forms of energy involved in the system. If the system components use the energy described by different types of Hamilton variables, then there must be a means of describing the conversion. There are two general types of elements suitable for describing energy conversion: those which do not store or dissipate energy (nonenergic), and those which are associated with energy storage or dissipation (energic). Both are described later in this chapter (Sections 2.5 and 2.6).

2.3 Circuit Models

As described above, there are two key tasks in the development of models of physical systems: determination of the ideal elements needed to model physical phenomena, and determination of the interconnections between elements. To this end, we have focused on determining the ideal elements which are needed to model physical phenomena in a system. In this section, we focus on one formalism for determining interconnections between elements, making it is possible to construct elementary models of physical systems. It is quite common when developing models to resort to an equivalent electrical circuit representation. The advantages of using circuit models are that they are universally understood, they immediately show how to construct electrical analogs for testing and simulation, and a number of computer simulation codes are widely available. We will begin with circuit models, but progress beyond them to more general modeling formalisms, in order to permit logical and consistent construction of transducer models.

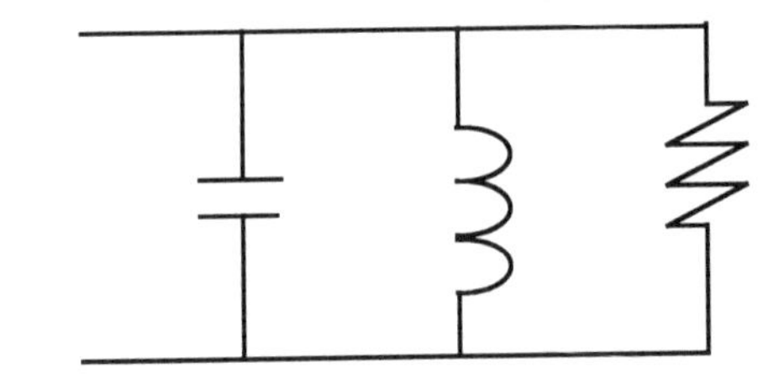

FIGURE 2.5. Electrical elements connected in parallel.

FIGURE 2.6. Electrical elements connected in series.

With elements defined as above, it remains to determine conventions which describe energetic relations when elements are connected by various ways. In circuits, it is Kirchhoff's laws which describe these conventions. Kirchhoff's voltage law says that the sum of the voltage around any closed-loop in a circuit is zero. As a result, elements connected in parallel, as in Fig. 2.5, have the same voltage. Kirchhoff's current law states that since charge is conserved, the sum of the input currents at any junction must equal the sum of the output currents at that junction. As a direct consequence, electrical elements connected in series, as in Fig. 2.6, have the same current.

We extend circuit models to nonelectrical systems by generalizing Kirchhoff's laws. Since voltage is an effort variable, the voltage law becomes a generalized effort law saying that the sum of the efforts around any closed-loop must be zero. Elements with the same effort, will appear in parallel in the system circuit model. The current law becomes a generalized flow law stating that the sum of the flows at any junction must be zero. Elements with the same flow will appear in series in the system circuit model.

Using the generalized circuit laws and the analogies developed previously we can now develop circuit models of elementary physical systems. For example, consider the simple harmonic oscillator shown in Fig. 2.7. The spring velocity is the difference in the velocities of the two ends, which is V. Similarly the dashpot velocity is the difference in the velocities of the two ends, which is also V. Hence the three elements all have the same velocity. A circuit model of the harmonic oscillator is thus the system shown in Fig. 2.8. The mass analog (inductor), spring analog (capacitor), and dashpot analog (resistor) are all electrically in series. The force source analog (voltage source) is also shown in the circuit.

If we alter the simple harmonic oscillator so that the spring and dashpot are connected end-to-end as in Fig. 2.9, then the three elements no longer have the same velocity. The velocity of the spring is now $V - V_1$ and the velocity of the dashpot is V_1. Note that the sum of the spring and dashpot velocities equals the mass velocity. This statement is the mechanical equivalent of the Kirchoff current

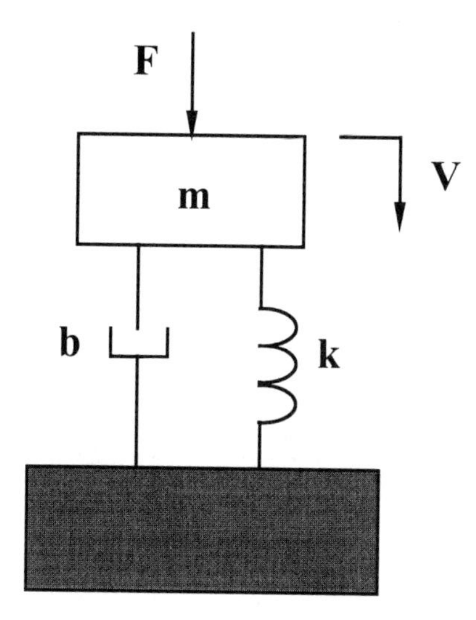

FIGURE 2.7. Simple harmonic oscillator.

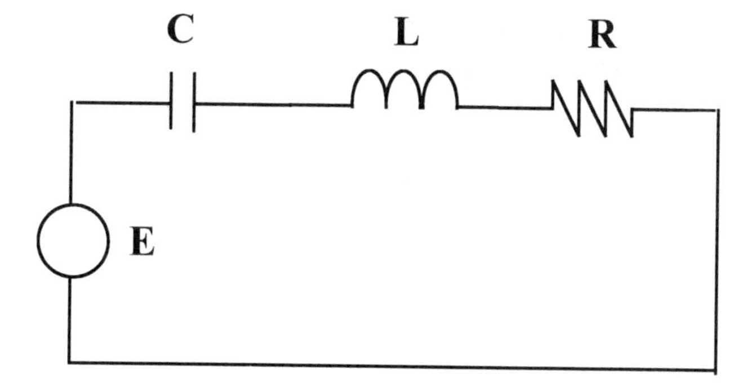

FIGURE 2.8. Electrical circuit model of a sinple harmonic oscillator.

law applied to junctions. It tells us that the analog electrical circuit is as shown in Fig. 2.10. Note that analyzing this circuit in terms of the effort, rather than the flow, the model suggests that the spring and dashpot are subject to the same force which is seen from Fig. 2.9 to be true.

We may develop elementary fluid system models in much the same way as was done above for mechanical systems. The main steps are to identify the elements needed to model the system, and to determine the connecting relationships. Once this is done it is a simple task to draw the circuit model using the analog of the ideal fluid elements. For example, consider the system shown in Fig. 2.11. In the simplest model, the valve is a 1-port resistance as it is the major source of head loss in the system. The pipe is modeled with a 1-port inertance describing the inertia of the flowing fluid. The water tank (whose fluid pressure is higher at the bottom thanks to gravity) is modeled with a 1-port capacitance describing the tank's storage of potential energy. It is purely a matter of personal preference whether we begin with

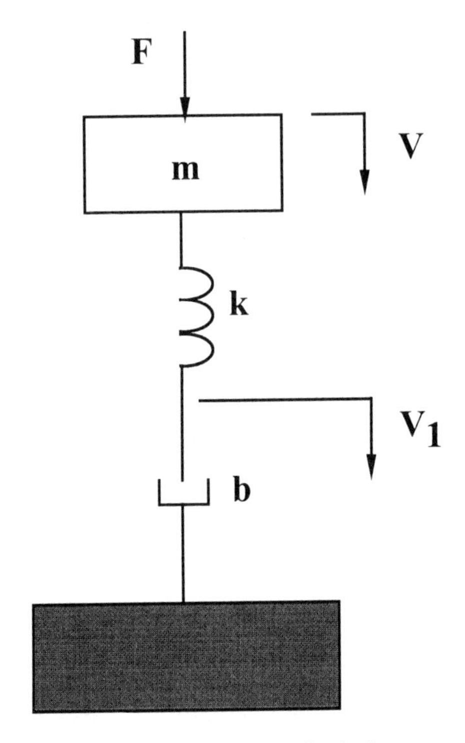

FIGURE 2.9. Simple mechanical system.

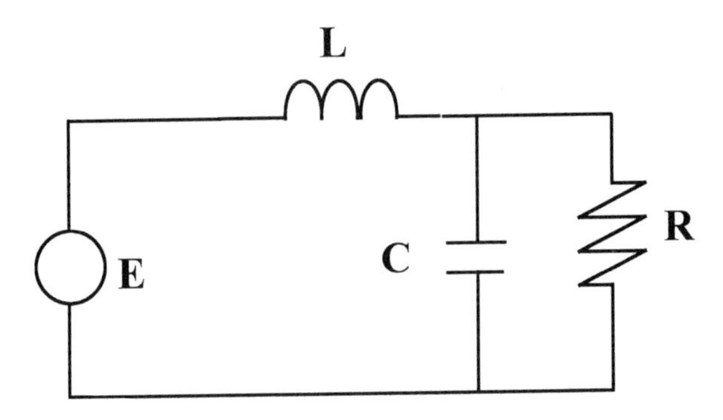

FIGURE 2.10. Circuit model of the simple mechanical system.

the flow or effort variable in determining connections between the elements. Here we choose to work with volume flow. If the elements have the same flow through them, they appear in series in the circuit model. If the flow separates, this is the fluid equivalent of an electrical T-junction. For example, the fluid system shown in Fig. 2.11 can be modeled by the circuit shown in Fig. 2.12.

As a final example of how we determine electrical circuit models of physical systems, let us consider again the nail gun shown in Fig 2.4. Our previous discussion of the nail gun led to the following list of ideal elements: an ideal source

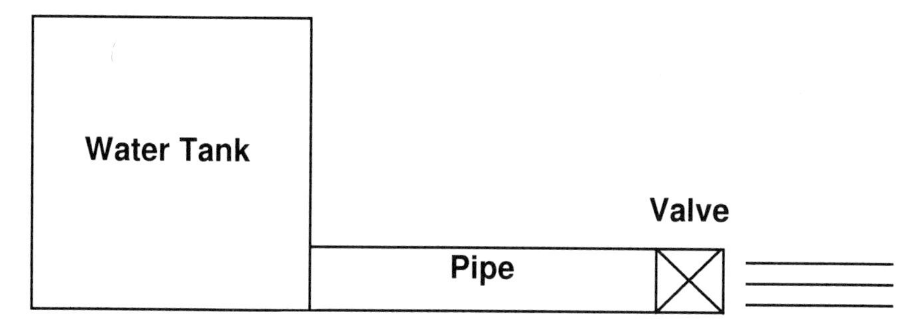

FIGURE 2.11. Simple fluid system (incompressible fluid).

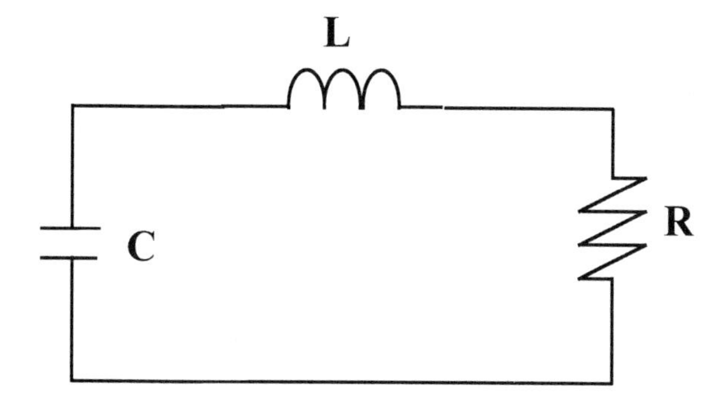

FIGURE 2.12. Circuit model of the simple fluid system.

of angular velocity (the motor), two ideal inertances (the plunger and the rotating piece), two ideal capacitances (the spring and the washer on the plunger), two ideal resistance elements (wheel-plunger friction and the washer on the plunger), and a nonenergic transforming element for rotational to translational energy conversion. We may determine the connections between these elements through consideration of the various element velocities. The spring is connected to a fixed object at one end, and to the plunger at the other end. Thus it has the same velocity as the plunger, so the plunger inductor and spring capacitor appear in series in the analog electrical circuit. The washer on the plunger has a single velocity associated with it, although we use two elements to describe its behavior. Thus the washer resistance and capacitance are associated with the same velocity. Then the electrical analog of the washer consists of a resistor and a capacitor in series. Because the washer is attached to the plunger, the spring and dashpot have a velocity at one end equal to that of the plunger. The stops are unmoving, so the other end of the spring and dashpot have zero velocity. The difference in velocities at the two ends is thus equal to the plunger velocity. Hence the washer capacitor and resistor are in series with the plunger inductor. Note that the washer is completely ineffective until the plunger has moved some distance. There are a number of ways to model this behavior but, for now, the simplest is to model the washer using a nonlinear spring and nonlinear dashpot, which have a thresh-

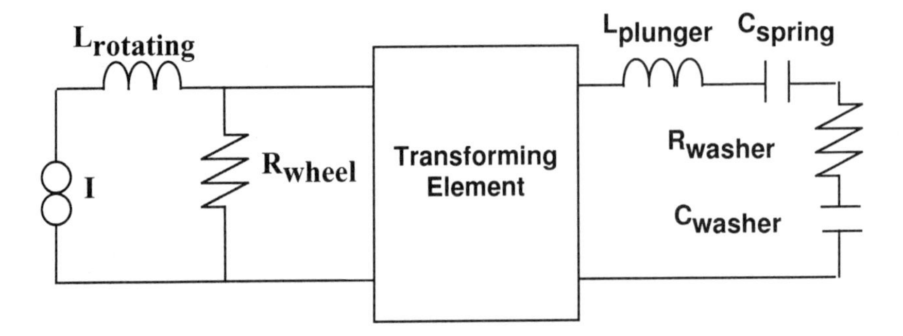

FIGURE 2.13. Circuit model of the nail gun.

old displacement below which they are inoperable. The motor is an ideal current source. The mass of the rotating piece is an inductor. The resistance associated with the wheel pushing on the plunger is associated with the velocity of the rotating part, so it is modeled with a resistor in series with the rotating part inductor. As in the case of the washer, the wheel is not in contact with the plunger at all times. Hence the friction resistance is also nonlinear. The conversion element for rotational to translational energy separates the translating elements from the rotating elements. The resulting electrical model is shown in Fig. 2.13, where we have simply used a box labeled "Transforming Element" to show that element of the model.

As with all models of real physical systems, there are a number of alternative models that could be developed for the nail gun. The weakest parts of the model shown here are the treatment of the washer, and the transforming element. In addition, we could easily make the model more sophisticated if the questions we are using the model to answer justify the added complexity.

2.4 Bond Graph Models

If electromechanical transducers were all as simple as harmonic oscillators, then the electrical circuit modeling formalism would be appropriate for them. Unfortunately, real systems, certainly including all electromechanical transducers, tend to be much more sophisticated than the simple harmonic oscillator, and the circuit formalism presents modeling problems which can be overcome by using a different approach. In this and the following sections of this chapter we describe the bond graph modeling formalism and its advantages (for transducers) over electrical circuit models.

Bond graph models were first developed in 1960 by Paynter [14] and have been used extensively since the early 1970s. The premise upon which bond graph models are based is simply that we may develop consistent system models by examining the flow of energy. In this sense they are similar to circuit models. However, as we will show, bond graph models are better suited for multiple energy domain systems and systems, in which there are multiport elements which store or dissipate energy. Because electromechanical transducers involve a minimum of two energy domains and at least one element with multiple ports, bond graph models provide a better platform for system modeling than do traditional circuit models. Hence, in this text, we will use bond graph modeling approaches extensively.

Some of the differences between bond graph models and circuit models are the results of historical bias; some are significant technical differences. The differences which are important in the modeling of transducers are discussed here.

A limitation of pure circuit models that results from historical bias is that they have been assumed traditionally to have elements described by linear constitutive relations. This restriction limits the models to linear systems, or systems which are weakly nonlinear and may be approximated by linear models valid in restricted regimes. By contrast, bond graph models are routinely applied to systems with nonlinear elements and do not suffer from the same historically imposed bias.

To understand the significant technical differences between bond graph models and pure circuit models, it is necessary to consider the general goals of system modeling: the ability to represent *all* physically realizable systems, and the insistence that *all* possible models should correspond to a physically realizable system. These two goals are almost mutually exclusive—the first arguing for generality and the second for restriction. No modeling technique is capable of meeting both goals, although some approaches are more successful than others. The modeling formalisms best able to meet both goals are those in which the only limitations imposed on models are explicit statements of fundamental physical restrictions, both conservation laws and causal relations. Any model consistent with these restrictions is permitted. Conventional circuit models explicitly show conservation laws but not causal relations. The major effect of this absense of causal information is the complication of computational procedures during model analysis. As described by Paynter [14] and Karnopp et al. [4], bond graph models which include causal descriptions serve to identify immediately the system order, state variables, and form needed for constitutive laws for nonlinear elements. This permits direct algorithmic determination of the system state equations.

A more serious shortcoming of pure circuit models is that they restrict models further than required by conservation laws and causal relations. In particular, pure circuit models do not permit the use of essential multiport energy storing or energy dissipating elements. Instead, essential multiport elements are modeled using artificial constructions of 1-port elements. The net result is a model which is a less accurate representation of reality, and which provides little physical insight, thus defeating the purpose of having a physical model. Since transducers necessarily involve the conversion of energy from one form to another, usually with accompanying dissipation and energy storage, this shortcoming of pure circuit models is quite a significant stumbling block.

In bond graph models a system port is represented by a line and may be augmented by listing the power variables associated with the energy at that port. If the ports of two subsystems are directly connected, as shown in Fig. 2.14, then the subsystems are said to be bonded. Bonds are assumed to be unable to store or dis-

A —— B

FIGURE 2.14. A system bond between subsystems A and B.

sipate energy so the energy leaving subsystem A must be received instantaneously by subsystem B and vice versa.

Just as in electrical circuits, the major components of bond graph models are ideal lumped elements and junctions between them. There are five ideal 1-port elements corresponding precisely to those in the electrical circuits: ideal effort source E, ideal flow source F, ideal capacitance C, ideal inductance I, and ideal resistance R. These elements are shown in Fig. 2.15, where the half-arrows indicate the direction of positive power flow. Note that the sources must show power flowing from them to the elements to which they are attached since they supply power to the system. The complementary structure indicates sinks. The resistance must show power flowing into it, since it represents energy taken from the system. The power flow half-arrows for the energy storing elements could be reversed without introducing any technical difficulties, but the convention shown, in which energy flow is positive if going into the element, seems to be the one most easily interpreted.

The junctions between elements correspond to the parallel and series electrical connections. The common effort (electrical parallel) junction is noted with a zero, and the common flow (electrical series) junction is noted with a one. Figure 2.16 shows examples of these junctions. In bond graph models we insist that junctions are themselves nonenergic, i.e., they can neither store nor dissipate energy. Hence, the input power at any junction must be instantaneously balanced by the output power at that junction. For a 0-junction, since every port has the same effort, this leads to a flow balance equation similar to the generalized Kirchhoff's current law. For the 1-junction, where every port has the same flow, the power balance leads to an effort equation corresponding to the generalized Kirchhoff voltage law. For the junction examples shown in Fig. 2.16 we get the following expressions:

$$f_1 = f_2 + f_3, \tag{18}$$

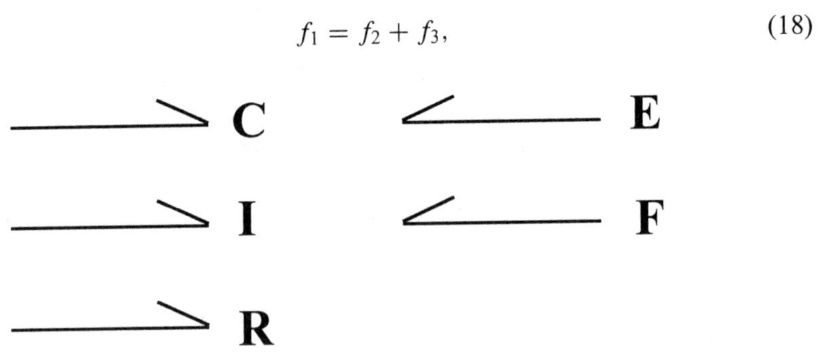

FIGURE 2.15. Ideal 1-port elements.

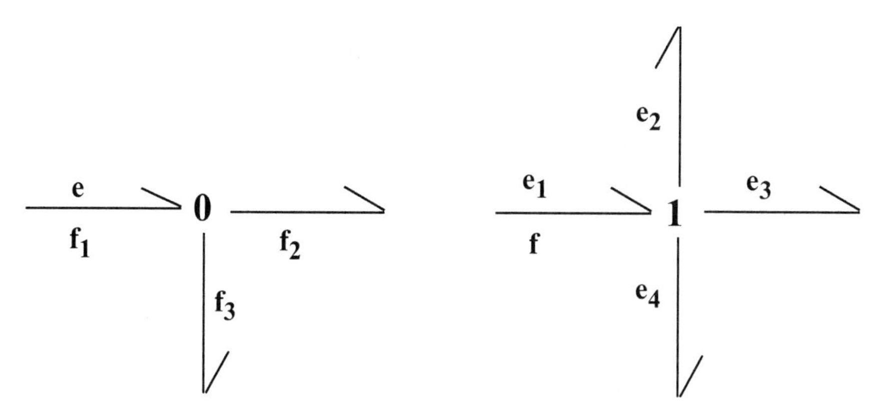

FIGURE 2.16. Ideal junctions; (a) common effort and (b) common flow.

for the 0-junction; and

$$e_1 = e_2 + e_3 + e_4 \tag{19}$$

for the 1-junction.

At this point there is a unique one-to-one mapping between elementary electrical circuits and elementary bond graph models. For example, the circuit models of the mass–spring–damper system given in Figs. 2.7 and 2.9 are shown in Figs. 2.17 and 2.18, respectively. Although it is certainly desirable that such a one-to-one correspondence can be made, the existence of this mapping is not sufficient motivation to justify using bond graphs rather than the more familiar circuit graphs. We begin to introduce the advantages of bond graph models by considering causal relations.

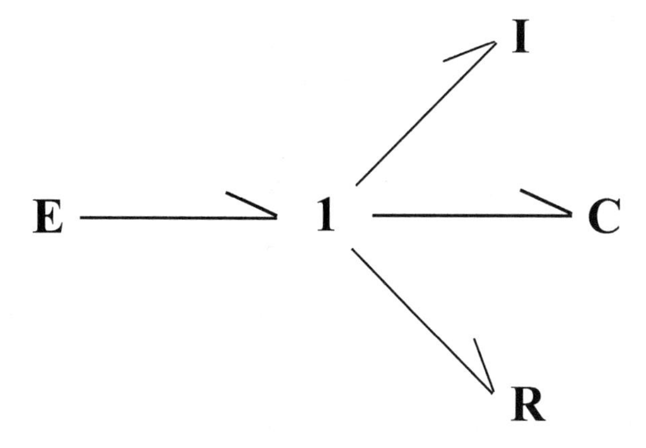

FIGURE 2.17. Bond graph model of a simple harmonic oscillator.

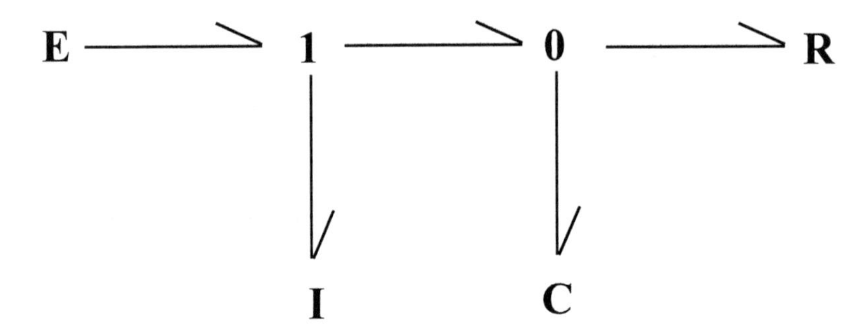

FIGURE 2.18. Bond graph model of the mechanical system shown in Fig. 2.9.

Consider two subsystems, A and B, which are bonded. The bond may be represented by a duplex pair of signals or oppositely directed branches. This signal duplex pair permits subsystem B to declare its presence in response to an input from subsystem A, and similarly allows subsystem A to respond with an output when supplied with an input from subsystem B. This existence of a signal duplex pair is one of the basic tenets of graph theory, and may be used to form the basis for causal arguments. It is convenient to imagine that the signal duplex pair is made up of the two port variables, i.e., that each signal indicates the value of one of the two port variables and from whence it has come. Thus, for a given bond, one of the subsystems will specify the effort and the other must be responsible for the flow. In bond graph notation, we indicate the signal direction by drawing a line perpendicular to the port line at the end of the bond toward which the effort signal is directed. The flow signal, then, is directed in the opposite direction. Note that the direction of power flow and signal flow are completely independent concepts.

Some of the primitive bond graph elements may be discussed easily in terms of signal flow and its direction. For example, consider the sources. Clearly an ideal flow source or sink must specify flow to the rest of the system, and an ideal effort source or sink must specify effort to the rest of the system. No other causal structure is permissible. Figure 2.19 shows the acceptable causal structures.

The causal structure of junctions is also easily discussed. At a common effort junction all ports have the same effort. Hence, exactly one port may specify the effort to that junction or a contradiction would be possible. The causal structure of a 0-junction is shown in Fig. 2.20, where the leftmost port is specifying the

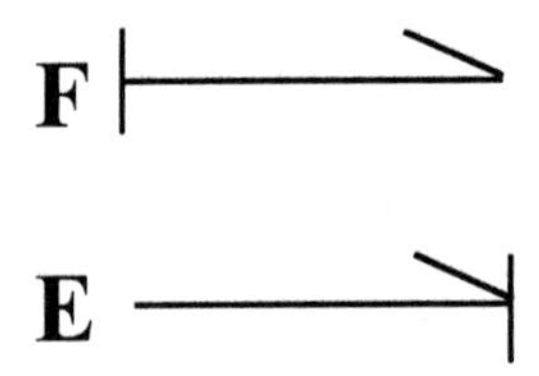

FIGURE 2.19. Causal structure for ideal sources.

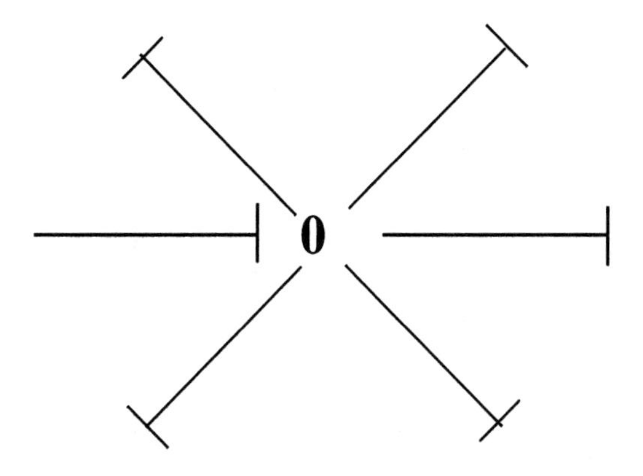

FIGURE 2.20. Causal structure for common effort junction.

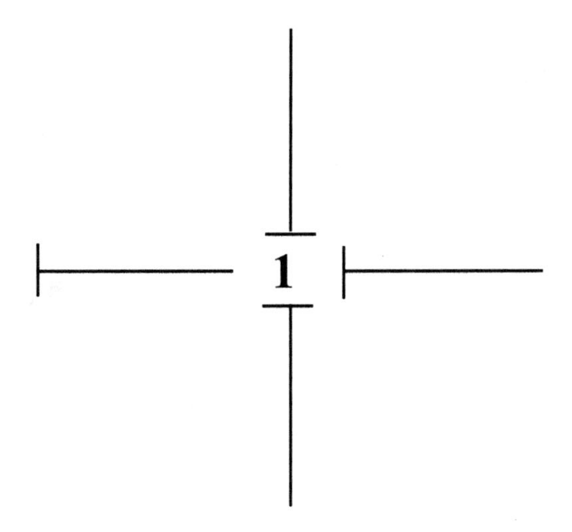

FIGURE 2.21. Causal structure for common flow junction.

effort to the junction so the remaining ports must specify a flow to the junction. Similarly, in a common flow junction exactly one port may specify the flow to the junction. Hence, the causal structure is as shown in Fig. 2.21, in which the leftmost port specifies the flow so the remaining ports must specify effort to the junction. Note that no power flow arrows have been shown in these figures since the concept of causality is independent of the direction of power flow.

Where causal structure becomes very useful, and somewhat more difficult to motivate, is in the energy storing 1-port elements. Let us begin with an ideal capacitance in which we have the constitutive relation $e = g(q)$. There are two possibilities: Either the effort is specified to the element by the remainder of the system, or the flow is specified to the element. Suppose first that the effort is specified. Then from the constitutive equation the displacement, q, is known precisely as long as g is an invertible function. Since the flow is simply the time derivative of q, f is known precisely. Hence, the power into the element is known at each instant in time. This causal structure is known as derivative causality and does not represent independent storage of energy. Now suppose that instead the system specifies the flow into the capacitance (e.g., the charging current of an electric capacitor). We may then integrate the flow to get the displacement, but q will not be known precisely until an initial condition is given. Hence, the power flowing into the element is not known precisely until an extra piece of information is given. This causal structure is referred to as integral causality and represents *independent* storage of energy by the capacitance. Each independent energy storage element increases the system order by one, since it represents one more initial condition that must be given to completely determine the system behavior.

Next we may consider the inertance element described by the constitutive equation $f = g(p)$. Since the inertance is the dual of the capacitance, we would expect that specifying the effort to the element will result in integral causality, and specifying the flow derivative causality. Indeed, if the effort is specified, we must temporally integrate it to obtain the momentum, but again an initial condition must be specified. Hence this causal structure represents independent energy storage. Conversely, if the flow were specified for an inertance, then the power into this element would be completely determined by the remainder of the system, and the element would not represent independent energy storage. Figures 2.22 and 2.23 show the independent and dependent energy storage configurations for the capacitance and inertance elements.

The remaining 1-port element is the resistance which is described by $e = g(f)$ or $f = g(e)$. In this element it does not matter which of the two variables is specified by the rest of the system. The power into the element is always known precisely as long as the functional relation is invertible. Hence a resistance element will never increase the system order. Note that for nonlinear resistances, the functional relation between the effort and flow variables may well not be invertible. In such cases, only one causal structure is permitted, but the system order is unaffected.

Using the concepts outlined here we may develop causally consistent models of elementary physical systems. We assign the causal strokes in a bond graph model

FIGURE 2.22. Independent causal structure for ideal capacitance and inertance elements.

FIGURE 2.23. Dependent causal structure for ideal capacitance and inertance elements.

using a logical progression from those elements which have only one causally consistent choice (e.g., sources), through those for which the choice reflects the system order (capacitances and inertances), to those elements whose causality does not reflect information about system order (e.g., resistances). As each causal stroke is assigned, it is keenly important to consider whether there are repercussions which immediately determine the causal strokes on other bonds.

The first step in causal stroke assignment, assignment to the sources and sinks, is always straightforward as these ideal elements have only one possible causal configuration. Of course, as source causality is assigned it is necessary to check that no other causal strokes in the model are determined. For example, if an effort source is connected to a common effort junction or a flow source to a common flow junction, then the causality associated with every other bond to the junction is established.

The second step in causal stroke assignment deals with the energy storage elements. The aim in this step is to assign as many of the ideal capacitances and inertances integral (independent) causality as is possible without a causal violation. Thus, we begin with any capacitance or inertance whose causality is not yet established, and assign independent causality. The repercussions of this assignment are propagated, should they exist. Then the process is repeated, moving on to any other energy storage element whose causality is not yet established, until all of the ideal capacitances and inertances have set causality. In general, there is an obvious explanation when an energy storage element has derivative (dependent) causality. For example, consider two flywheels which are rigidly linked together. They must have the same angular velocity, so the bond graph model would show both inertances attached to a 1-junction. As soon as one of the inertances is assigned integral causality, the other must be derivative, or a causal conflict arises at the junction. But this is not a surprise because the linked flywheels obviously could be modeled as a single flywheel with the same angular velocity. Once this step is complete, the system order is determined as the number of energy storage elements with independent causality.

Frequently, once the causalities of the energy storage elements are determined, the causal strokes for the entire system are set. If this is not the case, then the last step is to progress through the energy dissipation elements assigning a consistent set of causal strokes. The choice at this point has an impact in determination of the system equations, but not on the system order.

As an example, the bond graph model of a simple harmonic oscillator is shown in Fig. 2.24. Note that both the inertance and the capacitance are independent energy storage elements, leading to the conclusion that the system is second order. If we modify the system to put the spring and damper end-to-end as in Fig. 2.9, then the bond graph model is as shown in Fig. 2.25. Note that the system is still second order. In both of these cases, the assignment of independent causality to the energy storage elements determines the consistent causal structure for the resistance.

The mass, spring, damper examples above point out one of the advantages of causal assignment in a model, namely, the ability to determine the system order

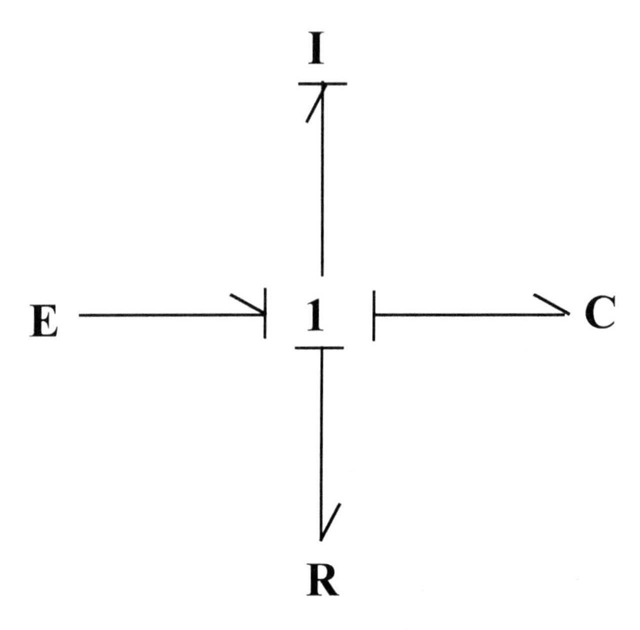

FIGURE 2.24. Bond graph model, including causality, for a simple harmonic oscillator.

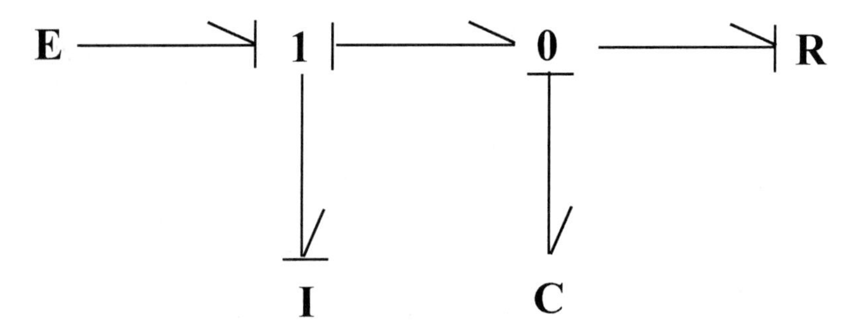

FIGURE 2.25. Bond graph model, including causality, for the simple mechanical system shown in Fig. 2.8.

without additional computation. This permits us to say much about system behavior prior to model evaluation. For example, a first-order system never exhibits resonant behavior. In addition, causal assignment provides a roadmap for derivation of the equations needed to describe the system behavior. Knowledge of the causal structure permits optimal equation derivation which can result in quite a savings in time for high-order systems.

Before leaving this description of causal structure and its relation to the system order, we should call attention to a frequently occurring pathological model in which the number of independent elements is not an indication of the system order. This typically arises when there are duplicates of a particular type of element. For instance, consider the simple harmonic oscillator whose bond graph model is shown in Fig. 2.24. If the spring were replaced by two springs connected mechanically in parallel, then the effect would be to add another capacitance element to the 1-junction. It would have the same causal assignment as the capacitance element in Fig. 2.24, suggesting that the system now has three independent energy storage elements. While it is true that there is an additional piece of information needed to fully define the system state, namely the initial condition of the new spring, it is not true that the system order has increased from second to third. This is because the two springs in parallel could be lumped into one effective spring. Hence the harmonic oscillator with two parallel springs would still be a second-order system. In general, when there are multiple occurrances of a particular type of element at a single junction, we must carefully examine the situation in order to determine system order. If the multiple elements could be lumped into a single effective element, then the system order is unaffected by the number of elements of that type present at that particular junction.

2.5 Nonenergic 2-Port Elements

In the previous sections of this chapter we have discussed only those systems which can be modeled using ideal 1-port elements and junctions. The systems which can be easily described using such elementary models do not include transducers because the models do not permit the conversion of energy from one form to another. To describe energy conversion it is necessary to use elements with two or more ports. In this section, we expand our elements to include 2-port elements which neither store nor dissipate energy. Section 2.6 addresses multiport elements which are associated with energy storage and dissipation. Thermal system elements, which are examples of multiport energic elements, are discussed in detail in Section 2.6.

Ideal, nonenergic 2-port elements are those elements which transform one set of power conjugate variables into another set without loss or gain of energy. In general such elements are described by the following constitutive relations:

$$\begin{aligned} e_2 &= g(e_1, f_1), \\ f_2 &= h(e_1, f_1), \end{aligned} \tag{20}$$

where the subscripts 1 and 2 simply label the two ports. If the relations are linear, then these equations reduce to

$$\begin{aligned} e_2 &= ae_1 + bf_1, \\ f_2 &= ce_1 + df_1, \end{aligned} \tag{21}$$

where a, b, c, and d are coefficients. The requirement that the elements be nonenergic (neither energy storing nor energy dissipating) imposes constraints on the coefficients in (21). This may be seen by multiplying e_2 and f_2:

$$\begin{aligned} e_2 f_2 = e_1 f_1 &= (ae_1 + bf_1)(ce_1 + df_1) \\ &= ace_1^2 + bdf_1^2 + (ad + bc)e_1 f_1. \end{aligned} \tag{22}$$

For these equalities to hold for arbitrary effort and flow the following conditions must bc mct:

$$\begin{aligned} ac &= 0, \\ bd &= 0, \\ ad + bc &= 1. \end{aligned} \tag{23}$$

Now consider the possible solutions of (23). From the first of the three equations either $a = 0$ or $c = 0$. Let us assume that $c = 0$. Then we must have $b = 0$ for the second equation to be true without causing a contradiction in the third. Further, we have $a = 1/d$ which yields the following:

$$\begin{aligned} e_2 &= e_1 a, \\ f_2 &= f_1/a. \end{aligned} \tag{24}$$

These equations relate like variables and are generally called the transformer equations. Typical examples of transformers include electrical transformers and gears. The rotational to translational transformation in the nail gun shown in Fig. 2.4 is also modeled with a transformer, as there is a correspondence between the rotational velocity and the translational velocity of the plunger.

Had we instead sought to solve (23) beginning with $a = 0$, we would insist that $d = 0$ and $b = 1/c$ which yields the gyrator relations:

$$\begin{aligned} e_2 &= f_1 b \\ f_2 &= e_1/b. \end{aligned} \tag{25}$$

Note that in these relations we relate an effort to a flow and vice versa. Examples of gyrators include the conversion from electrical to magnetic variables (to be discussed in Chapter 4) and gyroscopes.

Both electrical models and bond graph models recognize ideal 2-port nonenergic elements. Figure 2.26 shows the electrical notation that we will use, and Fig. 2.27 shows the bond graph representation. We may consider the causal structure of these 2-ports by considering the transformer and gyrator equations. In a transformer, like variables are related by the transformer modulus (a in (24)). If one port provides effort variable information to the transformer, then the effort at the other port is known as a consequence. Similarly, if one port specifies the flow, then the flow variable at the other port is known. Thus, the effort at both ports may not be specified; nor may the flow. In other words, of the four possible causal structures, only two are permissible. These are shown in Fig. 2.28. Similar arguments lead to the rejection of two of the four possible causal structures for gyrators. The two acceptable causal structures are shown in Fig. 2.29. Not surprisingly, these two structures are the two that were rejected for the transformer.

FIGURE 2.26. Electrical notation of (a) a transformer and (b) a gyrator.

FIGURE 2.27. Bond graph notation of (a) a transformer and (b) a gyrator.

FIGURE 2.28. Acceptable causal structures for a transformer.

2.6 Multiport Energic Elements

Transformers and gyrators are the only 2-port elements permitted in pure circuit models. This restriction imposes serious problems for transducer modeling since most transducers convert energy in a process involving energy storage or dissipation. Bond graph models permit description of the conversion of energy from one form to another with explicit energy storage or dissipation in the process. The description involves *energic* 2-port elements: 2-port capacitance, 2-port inertance, mixed 2-port inertance/capacitance, and 2-port resistance. These energic 2-ports describe the storage of potential energy, the storage of kinetic energy, and the irreversible conversion of energy to a form in which it cannot be used for work. They are 2-port elements because the energy stored or converted irreversibly is a function, not of one, but of two independent variables. For example, consider the cantilever beam shown in Fig. 2.30. The potential energy stored in the beam depends on both the linear displacement, x, and the bending angle, θ. Rather than model the cantilever as two separate 1-port capacitances, which assumes that there is no coupling of the bending and translational energies, we would model the beam using a single 2-port capacitance as shown in Fig. 2.31. Two constitutive equations relate the four variables F, τ, x, and θ in the following form:

$$F = F(x, \theta),$$

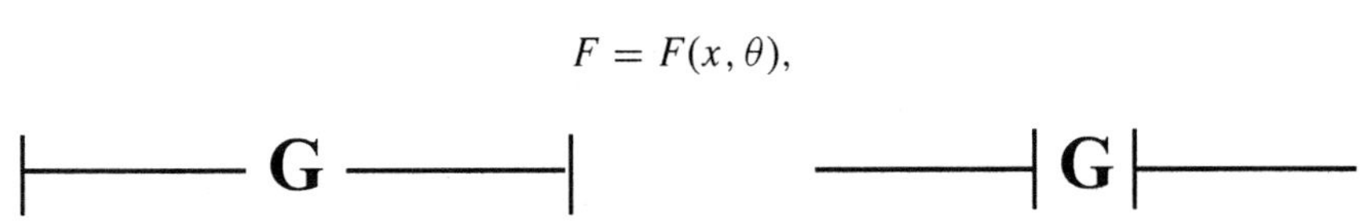

FIGURE 2.29. Acceptable causal structures for a gyrator.

FIGURE 2.30. Cantilever beam.

$$\overset{F}{\underset{V}{\text{———}}}\ C\ \overset{\tau}{\underset{\Omega}{\text{———}}}$$

FIGURE 2.31. Bond graph representation of the cantilever beam.

$$\tau = \tau(x, \theta). \tag{26}$$

The fact that we can model multiport energic elements is a key advantage of bond graphs in transducer modeling because it affords us the ability to develop more accurate models of transducers. The use of multiport elements also means that the thermal behavior of a system can be handled in a consistent and logical fashion. The difficulty with multiport elements in models is not computational complexity, but determining consistent constitutive equations. We will demonstrate with many transducer examples throughout the text, precisely how this determination can be made. Here we focus on thermal systems in order to define ideal thermal elements.

As previously mentioned, purely thermal systems exhibit potential energy storage and energy "dissipation" (if dissipation is taken to mean irreversible conversion of energy to an unusable form). The description of these two energetic mechanisms requires introducing essential 2-port elements.

First, consider the thermal potential energy storage of a medium. This is commonly described in terms of the specific internal energy of a substance, u. We are free to choose which two thermodynamic variables we will view as the independent variables. Let us choose the specific entropy, s, and the specific volume, v. Then the internal energy may be written as $u(s, v)$ and its total differential is

$$du = \frac{\partial u}{\partial s} ds + \frac{\partial u}{\partial v} dv. \tag{27}$$

We identify $\partial u/\partial s$ as the temperature T and $\partial u/\partial v$ as $-P$, where P is pressure. Hence we can rewrite Eq. (27) in the more recognizable form of

$$du = T ds - P dv. \tag{28}$$

Consider each term in (28). The form of the term $T\,ds$ is that of an effort variable multiplying a change in a displacement variable. In other words, it looks in form

FIGURE 2.32. Bond graph representation of thermal energy storage.

like the expression we would obtain for a small change in the energy stored in a spring. Thus we identify the term as being associated with a capacitance. The second term is similarly an effort times a change in a displacement variable. Thus it is also a capacitance term. Taken together, since there are two independent variables, the internal energy stored in a substance can thus be represented by a 2-port capacitance. The bond graph model for this is shown in Fig. 2.32. Given the signs in Eq. (28), one of the power flow arrows must point into the element symbol and the other out of the element symbol. Note also that since a capacitance represents energy storage, and the capacitance shown in Fig. 2.32 has two ports, the causality of the 2-port capacitance may add zero, one, or two orders to the system depending on whether none, one, or both of the ports are independent causality.

Now consider energy dissipation due to heat transfer. Suppose we have an insulated pipe and a thin membrane separating two media at different temperatures, as shown in Fig. 2.33. Then a certain amount of heat is transferred from one medium to the other. This rate of heat transfer yields the relation

$$T_1 \frac{dS_1}{dt} = T_2 \frac{dS_2}{dt}, \tag{29}$$

where S is the total entropy and the equation simply states that the heat leaving system 1 must enter system 2. This equation can be thought of as the defining constitutive equation for a 2-port resistance, such as shown in Fig. 2.34. Note that for this purely thermal resistance, the power flow arrows must both point in the same direction in order to define a consistent positive power flow direction.

A complete example of a system which may be modeled using multiport elements and multiple energy domains is shown in Fig. 2.35. The system considered here is a piston which moves in a compressible fluid. A thin membrane separates the pipe into two chambers, one of which has constant volume due to the presense of the membrane. The bond graph model, shown in Fig. 2.36, uses a transformer to convert from an applied force to a pressure on the fluid in chamber 2. The fluid

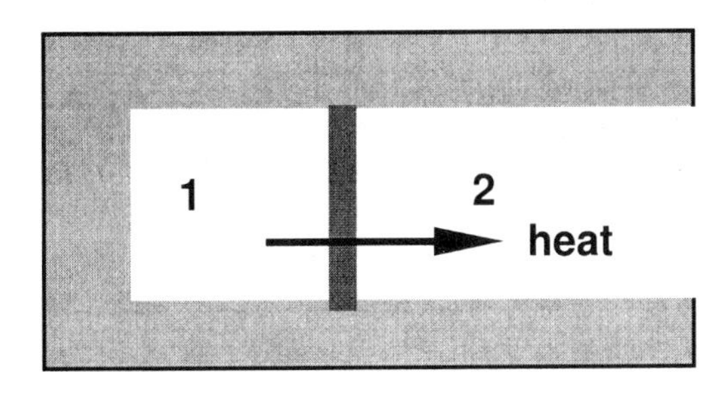

FIGURE 2.33. Heat transfer from one medium to another.

FIGURE 2.34. Bond graph representation of thermal resistance.

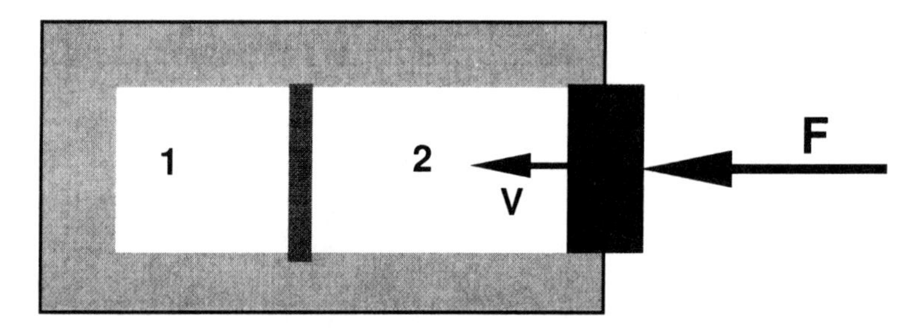

FIGURE 2.35. Thermal system with both heat transfer and energy storage.

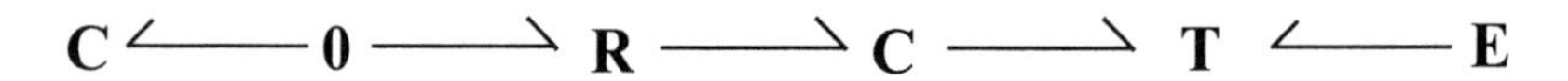

FIGURE 2.36. Bond graph representation of the thermal system.

in chamber 2 has internal energy which is modeled by a 2-port capacitance. The heat transfer from chamber 2 to chamber 1 is modeled using a 2-port resistance, and chamber 1 is modeled using a 1-port capacitance (since the volume is fixed). The 0-junction permits us to be consistent in defining positive energy as flowing into energy storage elements. The resulting model could be analyzed in a fashion identical to that used for systems modeled with 1-port elements only. A structured procedure for analysis of any model is described in the next section.

Another example of a system whose model includes a 2-port energic element is a piezoelectric accelerometer. An example of a piezoelectric compression accelerometer, sold by Dytran Instruments, Inc. [15], is shown in Fig. 2.37. In normal operation, the accelerometer is attached to a vibrating object. Acceleration of the base causes a compressional strain of plates of piezoelectric material, which gives rise to a voltage in response.

The accelerometer shown in Fig. 2.37 is a quite simple transducer. The inertial mass is modeled as a 1-port inertance. The springs attached to it may be modeled as a single lumped 1-port capacitance. The piezoelectric plates produce an output voltage that depends on strain and charge, so they may be modeled as a 2-port capacitance, with one mechanical port and one electrical port. The acceleration input to the base is usually modeled as an ideal flow source. The main dissipation is that which appears in connecting electronics, which we choose not to include here. Using these elements, the bond graph model which results is shown in Fig. 2.38. The causality shown in Fig. 2.38 indicates that the accelerometer, as modeled, is

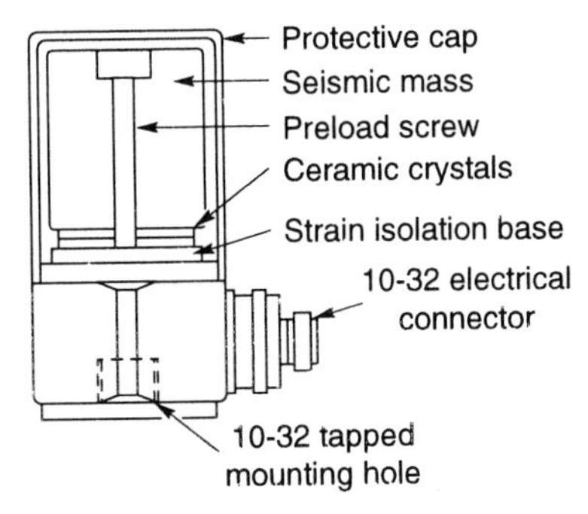

FIGURE 2.37. Piezoelectric compression accelerometer sold by Dytran Instruments, Inc. [15].

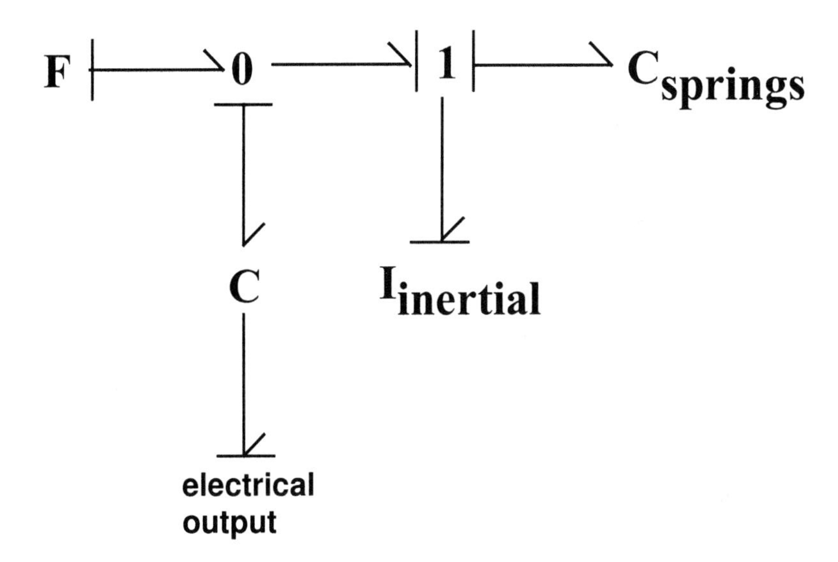

FIGURE 2.38. Bond graph representation of the piezoelectric compression accelerometer.

fourth order. Given that the resistance elements have been neglected, this means that the first two resonances of the accelerometer system can be predicted from this simple bond graph model.

Figure 2.38 is typical of transducer models in that it includes a 2-port energic element. It is this element that has no electrical analog, and that prompts our use of bond graphs throughout this book. As will be demonstrated in Chapter 5, which provides detailed information on piezoelectricity, if we try to develop a circuit model of the piezoelectric conversion of mechanical energy into electrical energy, then we are led to models which are demonstrably wrong, and which provide little or no physical insight.

2.7 Model Analysis

No discussion of modeling would be complete without mentioning how we can analyze the models developed. Generally the purpose of building models of physical systems is the desire to obtain a time or frequency response of the system. Such system characteristics can be determined by deriving the equations which describe the system performance, and solving them in the time or frequency domain.

There are many acceptable means of deriving equations which describe a physical model of a real system. For example, the classical approach to purely mechanical systems generates coupled second-order ordinary differential equations to describe the system. Modern system dynamics approaches, which can easily be applied across energy domains, instead opt for coupled first-order state equations. The advantages of the first-order state equations are that they are generally easier to derive, and they result in slightly less loss of information. (We can always generate second-order differential equations from first-order equations, but the reverse process cannot be used to generate a unique set of state equations.) The major disadvantage is that roughly twice the number of equations is used, so computation by hand is rendered less practical.

To begin to derive state equations we define the state variables of a physical system. The state variables are the minimum set of variables needed to completely describe the state of a system at any instant in time. Each independent energy storage element contributes a single state variable. For capacitance elements, the normal choice is the displacement variable. For inertance elements we normally choose the momentum variable.

For each state variable we seek an equation of the form

$$\frac{dx_i}{dt} = g_i(u_1, u_2, \ldots, x_1, x_2, \ldots, x_n), \tag{30}$$

where x_i are the n state variables and u_j are the system inputs. We determine these equations by manipulation of two types of equations for the system: constitutive equations for each element, and junction equations which describe the connections between elements. For linear systems we may write this form as

$$\frac{d}{dt}\mathbf{x} = [A]\mathbf{x} + [B]\mathbf{u}, \tag{31}$$

where we identify $[A]$ as the system matrix, $[B]$ as the input matrix, $\mathbf{x}$ as the state vector, and $\mathbf{u}$ as the input vector.

As an example of state equation derivation, consider the mechanical simple harmonic oscillator shown in Fig. 2.7. If we assume that the elements are described by linear constitutive equations, then we have:

$$p_m = mv_m, \tag{32}$$

$$x_k = F_k/k, \tag{33}$$

$$F_b = bv_b, \tag{34}$$

where the subscripts indicate the element the variable describes (m for mass, k for spring, and b for damper). In addition to these constitutive equations we have the following junction equations:

$$v_m = v_k = v_b, \tag{35}$$

$$F = F_m + F_k + F_b. \tag{36}$$

These five equations can be used to form the two coupled first-order state equations for the state variables p_m and x_k.

$$\begin{aligned}\frac{dp_m}{dt} &= F_m = F - F_k - F_b = F - kx_k - F_b = F - kx_k - bv_b \\ &= F - kx_k - bv_m = F - kx_k - \frac{b}{m}p_m,\end{aligned} \tag{37}$$

and

$$\frac{dx_k}{dt} = v_k = v_m = \frac{1}{m}p_m. \tag{38}$$

These equations now can be solved for a given force input F to determine the time or frequency response of the system.

The method outlined above for state equation derivation works for all physical system models regardless of whether the constitutive equations are linear or nonlinear, and whether the system uses 1-port or multiport elements. The only distinction is that multiport elements require more than one constitutive equation. To be precise, an element with n ports will always be described by n constitutive relations.

A reasonable means of checking that the state equations are likely to be correct is to ask two questions. First, has each of the constitutive and junction equations been used at least once? If not, there must be some compelling reason for not needing one of the equations. For example, we should not need any equation dealing with the complementary port variable of a source, e.g., the flow associated with an effort source. Second, does each input appear in at least one state equation? If the answer here is no, then one of the inputs has no effect on the system, a highly unlikely situation.

Finding the time response of the system from the state equations is a matter of solving the equations given a complete set of initial conditions, i.e., the initial value of each of the state variables. For most systems, this solution is most easily obtained numerically on a computer. Many commercial software packages are available to perform this task.

Finding the frequency response of the system from the state equations generally involves determining the transfer functions, i.e., the ratios of a system response variable to a system input. This is done by assuming that the inputs all have the following form of excitation:

$$\mathbf{u} = \mathbf{U}e^{st}, \tag{39}$$

where s here is the Laplacian operator. A linear system responds to a particular type of excitation with the same type of response, so we assume that the state vector is given by

$$\mathbf{x} = \mathbf{X}e^{st}. \tag{40}$$

Substitution of these assumed forms into the state equations permits derivation of the transfer functions. To obtain the frequency response, we substitute $s = j\omega$ into each transfer function. Typically, we then graph the magnitude and phase of the transfer function to permit interpretation of the response as a function of frequency. For example, for the simple harmonic oscillator, the state equations would become:

$$sp_m = F - kx_k - \frac{b}{m}p_m \tag{41}$$

$$sx_k = \frac{1}{m}p_m. \tag{42}$$

Manipulating these to find x_k/F we have

$$\frac{x_k}{F} = \frac{1}{ms^2 + bs + k}. \tag{43}$$

Substituting $s = j\omega$ and plotting the phase and magnitude as a function of frequency produces the results shown in Fig. 2.39.

While this description of the analysis of system models is necessarily brief, it captures the essense of what we can do with a good model of a physical system. For further information on the analysis of physical system models, the reader is referred to [4]–[8].

2.8 References

1. D. Hrovat and W. E. Tobler, *Bond Graph Modeling of Automotive Power Trains,* J. Franklin Inst., **328**, 623 (1991).
2. J. A. M. Remmerswaal and H. B. Pacejka, *A Bond Graph Computer Model to Simulate Vacuum Cleaner Dynamics for Design Purposes,* J. Franklin Inst., **319**, 83 (1985).
3. S. Hanish, *A Treatise on Acoustic Radiation*, Vol. 2, *Acoustic Transducers* (Naval Research Lab., Washington, DC, 1989).
4. D. C. Karnopp, D. L. Margolis, and R. C. Rosenberg, *System Dynamics: A Unified Approach* (Wiley, New York, 1990).
5. C. N. Dorny, *Understanding Dynamic Systems: Approaches to Modeling, Analysis, and Design* (Prentice Hall, Englewood Cliffs, NJ, 1993).
6. K. Ogata, *System Dynamics* (Prentice Hall, Englewood Cliffs, NJ, 1992).
7. W. J. Palm, III, **Modeling, Analysis, and Control of Dynamic Systems** (Wiley, New York, 1983).
8. J. L. Shearer and B. T. Kulakowski, **Dynamic Modeling and Control of Engineering Systems** (Macmillan, New York, 1990).
9. F. A. Firestone, *A New Analogy Between Mechanical and Electrical Systems,* J. Acoust. Soc. Am., **4**, 249 (1933).

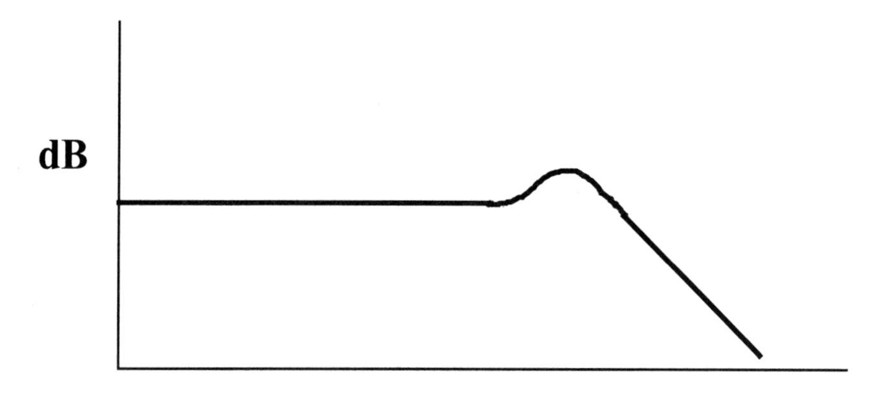

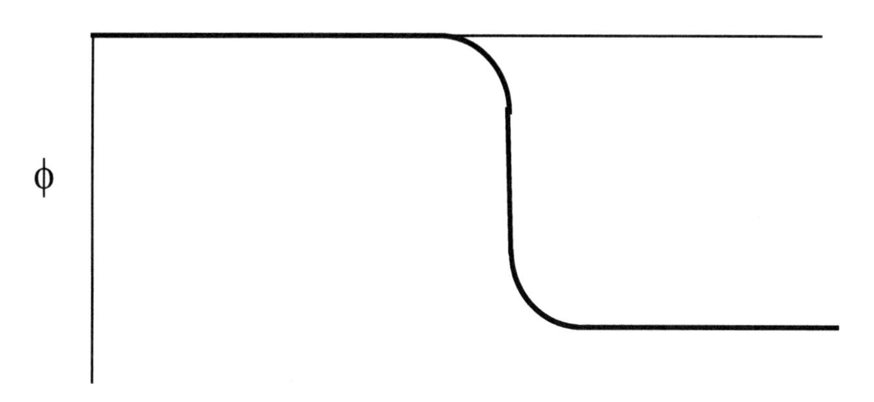

FIGURE 2.39. Frequency response of the simple harmonic oscillator: (a) magnitude and (b) phase versus frequency.

10. F. A. Firestone, *Twixt Earth and Sky with Rod and Tube; the Mobility and Classical Impedance Analogies,* J. Acoust. Soc. Am., **28**, 1117 (1956).

11. H. M. Trent, *Isomorphisms Between Oriented Linear Graphs and Lumped Physical Systems,* J. Acoust. Soc. Am., **27**, 500 (1955).

12. J. C. Maxwell, *A Dynamical Theory of the Electromagnetic Field,* Roy. Soc. Trans., *CLV* (1865); reprinted in *The Scientific Papers of James Clerk Maxwell* (Dover, New York, 1952).

13. K. T. Ulrich and S. D. Eppinger, *Product Design and Development* (McGraw-Hill, New York, 1995).

14. H. M. Paynter, **Analysis and Design of Engineering Systems**, (MIT, Cambridge, MA, 1960, 1961).

15. Dytran Instruments, Inc., 21592 Marilla St., Chatsworth, CA 91311.

Part II

Transduction Mechanisms

The introductory part of this book defined transducers and provided a common means for modeling many types of electromechanical transducers. In this part of the book, the focus shifts from the general to the specific, i.e., to the various mechanisms used in electromechanical sensors and actuators. Before beginning, a few words about the categorization scheme are in order.

As Chapter 1 explained, the definition of transducer used in this book is broader than that normally employed. In fact, it is not terribly common to find books which simultaneously consider sensors and actuators. In this book, sensors and actuators will be considered simultaneously, as they are frequently nearly identical devices which are simply operated by complementary means. The distinctions between sensors and actuators often result from nothing more than optimization of the specific numerical choice of parameters for operation. From a fundamental view, sensors and actuators may well be based on identical functional principles, making it advantageous to consider them together.

Traditionally transducers have been classed in three ways. First we may categorize and discuss them by their use. This results in classes such as microphones, accelerometers, and displacement measurement. This categorization is probably the most common found in textbooks on sensors. However, this approach is not helpful for a physical understanding of transduction processes because the fundamental principles may be quite similar in devices with very different applications. Further, by approaching a transduction discussion from the application viewpoint, we necessarily limit it to a survey of commercially available types of transducers. This stifles creativity and can leave questions about why certain types of transducers are not useful for a particular application.

The second method of classing transducers has been by basic energy type. Capacitive devices are those which rely on energy in an electric field in some manner. Inductive devices rely on magnetic fields. Resistive devices are those which clearly work through the irreversible conversion of energy. Although this classification scheme permits discussion of mechanisms of transduction rather than applications, the categories are so broad as to make comparison of transducers difficult. In other words, two capacitive devices might have nothing more in common than the fact that they rely on changes in an electric field.

The third alternative, and the one used in this book, is to categorize transducers by the material or structural behavior which leads to transduction. This effectively introduces subclasses into the capacitive, inductive, and resistive categories defined above. As we will discuss, there are many such subclasses. Using this categorization, it is possible to see similarities between various transducers, while understanding the details of the transduction mechanism associated with each class.

To demonstrate this last point more fully, consider an electromechanical transducer which works through changes in the capacitance of a parallel plate capacitor. For a parallel plate capacitor there are three variables which determine the capacitance: the area of the plates, the separation between the plates, and the electric permittivity of the material between the plates. A designer sees each of these three parameters as an opportunity for transduction since an electromechanical transducer can be constructed by causing any one of these three parameters to vary in response to a mechanical signal. Indeed, commercially available sensors exist which work through dynamic variation of each one of the three capacitance parameters. For instance, in many common microphones, the sound pressure causes the distance between the plates of a parallel plate capacitor to vary. In some angular displacement sensors, plates are rotated so that there is a change in the overlapping area of the plates and thus in the capacitance of the system. In some humidity sensors, the material between the plates changes its permittivity in response to changing moisture content in the surrounding air. In all three of these sensors, the electromechanical coupling is accomplished by changing the capacitance of a parallel plate capacitor. However, this basic similarity between the microphone, angular position sensor, and humidity sensor would be missed if we were to discuss them in terms of what is being measured. Thus, we choose in this book to use a categorization which emphasizes the similarities in various transducers and encourages a design perspective.

3

Transduction Based on Changes in the Energy Stored in an Electric Field

We begin our transduction mechanism discussion with transducers which are based on changes in the energy stored in an electric field. This class of mechanisms is capacitive. Some of these transducers accomplish energy conversion in a manner that is structurally dependent; some in a manner that is material dependent.

3.1 Electric Fields and Forces

To understand transduction based on changes in the energy stored in an electric field, it is necessary, of course, to understand electric fields and forces. This section is intended as a quick review of electric fields. For further information the reader should see any classical text on electricity and magnetism. A few of the better known texts are [1]–[5].

Suppose we have a charge q_1 fixed in space and we wish to know the electric field it establishes. Then we move an exploring charge toward the fixed charge and measure the force needed to keep the two charges at a given separation. The result of this experiment is known as Coulomb's law:

$$\mathbf{F} = \frac{q_1 q_2}{4\pi r^2 \epsilon} \hat{\mathbf{r}}, \tag{1}$$

where $\hat{\mathbf{r}}$ is the unit vector in the direction connecting the two charges and ϵ is the electric permittivity of the medium. The permittivity of free space is given by $\epsilon_0 = 8.85 \times 10^{-12}$ Farads/meter (F/m). The permittivities of some common materials are given in Table 3.1.

TABLE 3.1. Permittivities of some common materials.

Material	ϵ ($\times 10^{-12}$ F/m)
Air	8.85
Alcohol	228.33
Beeswax	25.67
Benzene	20.18
Carbon Dioxide	8.85
Diamond	48.68
Glass	32.75–88.50
Nylon	30.98
Paper	14.16–30.98
Polyamide	44.25
Polyethylene	20.36
Polystyrene	26.55
Polyvinyl Chloride	40.27
Rubber	22.13–309.75
Salt	53.10
Teflon	17.70
Water	694.72

The field established by q_1 should not depend on the nature of the exploring charge. Therefore we define the electric field as the force per charge, and it is given by

$$\mathbf{E} = \frac{q_1}{4\pi r^2 \epsilon}\hat{\mathbf{r}}. \tag{2}$$

The SI units of electric field are Volts/meter (V/m), which are the same as Newtons/Coulomb (N/C).

We may also discuss the ability of the electric field established by this charge to do work, i.e., the potential energy. In mechanical systems the potential energy is given by

$$\mathcal{E}_{pot} = \int \mathbf{F} \cdot \mathbf{dl}, \tag{3}$$

where l is the distance moved. Using this definition and the definition of the electric field as a force per charge, we can define a potential energy per unit charge as

$$e = \int \mathbf{E} \cdot \mathbf{dl}. \tag{4}$$

This potential energy per unit charge is referred to as the electrical potential, and the SI units are Volts. In this book we will often refer to this potential as voltage.

The above definitions suffice for a single, point charge. Now we consider a system of charges. Suppose we have a volume in space bounded by a surface S as shown in Fig. 3.1. The field established by the charge q interior to the surface is directed radially outward. At any point on the surface we may find the normal component of the electric field from

$$\mathbf{E} \cdot \mathbf{n} = \frac{q \cos(\theta)}{4\pi \epsilon r^2}, \tag{5}$$

where $\mathbf{n}$ is a unit vector normal to the surface. If we define a small segment of the surface as da, then we may also write that

$$\cos(\theta)\, da = r^2\, d\Omega, \tag{6}$$

where $d\Omega$ is the solid angle subtended by da as observed at the point occupied by the charge. If we integrate over the entire closed surface, we are integrating over 4π solid angles. Hence we get the result known as Gauss's law:

$$\int_S \mathbf{E} \cdot \mathbf{da} = \frac{q}{\epsilon}. \tag{7}$$

Equation (7) is the integral form of Gauss's law which states that the flux of the electric field through a closed surface is proportional to the total charge enclosed within the surface.

So far the only field parameter we have introduced is the electric field. Maxwell's equations in a vacuum use only two field parameters: the electric field $\mathbf{E}$ and the magnetic induction (or magnetic flux density field) $\mathbf{B}$. Because there is no such thing as a magnetic monopole, two of the four Maxwell equations are homogeneous. The other two are inhomogeneous equations. Maxwell's equations in a medium are given in two different forms: they may be expressed in terms of $\mathbf{E}$ and

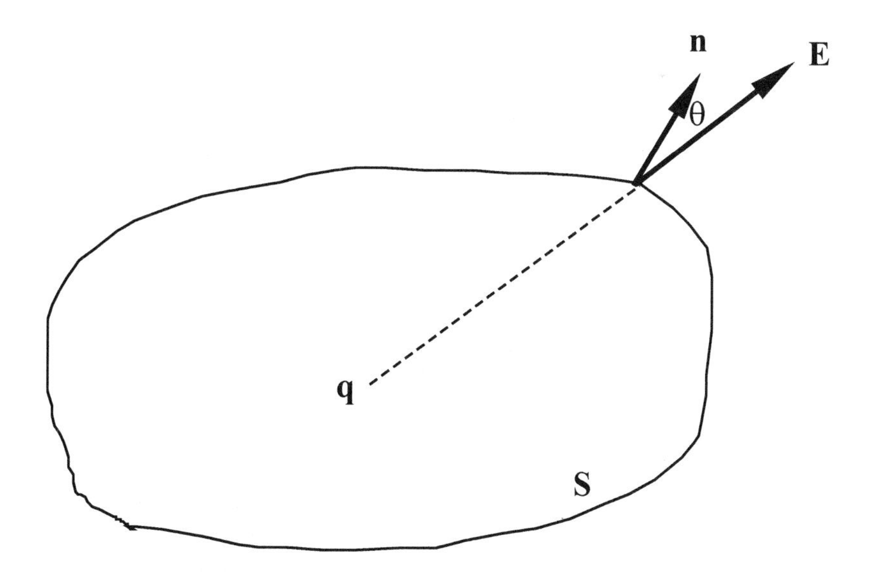

FIGURE 3.1. Arbitrary volume in space containing a charge.

B (this is the choice of most physicists), or two additional field parameters may be introduced for clarity, **D** and **H** (this is the choice of most engineers). **D** is referred to as the electric displacement (or charge density field) and **H** is the magnetic field. The main reason for adding two additional field variables is to permit ease in discussing macroscopic behavior. In this book, we choose to use all four field parameters as descriptors. Since **E** and **B** appear in the homogeneous Maxwell equations, it is traditional to consider them to be the independent field variables, while **D** and **H** are the dependent field variables.

We can describe the difference between the electric field and the charge density field by consideration of Gauss's law. Suppose the surface under consideration completely encloses a region or material with electric dipoles, i.e., closely coupled pairs of charges with opposite sense but equal magnitude. Then the total charge enclosed is nearly equal to the free (monopole) charge since the net charge of a dipole is zero. Under these conditions the electric field and electric displacement are equal. Now suppose that the surface splits the medium which contains dipoles. Then it is quite likely that some dipoles are partly inside and partly outside the volume bounded by the surface. We then correct macroscopically for this by defining the electric displacement as

$$\mathbf{D} = \mathbf{E} + 4\pi\mathbf{P}, \tag{8}$$

where **P** is the electric polarization field which arises due to the dipole moment. In nonferroelectric material exposed to weak fields, the polarization is proportional to the electric field. We may therefore rewrite (8) as

$$\mathbf{D} = [\epsilon]\mathbf{E}, \tag{9}$$

where $[\epsilon]$ is the matrix of electric permittivities. Indeed, (9) may be taken as the defining relation for the electric permittivity of a material. In many materials, the only nonzero elements of $[\epsilon]$ are the trace elements, and these are equal. Under these conditions, $[\epsilon]$ reduces to a scalar ϵ. For such isotropic materials, Gauss's law may be rewritten as

$$\int_S \mathbf{D} \cdot \mathbf{da} = q. \tag{10}$$

This explains why **D** is often called the charge density field.

While the material presented up to now on electric fields and forces is quite elementary, it is sufficient for developing an understanding of how capacitive transducers function. We focus this investigation by considering the behavior of the simplest capacitor type, namely that formed by two parallel plates separated by a small distance. As the following two sections will demonstrate, the parallel plate capacitor can be exploited in many ways to form electromechanical transducers. While most commercially available transducers use the parallel plate geometry, the methods presented here are generally applicable to capacitive structures, although specific results are geometry dependent.

3.2 Transducers Made with a Variable Gap Parallel Plate Capacitor

Now that we have reviewed some of the basic definitions and principles associated with electric fields, we can now use them to examine transduction mechanisms. Let us first consider a parallel plate capacitor with a dielectric separating the plates, as shown in Fig. 3.2. By definition, the capacitance is given by $q = Ce$. We may rewrite this using (10) as

$$q = Ce = DA, \tag{11}$$

where A is the area of each plate. Using (9) and assuming the material to be isotropic, we can rewrite this equation as

$$q = \epsilon E A. \tag{12}$$

If we further assume that the voltage establishes a uniform field between the plates, then from the definition of the electric potential we may write

$$q = \frac{\epsilon A e}{d}, \tag{13}$$

where d is the separation between the electrodes. The assumption of a uniform field is the same as requiring that the plate dimensions are large relative to the separation between the plates, thus making it reasonable to ignor the small fringing field at the periphery of the capacitor plates. Equation (13) permits us to define the capacitance as

$$C = \frac{\epsilon A}{d}. \tag{14}$$

Equation (13) may be thought of as the constitutive equation of the parallel plate capacitor when edge effects may be neglected.

Equation (14) applies to a specific geometry, namely the parallel plate capacitor. However, we note that for an arbitrary capacitor geometry, the capacitance will always be a function of the permittivity of the material between the electrodes and

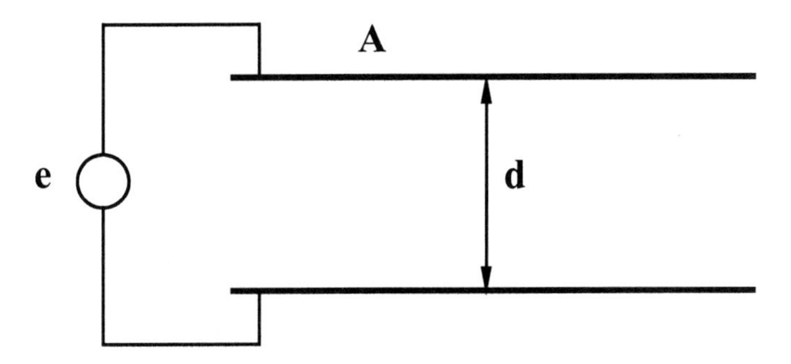

FIGURE 3.2. Parallel plate capacitor.

the geometrical parameters. It is this which permits the extension of the results obtained here to similar results in other geometries.

We can rewrite (13) to express the voltage in terms of the charge and the capacitor material and geometric parameters

$$e = \frac{dq}{\epsilon A}. \tag{15}$$

This equation has the form $e = g(q)$ which we used to define an ideal capacitance in Chapter 2, a fact that comes as no surprise.

Equation (15) provides valuable insight into how we can make use of a parallel plate capacitor in electromechanical transduction. Clearly, it is possible to change the voltage dynamically by using a mechanical means to temporally vary the electrode separation, the electric permittivity of the material between the plates, or the plate area. Each of these approaches has been used successfully in the development of electromechanical transducers. In what follows, we consider these various transduction mechanisms one at a time.

First suppose that we permit one plate of the capacitor to move so that the plate separation remains spatially uniform but varies in time. Because it takes mechanical energy to move the plate, and the capacitor stores electrical energy, there are now clearly two energy types associated with this device. Hence we model it as a 2-port capacitor. One port is associated with electrical power conjugate variables and the other with mechanical power conjugate variables. If the analysis shows that these two energy domains remain uncoupled, we can modify our model to two independent 1-port elements.

Given that the model of the variable gap parallel plate capacitor is as shown in Fig. 3.3, the challenge is to determine the constitutive relations. We do this in a manner that is guaranteed to produce expressions consistent with a statement of the energy stored in the device. First we note that the electrical energy stored by the capacitor is given by

$$\mathcal{E} = \frac{q^2}{2C} = \frac{q^2(d + x)}{2\epsilon A}, \tag{16}$$

where x is the displacement of the electrode from equilibrium as shown in Fig. 3.4. Note that this expression is a function of the two Hamiltonian variables q and x, confirming the need to use a 2-port capacitance in the model.

Next we note that the electrical power stored by the variable gap parallel plate capacitor is given by the time rate of change of the energy. We calculate this power,

FIGURE 3.3. Parallel plate capacitor.

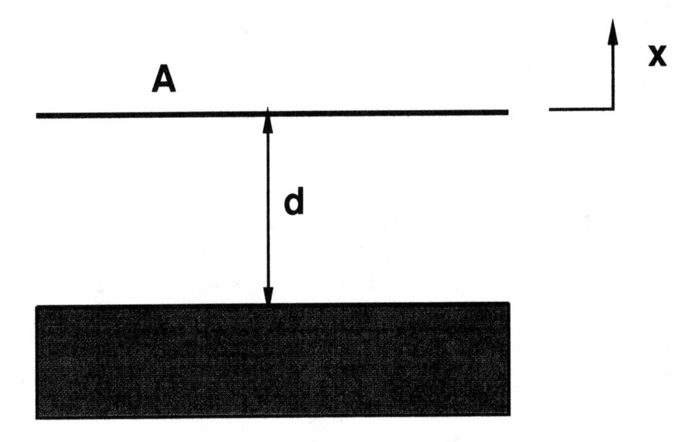

FIGURE 3.4. Parallel plate capacitor with varying gap dimension.

$\mathcal{P}$, by simple application of the chain rule:

$$\mathcal{P} = \frac{d\mathcal{E}}{dt} = \frac{\partial \mathcal{E}}{\partial q}\frac{dq}{dt} + \frac{\partial \mathcal{E}}{\partial x}\frac{dx}{dt}. \tag{17}$$

Since dq/dt is the current flow associated with the capacitor, $\partial\mathcal{E}/\partial q$ must be the voltage, because the product of these two yields the power. Similarly, dx/dt is the velocity of the plate and $\partial\mathcal{E}/\partial x$ is the force associated with it. From these identifications and (16) we may determine the constitutive equations for F and e:

$$e = \frac{q(d+x)}{\epsilon A}, \tag{18}$$

$$F = \frac{q^2}{2\epsilon A}. \tag{19}$$

There are a few items of interest in (18) and (19). First, because they were determined from the energy expression, they are guaranteed to be consistent with it. Second, the results indicate that there is a fundamental coupling between the electrical and mechanical energy variables. This suggests that it is not possible to represent the variable gap capacitor accurately using 1-port energy storage elements. Third, the results obtained could have been predicted prior to calculation. The voltage constitutive statement is merely a restatement of $q = Ce$ where C is time-varying. The force constitutive equation expresses the electrostatic force of attraction between the two plates.

Armed with a thorough understanding of the variable gap parallel plate capacitor, we may now discuss its use as a transduction mechanism. There are three basic approaches to using a varying capacitor gap as a transduction mechanism: relating the voltage to the displacement of the plate, relating the force and the voltage, and relating the force and the charge. Each of these approaches has been used with success. Below we discuss them one at a time. Following this presentation we will return to the other electromechanical transduction mechanisms associated with a parallel plate capacitance.

3.2.1 *Relating Displacement and Voltage*

The simplest means of using a variable gap parallel plate capacitor as a transduction mechanism is to relate the plate displacement to the voltage. Clearly, this may be done using (18). Note that this equation includes not only the voltage and displacement, but also depends on the charge on the plates. We may proceed in two ways to handle this complication: by assuming that the charge is constant and then checking the validity of this assumption; or by performing an analysis with the pertinent variables (e, x, and q) each written as a static portion and a time-varying portion. Here we choose the latter approach, as it applies more generally than the assumption of a constant charge on the plates. This approach also permits us to discuss the necessary conditions for producing a linear transducer, i.e., one in which there is a linear relation between the time-varying portion of the voltage and the time-varying displacement.

Suppose we have a dc bias voltage e_0 on the capacitor, and we either apply an ac voltage $\hat{e}$ to generate a displacement (actuator) or sense an ac voltage $\hat{e}$ in response to motion of the plate (sensor). Then

$$e = e_0 + \hat{e}. \tag{20}$$

The charge on the plates may then be modeled as a static charge q_0 and a time-varying charge $\hat{q}$, so

$$q = q_0 + \hat{q}. \tag{21}$$

Now we use (18) to relate the voltage to the displacement and charge

$$e_0 + \hat{e} = \frac{q_0 d}{\epsilon A} + \frac{q_0 x}{\epsilon A} + \frac{\hat{q} d}{\epsilon A} + \frac{\hat{q} x}{\epsilon A}. \tag{22}$$

Two of the six terms in (22) are static terms and the remaining terms are time-varying. Given that the static terms must balance even when the transducer is at rest, we can clearly divide (22) into two separate equations—a static equation and a dynamic equation. No further simplification is possible until additional assumptions are made. Further, the dynamic equation leaves us with three variables related: $\hat{e}$, $\hat{q}$, and x. Thus, the problem of desiring to eliminate charge from the picture still remains, and additional assumptions (conditions) must be imposed.

In (22), if we assume that e_0 is much larger than $\hat{e}$, that q_0 is much larger than $\hat{q}$, and that d is much larger than x (which might also be written as $\hat{d}$), then we can identify terms as being zeroth-order (dc), first-order, or second-order small. The zeroth-order terms provide us with the static equation:

$$e_0 = \frac{q_0 d}{\epsilon A}, \tag{23}$$

which is simply the relation previously discussed for a parallel plate capacitor with fixed plates. The first-order terms provide the following expression:

$$\hat{e} = \frac{q_0 x}{\epsilon A} + \frac{\hat{q} d}{\epsilon A}. \tag{24}$$

Note that there are two first-order terms on the right-hand side. The first relates the change in the capacitor gap to the voltage change, and the second relates the change in the charge on the plates to the voltage change. In transducers, in which the goal is to relate the gap change to the voltage change, we seek to minimize the effect of the changing charge. We accomplish this by electrically isolating the ac voltage and charge from the dc voltage and charge, so that the charge on the plates is essentially constant. Under these conditions, (24) can be approximated by

$$\hat{e} = \frac{q_0 x}{\epsilon A}. \tag{25}$$

It is this expression that is commonly used to provide the leading order, linear relationship between the motion of a plate in a parallel plate capacitor and the ac voltage. Note that this expression is written as though the plate motion results in the electrical signal. However, it is equally valid for the actuation in which application of an ac voltage is used to cause motion of one of the plates of a parallel plate capacitor.

One interesting way to view Eq. (25) is to rewrite it in terms of a sensitivity, i.e., as the ratio of an output variable to an input variable. As a sensor of the plate motion, the sensitivity is given by

$$\psi = \frac{\hat{e}}{x} = \frac{q_0}{\epsilon A} = \frac{e_0}{d}, \tag{26}$$

which suggests that sensitivity is improved as the dc bias voltage is increased or the plate separation is decreased. As an actuator of plate motion, the sensitivity is the inverse of the previous equation:

$$\psi = \frac{x}{\hat{e}} = \frac{q_0}{\epsilon A} = \frac{d}{e_0}, \tag{27}$$

which suggests that sensitivity is enhanced by increasing the plate separation or decreasing the dc bias voltage. However, recall that we began this discussion with the assumption that $\hat{e}$ is small relative to e_0. Thus, although the actuator sensitivity theoretically increases as e_0 decreases, there are strict limits to the validity of this sensitivity expression.

The second-order term in (22), and the neglected first-order term in (24) may be thought of as producing a perturbation to the expression given in (25). Effectively, these terms produce distortion of the signal, but it is generally quite small as long as the time-varying quantities are small relative to the static quantities. For the unperturbed system, since it is linear, we know that an ac excitation at frequency ω will result in response at the same frequency. Hence, imagine that x and $\hat{q}$ are sinusoidal functions of time at frequency ω. Then the second-order term will be a perturbation that is described by the square of a sinusoidal function at frequency ω. Using trigonometric identities, we can rewrite either $\cos^2(\omega t)$ or $\sin^2(\omega t)$ as the sum of two terms: one which is dc and one which varies as $\cos(2\omega t)$. The dc term will introduce a small change in (23), while the ac term introduces a new frequency component to the signal, namely, one at twice the excitation frequency. This small perturbation to the signal is referred to as the second harmonic distortion. One means of eliminating this distortion is to build a transducer which uses three plates with the center one moving. This transducer then resembles two working out of phase with each other. It is referred to as a push–pull configuration, and effectively doubles the sensitivity.

The discussion above has assumed that the static voltage (dc bias) is applied through use of a dc voltage source such as a battery. However, an interesting and useful alternative to this approach exists in the form of electrets. Electrets are materials which retain a polarization stably over long periods of time [6]–[9]. Effectively, these materials have an embedded charge, so if they are used as one plate of a parallel plate capacitor, they present a charge bias which is related to the effective voltage bias through (23). Originally electret devices used carnuba wax which had been melted and allowed to solidify in the presence of a high electric field. In 1962 Sessler and West [6] found that polytetraflouroethylene (PTFE) and fluroethylene propylene (FEP), both known as Teflon®, are electret materials. Since then virtually all electret transducers have been made using thin-film polymers. Compared to the sensors and actuators using a battery or power supply to provide a dc voltage bias, electret devices offer the advantage of not requiring an external power supply. They also tend to be more linear, because there are fewer means of generating distortion, and are typically lower in cost. By 1980 over 100 million electret microphones were manufactured every year worldwide. In 1994, the estimate was that the worldwide production of electret transducers would number nearly 800 million. They may be found as standard items in telephones, tape recorders, and hearing aids. For more information on electrets, see Sessler[8].

There are a very large number of electromechanical sensors and actuators which exploit the relationship between the voltage and variable gap dimension. Some examples of these devices are presented here, but by no means is the set of examples comprehensive. The examples have been chosen to demonstrate the breadth of scale and application of this transduction mechanism.

Shown in Fig. 3.5 is an example of a variable gap displacement sensor which takes advantage of the x to $\hat{e}$ transduction mechanism. This sensor is sold commercially by Mechanical Technology Inc. [10] and its operation was described in detail by Philips and Hirschfeld [11]. In this device a probe is maintained at a high potential. As the probe is moved toward another metal object (target), the change in the potential is sensed in order to determine the distance to the target. This sensor is used in the frequency range of dc to 20,000 Hz, and can measure distances as small as 10^{-7} in.

Shown in Fig. 3.6, taken from Neuman and Liu[12], is an array of force sensors fabricated using techniques developed for microelectronics. In this array the sensors are formed from two separate pieces which are joined in the final process. In one piece, a substrate is coated with a continuous layer of a metal that forms one plate of the set of parallel plate capacitors. An elastomer is printed onto the metal. In the second piece, the metal layer is patterned so that there are a multitude of plates each of which defines a separate force sensor. The elastomer is printed onto this piece as well. By assembling the two pieces, elastomer to elastomer,

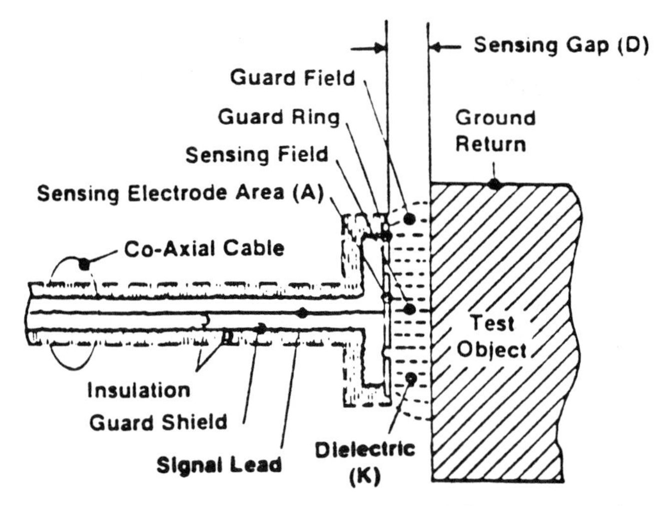

FIGURE 3.5. Mechanical Technology Inc. [10] displacement probe.

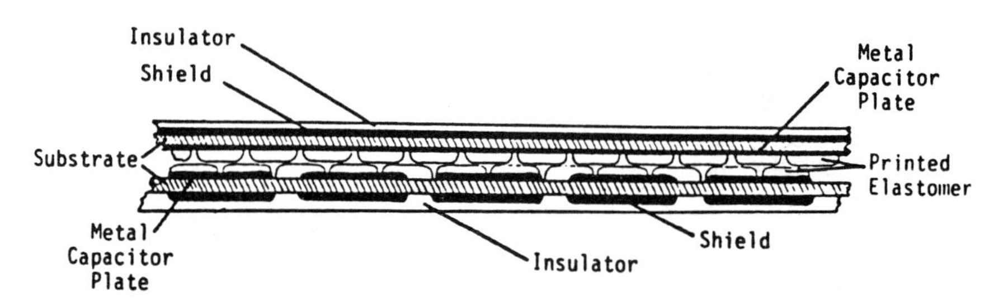

FIGURE 3.6. Array of force sensors (taken from Neuman and Liu[12]).

the force sensor array is formed. The elastomer provides stiffness and defines the permittivity of the medium between the two plates of the capacitors. Application of a force reduces the gap between the plates and produces an output voltage.

Another very small force sensor has been described by Despont et al. [13] and is shown in Fig. 3.7. This sensor is micromachined from silicon, and employs a differential approach to capacitive sensing. It uses a moving part with two arms located on opposite sides of a pair of fixed electrodes. Thus, as the capacitance increases on one side, it decreases on the other. Through judicious handling of the electronics, this can be used to increase the force sensor sensitivity while reducing its response to thermal variations.

Shown in Fig. 3.8 is another example of an electromechanical sensor based on varying the parallel plate capacitor gap and relating the voltage to the gap change. This device is sold by Setra Systems, Inc.[14], and senses pressure. The unit shown in Fig. 3.8(a) is an absolute or gage pressure sensor, and that shown in Fig. 3.8(b) is a differential pressure sensor, obtained by introducing a second port

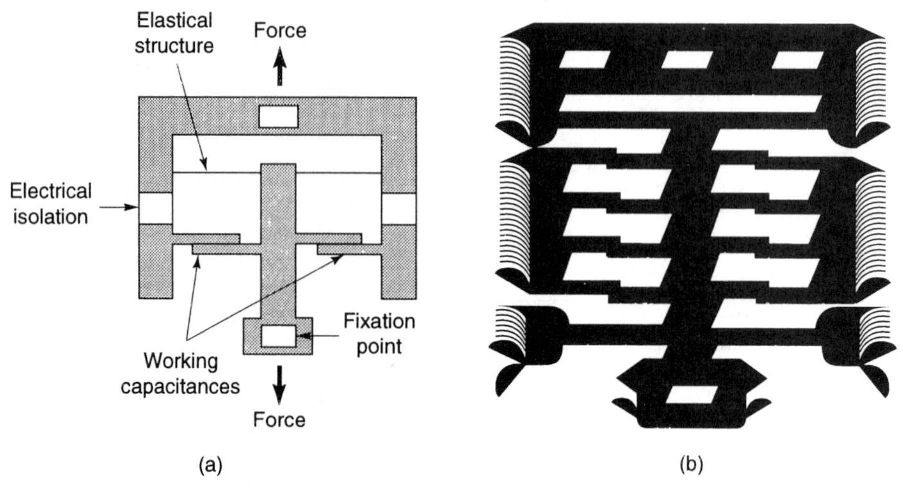

FIGURE 3.7. A force microsensor presented by Despont et al. [13].

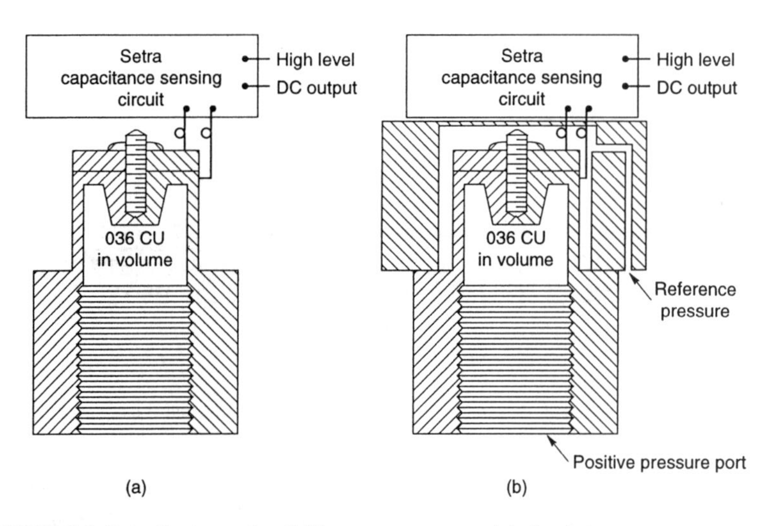

FIGURE 3.8. Setra Systems, Inc.[14] pressure sensors: (a) absolute or gauge pressure and (b) differential pressure.

for a reference pressure on the top surface. Units of these types are manufactured by Setra Systems, Inc. for pressure ranges of 0–25 psi to 0–10,000 psi.

Another pressure transducer which relates displacement and voltage in a parallel plate capacitor structure is shown in Fig. 3.9. In this barometric pressure sensor described in Whitaker and Call [15] and sold by Atmospheric Instrumentation Research, Inc. [16], two metal diaphragms are bonded to opposite sides of a ceramic disk. The space between the diaphragms and the disk is evacuated. This effectively results in two variable capacitance devices in which each responds to pressure changes with capacitance changes. The use of two capacitors hooked in serial electronically, means that this pressure sensor can be made relatively insensitive

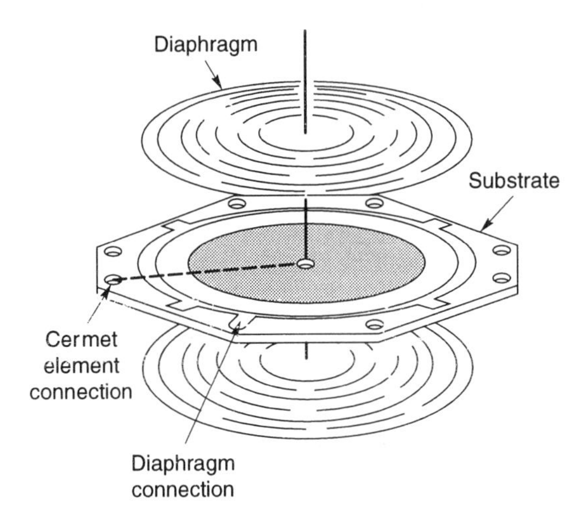

FIGURE 3.9. Barometric pressure sensor made by Atmospheric Instrumentation Research, Inc. [16].

to influence factors that would tend to produce oppositely directed motion of the two diaphragms, such as shock and vibration.

Yet another example of this type of transducer is a condenser microphone of the sort shown in Fig. 3.10, and described in detail in References [17]–[19]. In this sensor a thin metal diaphragm is held under tension above a fixed metal plate. The membrane responds to a sound pressure by moving (slightly) to yield an output ac voltage that is nearly proportional to the sound pressure. In the microphone shown in Fig. 3.10, the fixed plate is perforated so that the air trapped between the two plates of the capacitor may communicate with another volume of air behind the fixed plate. This lowers the stiffness of the air cavity and increases the system sensitivity although increases in motion also result in greater distortion to the signal. This sort of microphone has been duplicated extensively using silicon fabrication processes. For instance, Fig. 3.11 shows an underwater condenser microphone (a hydrophone) described by Bernstein [20]. In this sensor the moving electrode is perforated.

Figure 3.12 shows an elementary bond graph model of the condenser microphone. The membrane is modeled by a mass, I, and a stiffness, C_T, to account for the tension. These elements are connected to a 1-junction because there is a single velocity associated with them. The air cavities are a lumped stiffness dominated by the size of the back cavity, C_b. The back cavity stiffness operates on the difference in the amount of air moving through the perforations and that moving out of the pressure equalization tube. Thus C_b is attached to a 0-junction connecting between the perforations and the pressure equalization tube. Motion of the air through the perforations and through the air equalization tube are modeled by resistances, R_p and R_t. respectively. The conversion of energy in the transduction process is mod-

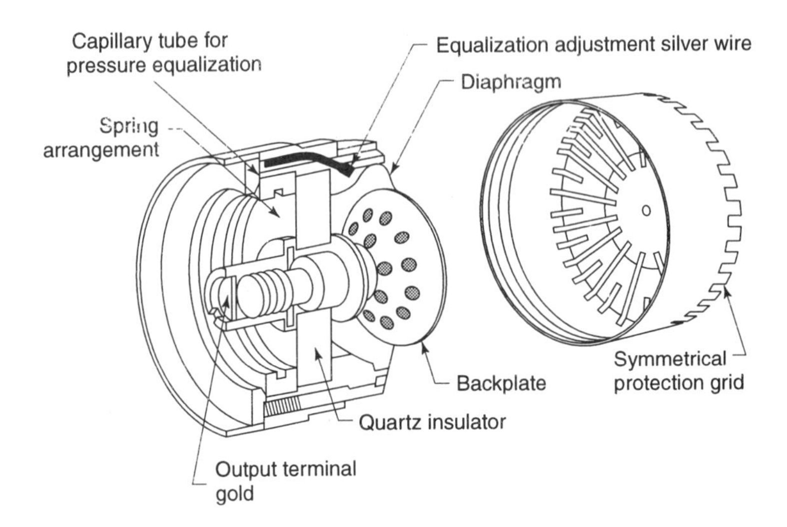

FIGURE 3.10. Condenser microphone. Figure from Rasmussen et al. [18].

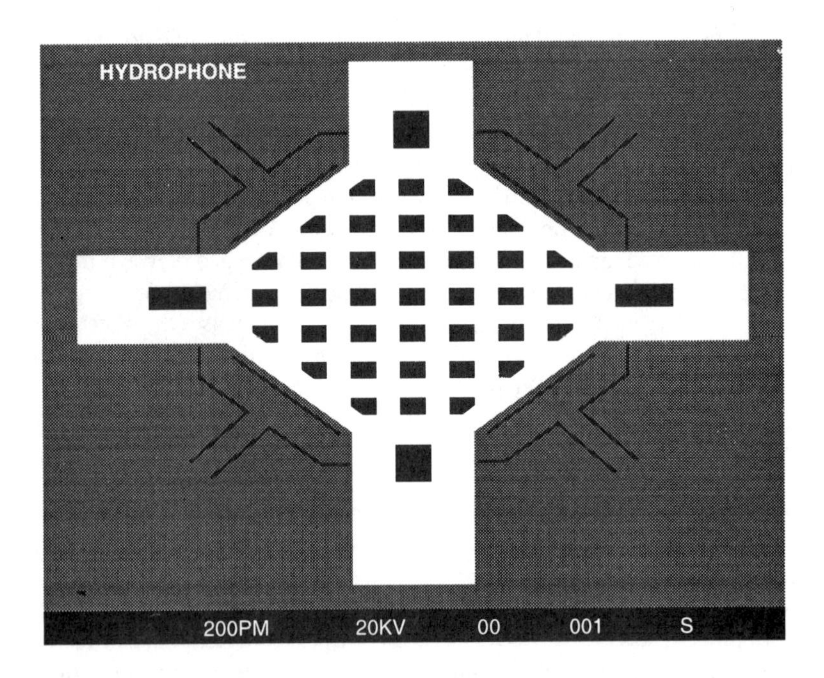

FIGURE 3.11. Example of a miniature hydrophone made using VLSI fabrication techniques. This hydrophone was described by Bernstein [20].

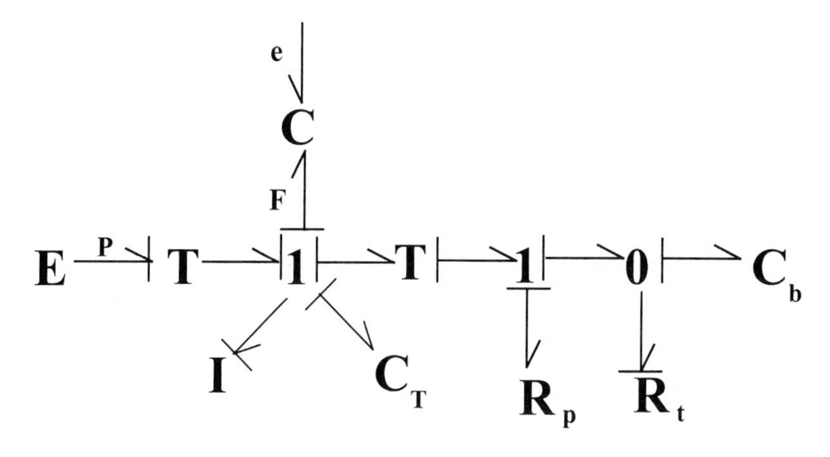

FIGURE 3.12. Bond graph of a condenser microphone.

eled using a 2-port capacitance with one electrical and one mechanical port. A transformer describes the mechanical to fluid coupling of the membrane and the surrounding fluid. The sound pressure is modeled using an ideal effort source.

Note that in this model the dc bias voltage and the signal voltage are lumped into one port. This introduces a problem in terms of determination of the causal structure of that particular port, since the dc bias clearly supplies effort, but the output signal is clearly determined by the system attached to the electrical port. It is possible to separate the dc bias and the ac signal into two electrical ports on the multiport C. This is shown in Fig. 3.13. Note that with the separation of the dc bias and the ac signal, the causal ambiguity is removed. The input dc voltage is bonded to an ideal effort source, with the causality appropriate to that. The ac signal port is bonded to processing electronics, and has causality appropriate to it being an output signal from the microphone. This bond graph model indicates that the condenser microphone is a fifth-order system. The state variables are the momentum of the membrane, the displacement of the membrane, the volume of air between the plates, the volume of air in the back cavity, and the charge output. At best, this model can serve to predict the first two system resonances, since resonance is a second-order phenomenon.

Using this model it is possible to determine the theoretical response of the condenser microphone shown in Fig. 3.10. To do this we would express the constitutive relations defining each of the elements in the model, and use these, plus the junction equations, to derive the state equations. From the state equations we could simulate the system time or frequency response, or determine transfer functions such as the sensor sensitivity ($\psi = \hat{e}/P$). We leave the details of this analysis to the interested reader, and focus on limiting behavior here.

Consider the low-frequency response of the microphone. The equalization port modeled by R_t is frequency dependent, with its resistance reducing to zero for very low frequencies. Effectively this means that the 0-junction at the right-hand end of the bond graph model disappears, i.e., that the back cavity is infinite. Then

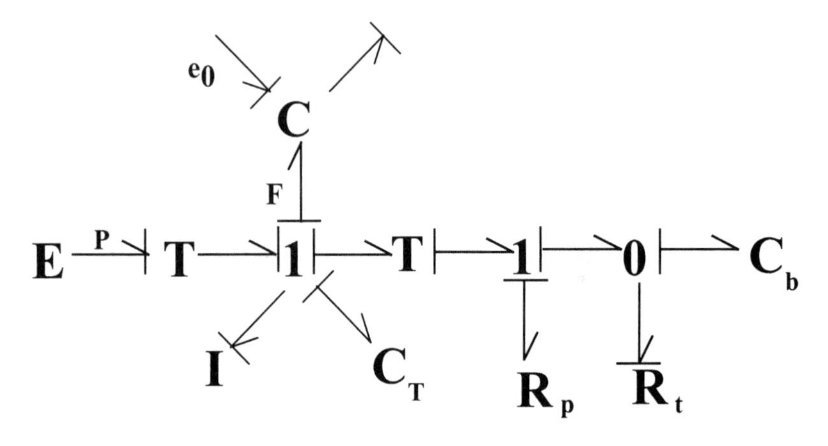

FIGURE 3.13. Another bond graph of a condenser microphone—with a 3-port capacitance.

the perforation resistance is the only element remaining to the right of the second transformer in the model, and it could be reflected through the transformer, which will simply scale the resistance. At the remaining 1-junction the impedances add. Since the impedance of an inertance is proportional to ω, that of a resistance is constant, and that of a stiffness is proportional to $1/\omega$, the capacitance elements will dominate for low frequencies. In other words, at low frequencies, the condenser microphone looks like a spring, or more precisely, two springs mechanically in parallel: the spring tension and the 2-port spring. The force divides between the two springs in a frequency independent manner determined by the relative stiffnesses of the springs. Since the output of the multiport C responds to force, at low frequencies, the response of the microphone is a flat function of frequency. At high frequencies, the inertance dominates over the capacitance elements at the main 1-junction, leading to a frequency response which decreases with increasing frequency. At resonance, in which the inertance and capacitance terms cancel, the resistance dominates.

Figure 3.14, taken from Sessler[8], shows a typical electret microphone. It is very similar in construction to a condenser microphone.

Finally, Fig. 3.15, taken from Fraden [21], shows the structure of a miniature flow rate sensor built from silicon. In this device, fluid flow disturbs a membrane which forms one plate of the parallel plate capacitor. The change in capacitance can be uniquely correlated with the flow rate for a given material. A device of this sort has been described by Cho and Wise [22].

3.2.2 *Relating Force and Voltage*

A second common means of using a variable gap parallel plate capacitor in transduction is to couple the force and the voltage relations. This mechanism is typically exploited in actuators where an input voltage is used to generate a force which subsequently produces a motion. By examining (18) and (19) we can see that it is

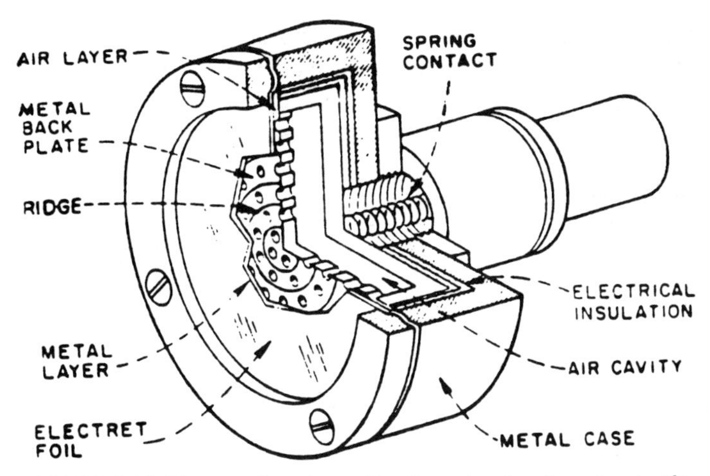

FIGURE 3.14. Electret microphone. The figure is taken from Sessler[8].

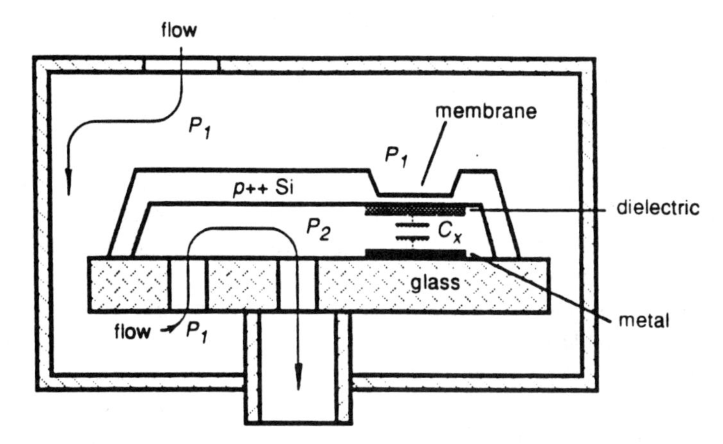

FIGURE 3.15. Structure of a miniature flow rate sensor. Figure taken from Fraden[21].

possible to write the following:

$$e = \frac{2F(d + x)}{q}. \tag{28}$$

This equation is not quite in the desired form because we have four variables rather than simply the two we desire. We can eliminate either x or q but not both. We eliminate q by noting that if we assume that the electric displacement is uniform in the gap, $q = DA = \epsilon EA = \epsilon eA/(d + x)$. Substituting this into the equation above yields

$$e^2 = \frac{2F(d + x)^2}{\epsilon A}. \tag{29}$$

Note that (29) shows that the relationship between the force and voltage is not linear, but quadratic. It also shows that the motion of the plate tends to complicate this mechanism. For many applications, the fact that the force/voltage relationship is quadratic is immaterial. We simply use the quadratic relations to develop a desired force level. Examples of this are presented shortly. Additionally, one can nominally linearize the transduction relation as shown below.

If we assume that the voltage may be divided into a large bias voltage e_0 and a small signal voltage $\hat{e}$, and that the force similarly has a dc component F_0 and a time varying portion $\hat{F}$, then we may write

$$e_0^2+2e_0\hat{e}+\hat{e}^2 = \frac{2d^2}{\epsilon A}\left[F_0 + \left(\frac{2F_0x}{d}+\hat{F}\right)+\left(\frac{F_0x^2}{d^2}+\frac{2\hat{F}x}{d}\right)+\frac{\hat{F}x^2}{d^2}\right], \quad (30)$$

where we have grouped terms of the same order on the right-hand side. With this, we may make the following assignments, accurate only to first order:

$$e_0^2 = \frac{2d^2F_0}{\epsilon A}, \quad (31)$$

$$2e_0\hat{e} = \frac{2d^2}{\epsilon A}\left(\frac{2F_0x}{d}+\hat{F}\right). \quad (32)$$

The dc terms describe the static force of attraction between the plates. There are two linear ac terms, one varying with displacement and one with force. In order for the transducer to respond to force rather than to displacement, it is necessary that $|2F_0x/d| << |\hat{F}|$. Since $\hat{F}$ is assumed to be small compared to F_0, we meet the inequality restriction by insisting that x be very small compared to d. In other words, we make the system very stiff mechanically so that there is little relative motion of the plates.

In (30) we have ignored the quadratic and cubic driving terms. We can consider these terms as small perturbations to the system described by the zeroth- and first-order terms. The second-order terms give rise to a small perturbation of the dc balance in (31) and a second harmonic distortion term. The effect of the third-order terms can be determined by expanding them. For example, suppose that the excitation is described by a $\cos(\omega t)$ function. Then the third-order terms will vary as $\cos^3(\omega t)$, which can be rewritten as

$$\cos^3(\omega t) = \frac{1}{8}[e^{j\omega t}+e^{-j\omega t}]^3. \quad (33)$$

This can be expanded and rewritten as follows:

$$\cos^3(\omega t) = \frac{1}{4}[\cos(3\omega t)+3\cos(\omega t)]. \quad (34)$$

From this result, we conclude that the third-order terms will have the effect of introducing a small perturbation to the linear signal and a small component of third harmonic distortion.

Because the linearized version of this force to voltage transduction mechanism is quite convoluted, it is rarely exploited. However, the quadratic form of this transduction mechanism is used in actuators. For instance, much of the recent work on micron-scale actuators uses electrostatic forces generated between two

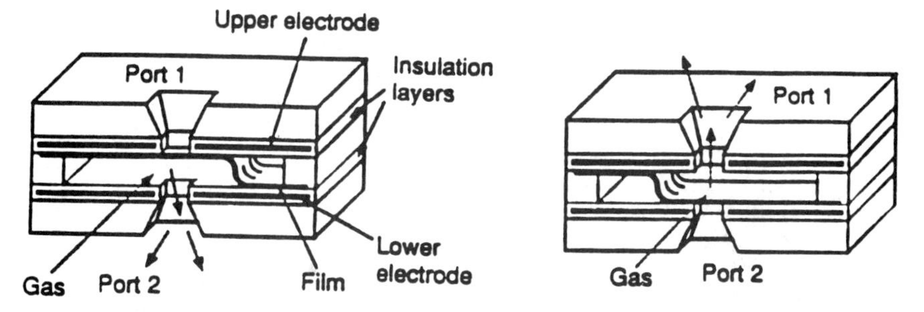

FIGURE 3.16. Electrostatic microactuator proposed by Sato and Shikida [23].

capacitor plates. Figure 3.16, taken from Sato and Shikida [23] shows one such example. In this proposed actuator, a conductive film between two metal electrodes is moved in response to an applied voltage. The film serves as one plate in a parallel plate capacitor formed either with the upper electrode as the second plate, or with the bottom electrode as the second plate. In response to the applied voltage, the film experiences a force and collapses onto the excited electrode plate. The intent in this transducer is to use the film position to open and close a microvalve. Because the actuator is bistable, i.e., has two operating points only, the nonlinear force/voltage relationship is completely irrelevant to operation.

Another example of an actuator exploiting the nonlinear form of the force/voltage relationship in a parallel plate capacitor serves as the drive mechanism in a surface profiling transducer as pictured in Fig. 3.17, taken from Kenny et al. [24]. Here, very small deflections are required, and the frequency of operation of the actuation is quite low.

Yet another example of an actuator exploiting the force/voltage relationship is a linear actuator fabricated using VLSI techniques and shown schematically in

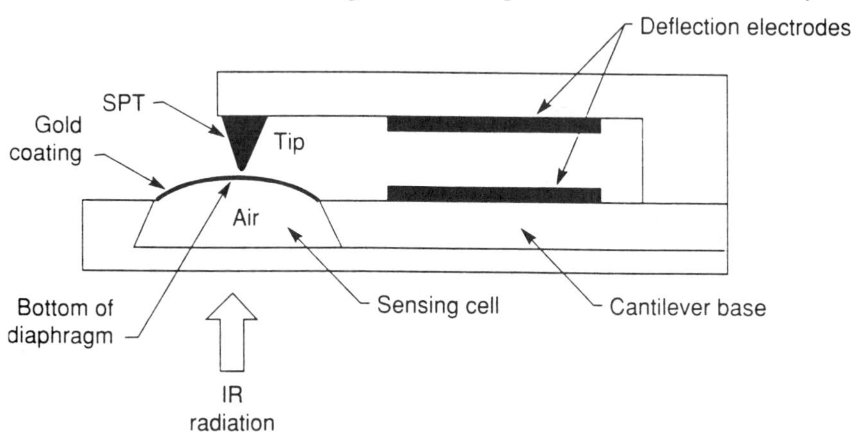

FIGURE 3.17. Electrostatic force actuator for surface profiling microscope. Reprinted with permission from T. W. Kenny, W. J. Kaiser, S. B. Waltman, and J. K. Reynolds, *Novel Infrared Detector Based on a tunneling Displacement Transducer*, Appl. Phys., Lett. **59**, 1820 (1991). Copyright 1991 American Institute of Physics.

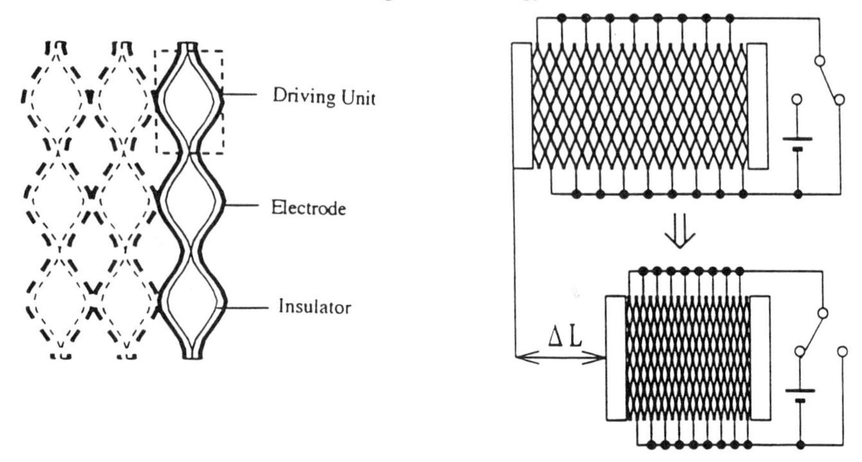

FIGURE 3.18. Microlinear actuator described by Minami et al.[25]. This transducer was made using VLSI fabrication techniques.

Fig. 3.18, taken from Minami et al.[25]. This actuator uses a honeycomb structure of electrodes rather than parallel plates. Application of a voltage compresses the honeycomb, producing a net linear motion. Honeycomb structures for linear actuation have been used on macroscopic scales as well as microscopic scales. Generally the resulting actuator produces high strains in the structure. Thus, the device certainly does not meet the small displacement criterion above for linear operation. This does not pose a problem as long as the nonlinearities are known and taken into account in operation.

Finally, an example which does take advantage of the linearized force to voltage transduction mechanism in a parallel plate capacitor structure is the electrostatic loudspeaker. In this device, the speaker cone is one of the plates in the capacitor, the other being fixed. A dc bias voltage gives the cone an equilibrium position which depends on the stiffness resisting the attraction force between the plates. An ac voltage signal gives rise to a force which moves the speaker cone to generate sound. Typical motions are on the order of microns to millimeters for a membrane whose size is on the order of tens of centimeters to meters.

3.2.3 Relating Force and Charge

A third means of potentially using a variable gap capacitor as a transducer is to relate the force and charge. This may be done directly using (19). Note that the force is related to the square of the charge, so the relationship is nonlinear. If we assume that there is a bias charge q_0 and a small time-varying charge $\hat{q}$, then we have

$$F_0 + \hat{F} = \frac{q_0^2}{2\epsilon A} + \frac{q_0\hat{q}}{\epsilon A} + \frac{\hat{q}^2}{2\epsilon A}. \tag{35}$$

We can identify the first and third terms as the static force balance. The second and fourth terms provide the linear relationship between the time-varying charge and the time-varying force. The remaining term is generally regarded as a perturbation that leads to second harmonic distortion (due to its quadratic nature).

This transduction mechanism is hardly ever exploited, and is presented simply for the purpose of completeness. Equation (35) suggests one could use the the varying gap parallel plate capacitor as a force sensor or a force actuator. As a sensor, an input force would produce a charge perturbation. However, this transduction approach is nearly impossible to employ because other forces associated with the transducer tend to be large compared to the electrostatic force. Thus, application of a force to one plate of a capacitor generally will not produce the charge perturbation predicted by Eq. (35). As an actuator, Eq. (35) correctly predicts that an input charge perturbation can produce an electrostatic force. However, one normally chooses to produce the charge perturbation by application of a voltage as was described in Section 3.2.2.

3.3 Transducers Using Other Means of Varying the Capacitance in a Parallel Plate Capacitor

The transduction mechanisms discussed above involve changing the capacitance of a parallel plate capacitor by changing the distance between the electrodes, all other parameters remaining fixed. If we consider the basic definition of the capacitance as $C = \epsilon A/d$, it is clear that we have focused on only one of three parameters that could be varied. Let us now consider what would happen if we chose instead to vary A or ϵ in order to produce a viable transducer.

3.3.1 Variable Area

First consider a capacitor in which A varies. It is difficult to imagine actually changing the area of a transducer in operation. However, we could change the *effective area* of the capacitor by permitting one of the plates to move normal to the direction of the electric field. Then, ignoring fringing fields, the effective area of the capacitor would be given by the area in which the two plates overlap. In the simplest case of this type, the plates are maintained at a fixed separation distance, and one of the plates moves in one of the two principal transducer axes normal to the field. This is shown graphically in Fig. 3.19. Now the capacitance may be written as

$$C = \frac{\epsilon w y}{d}, \tag{36}$$

where w is the width of the plates (into the page) and y is the length along which they overlap. The energy stored by the capacitor is given by

$$\mathcal{E} = \frac{q^2 d}{2\epsilon w y}, \tag{37}$$

which clearly shows that the energy varies inversely with the region of overlap.

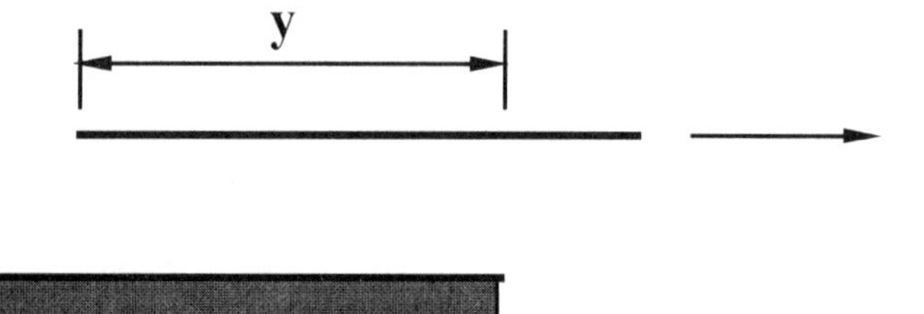

FIGURE 3.19. Capacitive transducer in which the effective area varies dynamically.

We may use (37) to determine the constitutive equations of this 2-port capacitance in much the same manner we used for the variable gap capacitor. We identify $\partial \mathcal{E}/\partial q$ as the voltage, which is given by

$$e = \frac{qd}{\epsilon w y}. \tag{38}$$

We identify $\partial \mathcal{E}/\partial y$ as the lateral force:

$$F = -\frac{q^2 d}{2\epsilon w y^2}. \tag{39}$$

Note that since y is varying in time, these constitutive relations are nonlinear. Note also that the lateral force decreases as the overlap between the plates increases. The negative sign in (39) indicates that the lateral force is a restoring force such as we would find in a spring.

Just as in the variable gap capacitor, there are a number of ways we might choose to use the variable effective area capacitor in transduction: by relating y and e, by relating F and e, or by relating F and q. In each of these cases, a choice must be made to work with the nonlinear relations or to produce a nominally linear transduction relation for restricted ranges.

First consider transduction relating e and y. This is analogous to the variable gap transduction mechanism in which the normal displacement and voltage are related. Using (38) we see that there is now a nonlinear relationship between the voltage and lateral displacement. As we have done previously, we can linearize this relation by assuming that the motion (lateral position) of the moving plate can be written as

$$y = y_0 + \hat{y}. \tag{40}$$

Then (38) can be rewritten as

$$e_0 + \hat{e} = \frac{d}{\epsilon w y_0}\left[q_0 + \left(\hat{q} - \frac{q_0 \hat{y}}{y_0}\right) + \left(\frac{q_0 \hat{y}^2}{y_0^2} - \frac{\hat{q}\hat{y}}{y_0}\right) + \text{higher order terms}\right]. \tag{41}$$

If we assume that $|\hat{q}/q_0|$ and $|\hat{y}/y_0|$ are small, then we can ignore terms higher than first order. Under these conditions, the zeroth-order term is once again a definition of the capacitance of a parallel plate capacitor with fixed plates. The linear terms yield a relation between the ac voltage and both the ac charge on the plates and the ac motion:

$$\hat{e} = \frac{d\hat{q}}{\epsilon w y_0} - \frac{d q_0 \hat{y}}{\epsilon w y_0^2}. \tag{42}$$

We wish to characterize this transduction mechanism in terms of a simple relationship between $\hat{e}$ and $\hat{y}$. Hence we again resort to electrical isolation of the ac system portion from the dc bias. Under this condition, (42) can be approximated as follows:

$$\hat{e} = -\frac{d q_0 \hat{y}}{\epsilon w y_0^2}. \tag{43}$$

As a practical matter, the harmonic distortion present in this transduction mechanism could be substantial because there are an infinite number of higher-order terms being neglected. However, the distortion introduced given the conditions assumed in deriving (43) is probably small.

An interesting example of a transducer exploiting the variable area transduction mechanism using the position/voltage relationship is the rotational position sensor shown in Fig. 3.20, taken from Fraden [21], and sold by GMC Instruments Inc. [26]. This sensor uses a pair of parallel plate capacitors in an electronically differential arrangement. As the wedge-shaped rotor moves, the effective capacitor area in one set of plates increases while that for the other decreases. The spiral shape of the electrodes is chosen because it produces a linearly increasing or decreasing electrode effective area/rotational position relationship.

Another example of a variable area sensor which exploits the voltage/position relationship is the slip sensor shown in Fig. 3.21 and described by Fan et al. [27]. In this transducer a normal force changes the gap in a pair of parallel plate

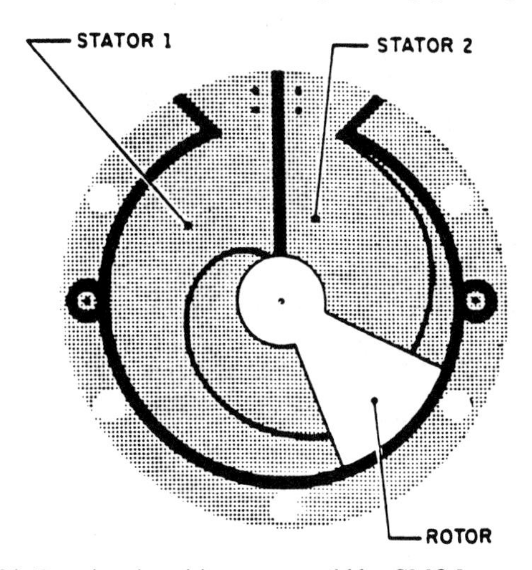

FIGURE 3.20. Rotational position sensor sold by GMC Instruments Inc. [26].

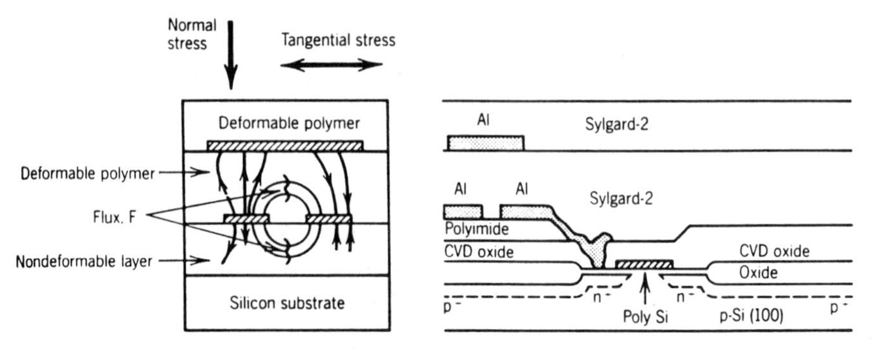

FIGURE 3.21. Schematic of a slip detector from Fan et al. [27].

capacitors. Slip, or shear force, moves the upper electrode from a position in which it is balanced between the two lower electrodes, to a position in which it only partly lines up with one of the bottom electrodes. This means that there is an effective area decrease for one of the capacitive sensors which may be detected through subtraction of the signal voltages produced by the capacitor pair.

We also could use the lateral force in a manner relating e and F. From the two constitutive equations, (38) and (39), we can write

$$F = -\frac{eq}{2\epsilon y}. \tag{44}$$

We note that this equation relates all four time-varying parameters, much as was the case for the variable gap transduction mechanism relying on the relation between F and e. We eliminate the q and y dependence by assuming that the electric displacement is constant over the active area of the capacitor, although this is clearly a poor assumption for cases in which the active area is small compared to the surface area of the edges of the effective capacitance region. With the constant electric displacement assumption we have $q = Ce = \epsilon wye/d$. Thus we can rewrite (44) as

$$F = -\frac{e^2 w}{2d}, \tag{45}$$

which yields a quadratic relation between the force and the voltage. Note that this transduction relation is much less complex than its variable gap analog. As in all of the other transduction methods discussed so far, we could linearize this relationship by assuming the force to consist of a large dc and a small ac component, while the voltage similarly consists of a large dc and small ac component. The resulting zeroth-order and first-order relations would be:

$$F_0 = -\frac{e_0^2 w}{2d},$$
$$\hat{F} = -\frac{e_0 \hat{e} w}{d}. \tag{46}$$

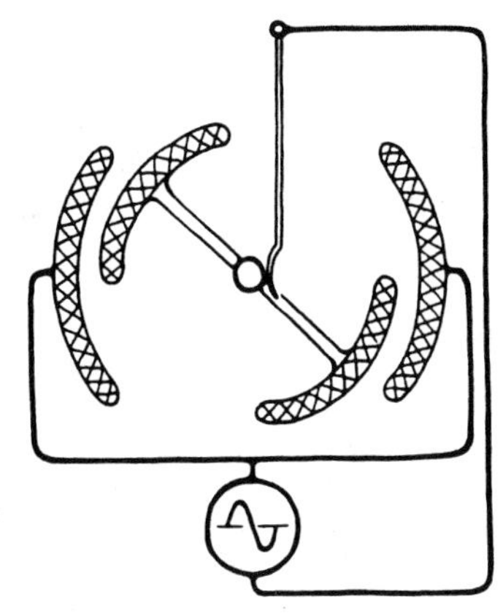

FIGURE 3.22. Electrostatic motor in which torque results from variable area capacitors. Figure taken from Jefimenko[28].

The remaining term which is quadratic in $\hat{e}$ would produce second harmonic distortion. It should be noted that because this transduction approach is generally used for actuation, where efficiency tends to matter more than linearity, the linearization technique is not often applied.

An example of a transducer which uses the voltage/force relationship in a variable area capacitive arrangement is the electrostatic motor shown in Fig. 3.22 and described in Jefimenko [28]. The turning torque in the motor is the direct result of the varying area between the sets of capacitors and exploitation of the force/voltage relationship. The voltage applied is at a frequency that matches the rotational speed.

Electrostatic motors also have become very popular on microscopic scales because of the relative ease with which they may be produced in silicon. Figure 3.23, taken from Brysek et al.[29], shows a typical example of a micromotor in which the torque results from application of a voltage sequentially to fixed electrodes in the stator].

A final example of an actuator which uses this voltage to force transduction mechanism is the solid-state linear micromotor shown in Fig. 3.24 and discussed in Daneman et al.[30]. Here a comb capacitor arrangement is used to apply a lateral force to a structure. The structure moves in response until the spring force supplied by cantilever beams balances the static force resulting from application of the dc voltage. Through use of impact arms and four sets of comb capacitors, a slider mechanism can be moved linearly. The comb capacitor has become a very popular lateral driving mechanism for small solid-state devices since its introduction by Lim et al. in a device to measure static friction coefficients on microscopic scales [31].

A third transduction mechanism using lateral capacitor plate motion relates the force and charge. However, this mechanism is not straightforward because the lateral motion influences the force and charge relation, as is obvious from (39). Because of these complications, this transduction mechanism typically is not ex-

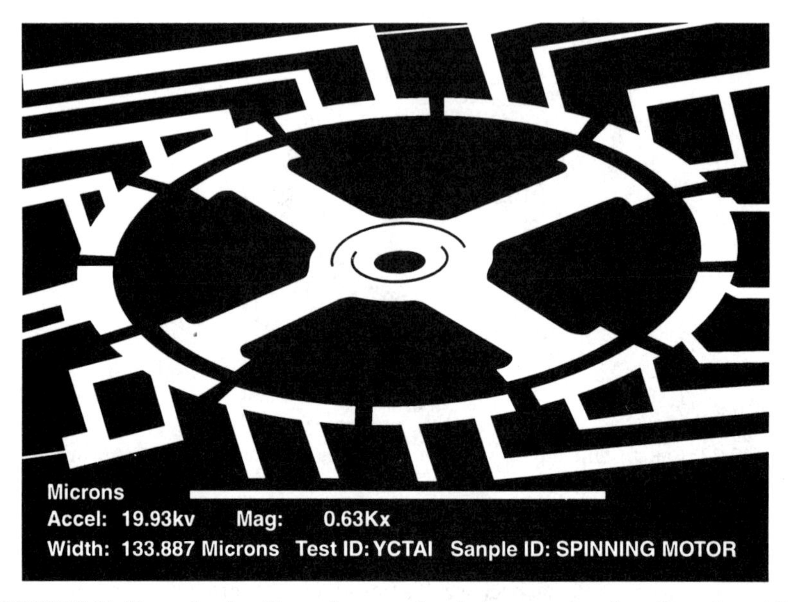

FIGURE 3.23. Example of a silicon electrostatic micromotor taken from Brysek et al.[29].

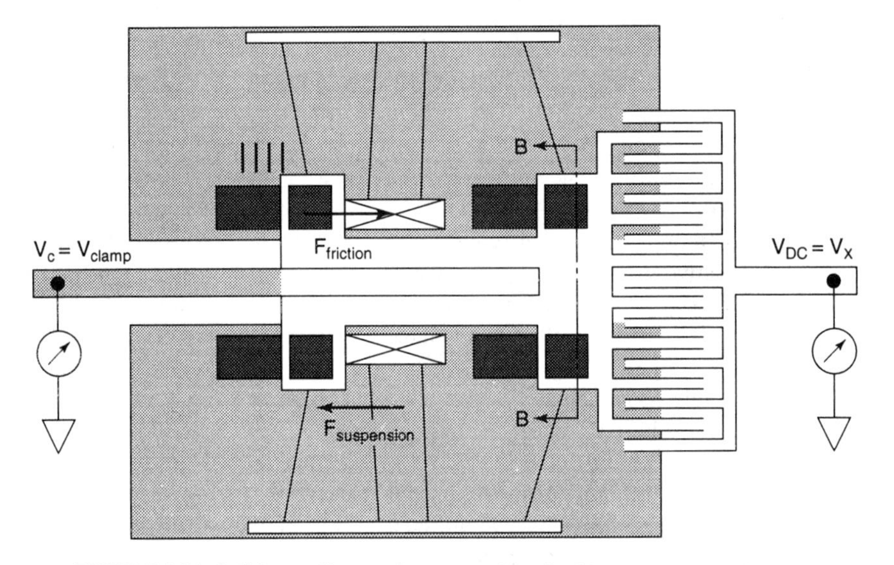

FIGURE 3.24. Solid-state linear micromotor described by Daneman et al.[30].

ploited in electromechanical sensors and actuators. It is included here primarily for completeness. To demonstrate how convoluted this force/charge relationship can be, we show below the effect of attempts to nominally produce a linear force/charge relationship.

If we assume that all three quantities can be written as the sum of a large static value and a small perturbation, then we can use (39) to write, correct to second order,

$$F = -\frac{d}{2\epsilon w y_0^2}\left[q_0^2 + \left(2q_0\hat{q} - \frac{2q_0^2\hat{y}}{y_0}\right) + \left(\hat{q}^2 - \frac{4q_0\hat{q}\hat{y}}{y_0} + \frac{3\hat{y}^2 q_0^2}{y_0^2}\right)\right]. \quad (47)$$

From this relation it is clear that there are three sources of second harmonic distortion, each of which is small if $|\hat{y}/y_0| << 1$ and $|\hat{q}/q_0| << 1$. There are also two terms which are linear in the perturbation variables, one of which depends on charge only, and the other which depends on the dc charge and the plate movement. In order for the charge and force linear relation to be dominant, the second of these two linear terms must be small, i.e., we must have $|\hat{q}/q_0| > |\hat{y}/y_0|$. As a practical matter, this could be achieved by using large plates so that the lateral motion range is kept to a small fraction of the plate dimension. However, the bottom line is clearly that this transduction mechanism can be linearized only for very restricted ranges of operation.

3.3.2 *Varying Permittivity*

Now consider the possibility of varying the electric permittivity of the material between the plates of the capacitor. This is normally done by laterally moving a dielectric material into and out of the space between the electrodes as shown in Fig. 3.25. This changes the effective value of the electric permittivity of the entire medium between the plates of the capacitor and can be used to detect motion of the dielectric.

To determine the effective electric permitivity, ϵ_{eff}, consider an instant in which the moving dielectric is neither totally removed from nor totally occupying the plate gap. Then the capacitor can be divided into two pieces. One piece has area A_1 and a dielectric with permittivity ϵ_1. The other portion has area A_2 and electric permittivity ϵ_2. The active area of the capacitor is fixed at A so we must have $A = A_1 + A_2$, and the length of the capacitor plates in the direction of motion of the dielectric is l. Since there is a single voltage difference between the plates we

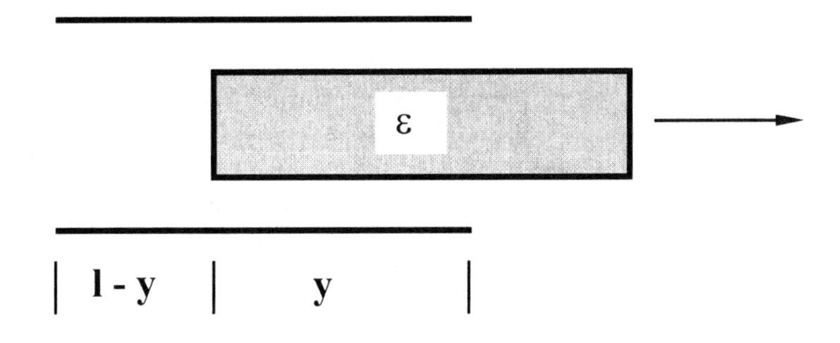

FIGURE 3.25. Capacitive transducer in which the permittivity varies spatially.

force the two capacitor segments to have the same voltage. Thus we may write

$$q_i = \frac{\epsilon_i A_i e}{d} \tag{48}$$

for each portion of the segment. We find the total charge by summing over the two portions. This yields

$$q = \frac{(\epsilon_1 A_1 + \epsilon_2 A_2)e}{d}. \tag{49}$$

Comparing this equation to the standard $q = \epsilon Ae/d$ we see that the effective electric permittivity is given by

$$\epsilon_{\text{eff}} = \frac{\epsilon_1 A_1 + \epsilon_2 A_2}{A} = \frac{\epsilon_1 y + \epsilon_2(l - y)}{l}, \tag{50}$$

where y is the length of the capacitor between which the dielectric is described by ϵ_1. Combining these equations we can finally write the voltage constitutive relation

$$e = \frac{qd}{[\epsilon_1 y + \epsilon_2(l - y)]w}. \tag{51}$$

Clearly this equation provides a nonlinear relation between the voltage and the motion of the dielectric in the gap that is quite similar to that seen in the varying area cases above.

The standard means of using this equation in transduction is to relate the voltage nonlinearly to the displacement y. This is typically done using a dielectric with a high permittivity moving in an air gap. Under these conditions, the denominator of (51) is dominated by the moving dielectric term, i.e., the $\epsilon_1 yw$ term. Then the voltage displacement relation is almost identical to (38) and the same conclusions follow. We can enhance the analysis of this system by expanding (51) in a Taylor series to find the zeroth-, first- and second-order terms without making the assumption that the moving dielectric term dominates. This makes it possible to determine the sources of distortion in the system.

Figure 3.26, taken from Fraden [21], shows an example of a cylindrical position sensor which exploits this transduction mechanism, in a device there are two fixed electrodes forming a parallel plate capacitor. A dielectric shaft which moves is threaded between the plates. Thus the shaft motion affects the permittivity of the medium between the plates.

Another example of a transducer that uses variable permittivity in a parallel plate capacitor is the water level gauge shown in Fig. 3.27 and described in Fraden [21]. Here the gauge consists of parallel plates typically separated by air and aligned vertically. Rain introduces water between the plates, causing a change in the effective permittivity and capacitance of the system. While this can be detected in the method described by (51), the time scale of the capacitance change is so long that an alternative method is often sought. One popular alternative is to use the transducer as the capacitor in a resonant circuit. Then the frequency of resonance is determined by the capacitance, which is related to the amount of water between the plates.

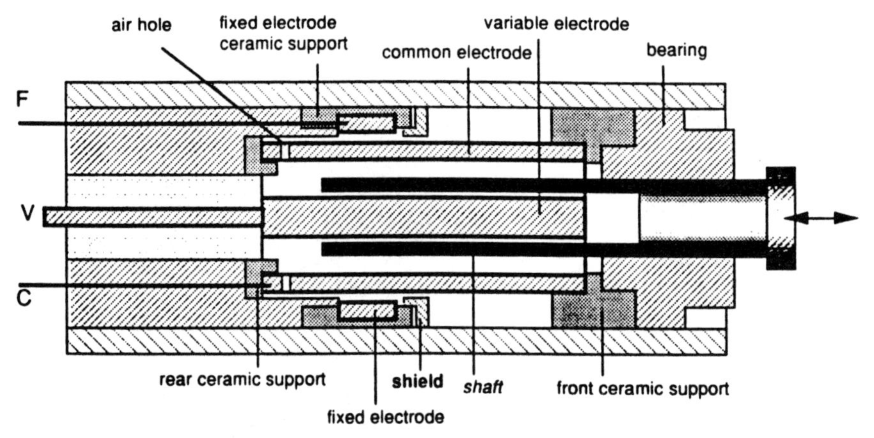

FIGURE 3.26. Cylindrical capacitive position sensor (from [21]).

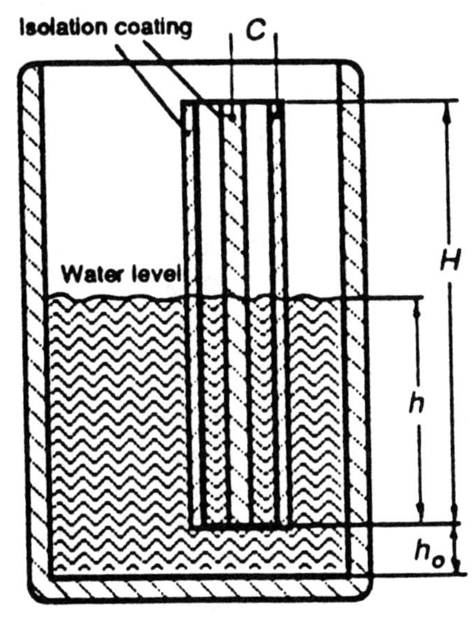

FIGURE 3.27. Water level gauge based on variable permittivity in a parallel plate capacitor (from [21]).

Another example of a transducer using variable permittivity in a capacitor is shown in Fig. 3.28. This position sensor is manufactured by Turck, Inc. [32] and uses a very interesting geometry. Instead of two parallel plates separated by a gap, the Turck sensor places the two electrodes on the same plane. The electric field lines go from the inner electrode to the outer electrode in arcs through the air in front of the electrodes. By moving an object close to the sensor, we can increase the capacitance by forcing the field lines to go through the object, which presumably has a much higher permittivity than air. This change in capacitance produces a signal if a dc bias voltage is present. It can also be sensed by using

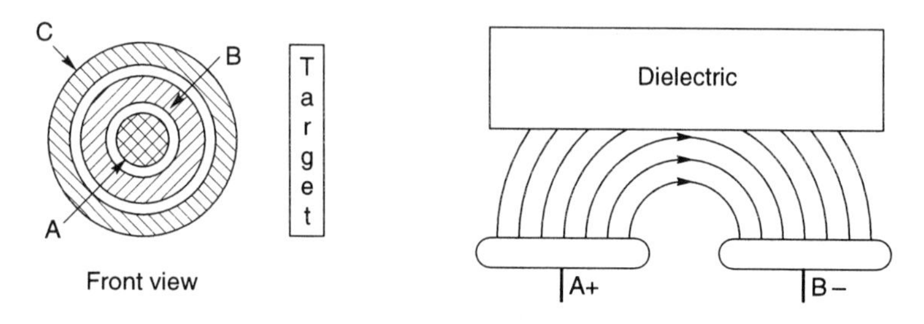

FIGURE 3.28. Turck, Inc. [32] position sensor for nonconducting targets.

the transducer as the capacitor in a resonant circuit and detecting the resonant frequency. Given that this transducer monitors how close an object has come to the device, it falls under the category of transducers generally known as proximity sensors. This particular proximity sensor works for nonconducting targets. Other types of proximity sensors are specifically designed for conducting targets.

Still another example of a transducer taking advantage of variable permittivity in a capacitor is the humidity sensor shown in Fig. 3.29. This device, manufactured by Hy-Cal Sensing Products, Division of Micro Switch [33], requires no moving parts, choosing to monitor relative humidity on the basis of changes in the permittivity of the dielectric material between the two electrodes. As shown in Fig. 3.29, a thin polymer layer is deposited on top of another thin polymer layer (polyimide) with a porous platinum layer in between. The capacitance pertinent to the transduction is between the platinum and the silicon substrate, which is rendered conducting through boron diffusion. In this device, the water content of the polyimide causes variations of up to 30% of the permittivity. Thus, changes in permittivity can be related directly to relative humidity. Given the small size of the sensor, it is possible to make its response time far faster than conventional humidity sensors. The devices sold by Hy Cal Sensing Products report response times as low as 15 s.

Although it is not as common as the voltage/position relationship, we could choose to use a varying electric permittivity to measure or produce a lateral force

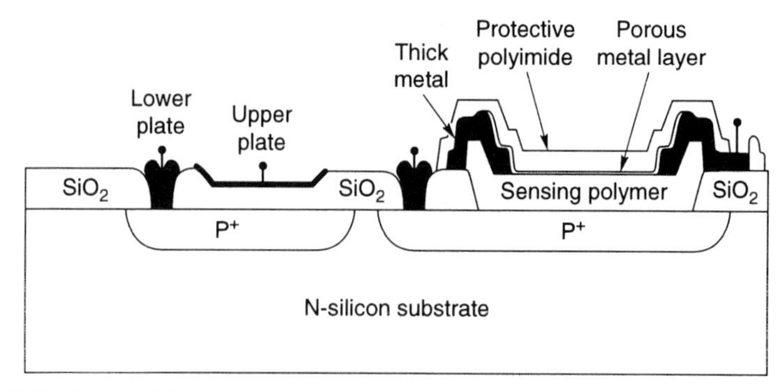

FIGURE 3.29. Humidity sensor sold by Hy Cal Sensing Products, Division of Micro Switch [33].

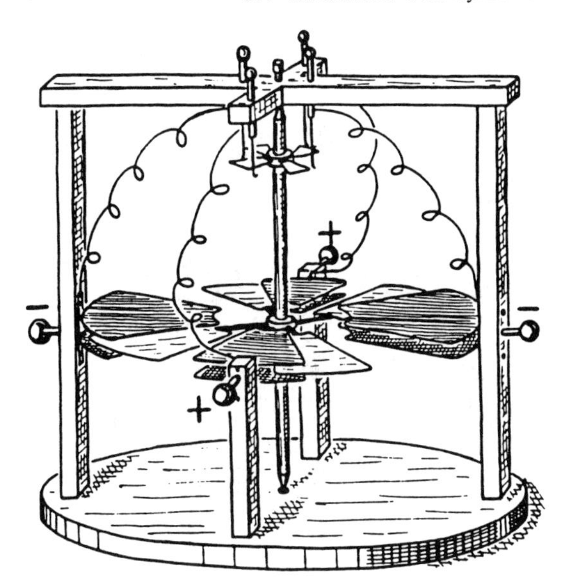

FIGURE 3.30. Electrostatic motor using a nonconducting rotor. Figure taken from Jefimenko [28].

similar to that described in Section 3.3.1. Analytical description of this particular transduction mechanism is left as an exercise for the reader, with a note that in the limit of $\epsilon_1 >> \epsilon_2$ this is the same as the analysis presented for the moving plate (variable area) case. Examples of use of this sort of transduction mechanism generally are limited to actuators. For instance, Fig. 3.30, taken from Jefimenko [28], shows an electrostatic motor made with fixed rotor plates and moving dielectrics on the rotor. The same sort of approach has been used for electrostatic motors with electrets on the rotor, as shown in Fig. 3.31. Because of the polarization present in electrets, they can provide electrostatic attraction-induced torque in addition to the torque caused by the varying permittivity between the plates.

3.4 Transducers with Cylindrical Geometry

Up to now we have considered the parallel plate capacitor geometry only. While the methods demonstrated for the parallel plate capacitor apply generally for arbitrary capacitor geometries, it is useful to demonstrate this through consideration of another popular geometry—cylindrical geometry. Here we imagine that we have a cylinder of radius r_1 centered inside a cylindrical shell with an outer radius of r_2. This situation is shown in Fig. 3.32. For this geometry, the capacitance is given by

$$C = \frac{2\pi\epsilon l}{\ln(r_2/r_1)}, \tag{52}$$

where l is the cylinder length [1]–[5]. Note that this is similar to the expression for capacitance in the parallel plates, in that it depends on the permittivity of the

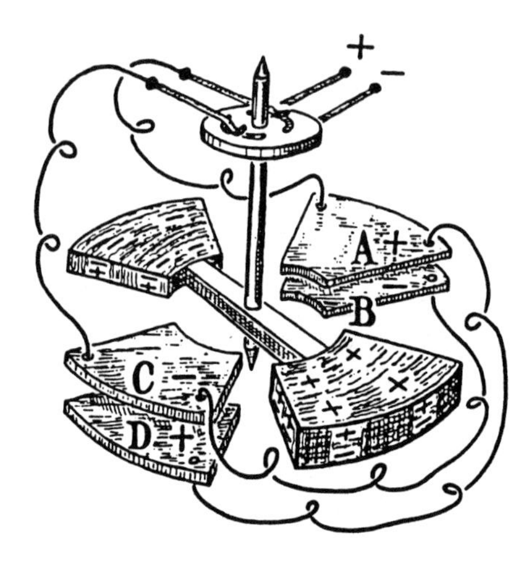

FIGURE 3.31. Electrostatic motor using electrets on the rotor. Figure taken from Jefimenko[28].

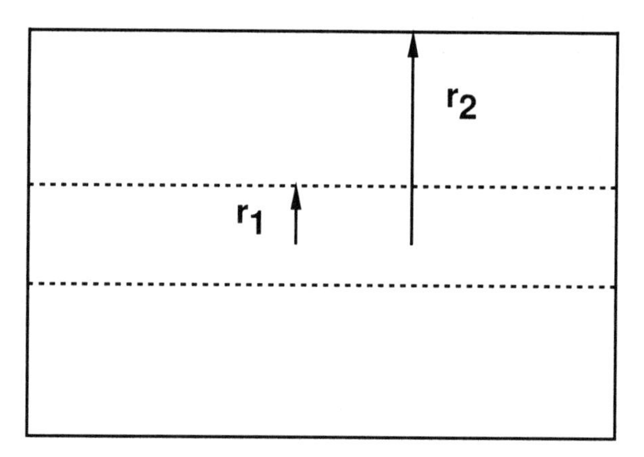

FIGURE 3.32. Cylindrical capacitor.

material between the electrodes and on geometrical parameters. From $q = Ce$ we may write

$$e = \frac{q \ln(r_2/r_1)}{2\pi \epsilon l}. \tag{53}$$

This expression suggests that there are three fundamental parameters which could be varied mechanically in order to induce a voltage, or which could be stimulated to change by application of a voltage: the permittivity, the cylinder length, and $\ln(r_2/r_1)$. The permittivity could be varied in much the same manner as was described for the parallel plate geometry. A piece of material with high permittivity could be slid into and out of the fixed cylindrical electrodes to produce a voltage signal, or induced to move in response to application of a voltage. Alternatively, a material whose permittivity changes with the environmental characteristics could be introduced between the two cylinders. Similarly, the cylinder length could be made a function of time by moving one of the electrodes. Then the effective length is approximately the length of electrode overlap. The radius ratio could be varied by allowing the outer or inner electrode to be made of a flexible material which deforms. For example, we could certainly imagine making a pressure sensor using the pressure to cause the outside electrode (presumably made of a foil or a metal-coated polymer) to expand in a breathing mode.

Determining the various constitutive relations which can be exploited in the cylindrical geometry requires using precisely the same approaches as were used in previous sections for the parallel plate geometry. Rather than present all of these, we present simply one example—the case in which the inner electrode radius is fixed and the outer radius varies (as in the imagined pressure sensor above). We choose this particular example because it represents the greatest departure from the parallel plate geometry. We begin by expressing the energy as

$$\mathcal{E} = \frac{q^2}{2C} = \frac{q^2 \ln(r_2/r_1)}{4\pi\epsilon l}. \tag{54}$$

Differentiation with respect to time produces the power:

$$\frac{d\mathcal{E}}{dt} = \frac{\partial\mathcal{E}}{\partial q} i + \frac{\partial\mathcal{E}}{\partial r_2}\frac{dr_2}{dt}. \tag{55}$$

We thus derive the following fundamental constitutive relations for the voltage and the radial force, F:

$$e = \frac{\partial\mathcal{E}}{\partial q} = \frac{q \ln(r_2/r_1)}{2\pi\epsilon l}, \tag{56}$$

$$F = \frac{\partial\mathcal{E}}{\partial r_2} = \frac{q^2}{4\pi\epsilon l r_2}. \tag{57}$$

These constitutive equations are similar to those for the parallel plate capacitor, except for the form of their dependence on the radii. As (56) and (57) indicate, the transduction mechanism with a cylindrical geometry leads to nonlinear constitutive relations. We can certainly work with the nonlinear constitutive relations, but it is more common to use a linear approximation. Suppose we are interested in a pressure sensor, for example, in which the change in r_2 is used to produce a signal voltage. Then (56) is the expression with which we start. If we assume that the charge, voltage, and varying radius can each be expressed as a static term (subscript 0) and a small perturbation (quantities with a circumflex), then we have the following:

$$e_0 + \hat{e} = \frac{(q_0 + \hat{q})\left[\ln(\frac{r_{20}}{r_1}) + \ln(1 + \frac{\hat{r}_2}{r_{20}})\right]}{2\pi\epsilon l}. \tag{58}$$

We may rewrite this expression using a Taylor series expansion for $\ln(1 + \hat{r}_2/r_{20})$ with the assumption that $r_{20} >> \hat{r}_2$. This yields

$$e_0 + \hat{e} = \frac{1}{2\pi\epsilon l}\left[q_0 \ln\left(\frac{r_{20}}{r_1}\right) + \hat{q}\ln\left(\frac{r_{20}}{r_1}\right) + \frac{q_0\hat{r}_2}{r_{20}}\right], \tag{59}$$

correct to first order only.

Balancing the static terms, we have

$$e_0 = \frac{q_0 \ln(r_{20}/r_1)}{2\pi\epsilon l}, \tag{60}$$

which is nothing more than a restatement of (53). The terms which are first-order small produce

$$\hat{e} = \frac{1}{2\pi\epsilon l}\left[\hat{q}\ln\left(\frac{r_{20}}{r_1}\right) + \frac{q_0\hat{r}_2}{r_{20}}\right]. \tag{61}$$

It is the second of the two terms on the right-hand side that we wish to emphasize, so we use the same electrical isolation approach as described for the parallel plate capacitor. Under these conditions, we finally have an estimate of the sensitivity of the pressure sensor:

$$\psi = \frac{\hat{e}}{\hat{r}_2} = \frac{q_0}{2\pi\epsilon l r_{20}} = \frac{e_0}{r_{20}}\ln(r_{20}/r_1). \tag{62}$$

This sensitivity expression indicates that the pressure sensor response could be increased, for a fixed r_1, by increasing the bias voltage or by decreasing the static radial position of the flexible electrode. Note that this is the same sort of result we obtained when considering the sensitivity of the parallel plate capacitor. There we concluded that, as a sensor, the sensitivity increases with increasing bias voltage and decreasing gap between the plates.

In deriving the linear approximation above terms of order $(\hat{r}_2)^n$ for $n > 1$ were ignored. As these terms are nonzero for all values of n, there are many sources of distortion in the cylindrical transducer imagined above. This is a distinction from the parallel plate geometry which only has a second harmonic distortion term. For small radial motions relative to the static position, the distortion terms may be ignored relative to the signal.

3.5 Gradient Transduction Using Two Dielectrics

So far we have limited our discussion to situations involving a single solid dielectric material. Another set of transduction mechanisms which is based on changes in the energy stored in an electric field uses two solid dielectric materials with different permittivities. When such materials are placed in an electric field, stresses arise at the interface. There are two cases: normal gradient and tangential gradient. We present this mechanism of transduction here for completeness, but before beginning we note that this particular transduction mechanism is not often used in sensing. There are simply too many alternatives that offer better performance in terms of sensitivity or efficiency. However, as will be discussed, the use of gradient transduction could provide a means of noninvasively producing sound waves in solid materials. In other words, gradient transduction mechanisms could be very useful for actuation.

First consider a parallel plate capacitor in which there are two dielectrics filling the gap between the plates, but they are oriented so that the interface is perpendicular to the plates. This is referred to as the tangential gradient arrangement, and resembles the parallel plate capacitive transducer in which a solid dielectric is moved laterally between two fixed plates. If we apply a bias voltage, as shown in Fig. 3.33, the electric field magnitude must be the same in the two dielectrics. Thus we have two capacitors in parallel. Hence, the effective capacitance for the system shown in Fig. 3.33 is given by

$$C_{\text{eff}} = C_1 + C_2, \tag{63}$$

where $C_i = \epsilon_i w y_i / d$.

We would normally calculate the energy stored by the capacitor from $\mathcal{E} = q^2/(2C)$. However, in this case, it is more convenient to use the co-energy form $\mathcal{E}^* = Ce^2/2$ since the voltage is specified in the system. Since the capacitance is independent of q and e, and the constitutive relation defining a capacitor is linear, the energy and co-energy are the same value. Using the co-energy expression we have

$$\mathcal{E}^* = \mathcal{E} = \frac{(\epsilon_1 y_1 + \epsilon_2 y_2) w e^2}{2d}. \tag{64}$$

We can rewrite this expression, recognizing that y_1 and y_2 sum to the net length, l:

$$\mathcal{E} = \frac{[\epsilon_1 y_1 + \epsilon_2 (l - y_1)]\, w e^2}{2d}. \tag{65}$$

From this we calculate the net interface force using $F = \partial \mathcal{E} / \partial y_1$:

$$F = \frac{w e^2 (\epsilon_1 - \epsilon_2)}{2d}, \tag{66}$$

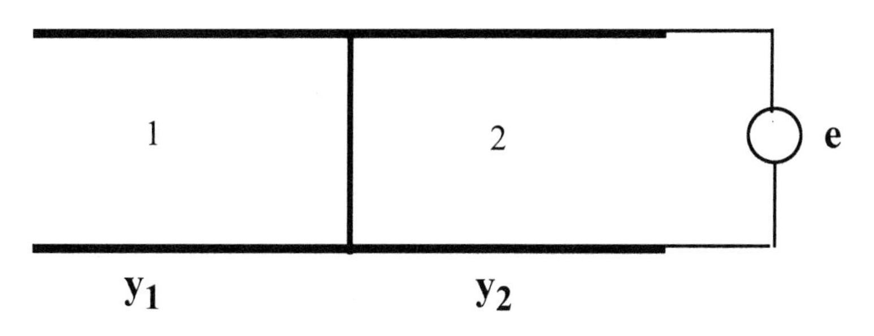

FIGURE 3.33. Tangential gradient capacitive transducer.

which directly relates the voltage and interface force. Once again the relationship is nonlinear and it depends on the values of permittivity of the dielectrics. Greater interface force is generated for cases of widely varying electric permittivities.

The tangential gradient capacitor can be used to noninvasively generate and monitor stresses at the interface between two solid materials. For instance, as an actuator, driving with a voltage at frequency ω will produce interface forces at frequency 2ω (due to the quadratic nonlinearity). These interfacial forces could be used to launch acoustic waves in the media. If desired, we could also linearize the force/voltage relationship by application of a large dc voltage and a small ac voltage. However, in actuation we normally put a higher premium on the magnitude of the imposed force than on the linearity of the transduction, and this would suggest simply working with the quadratic nonlinearity.

Now let us consider the normal gradient transduction mechanism. Figure 3.34 shows a parallel plate capacitor with two dielectric materials between the electrodes. The dielectrics are arranged so that there is a step change in material as we pass in a line between the plates. This is referred to as the normal gradient arrangement. Note that the normal gradient transducer is distinctly different from a push-pull parallel plate transducer, although they might resemble each other mechanically. In the push–pull transducer, there is an electrode separating the two media which is grounded. This effectively permits the push–pull device to act as though it were two independent transducers. In the normal gradient device, the separation between the two media is not grounded, and the media are thus tightly coupled electrically. In fact, the normal gradient device is nothing more than two capacitors in series electrically. Then the effective capacitance of this system is given by

$$C_{\text{eff}} = \frac{1}{1/C_1 + 1/C_2}, \tag{67}$$

where C_1 and C_2 are the capacitances of each section.

We next may write the energy stored by the capacitor as

$$\mathcal{E} = \frac{q^2}{2C_{\text{eff}}} = \frac{q^2(C_1 + C_2)}{2C_1C_2}. \tag{68}$$

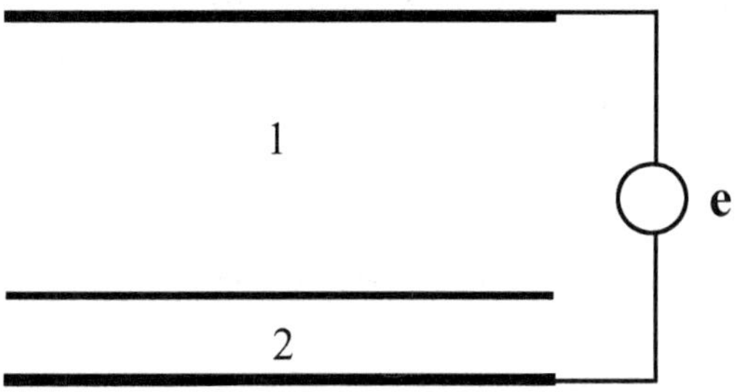

FIGURE 3.34. Normal gradient capacitive transducer.

We determine the capacitances for each section as $C_1 = \epsilon_1 A/d_1$ and $C_2 = \epsilon_2 A/d_2$, where d_1 and d_2 are the gap dimensions for each of the two capacitors, as shown in Fig. 3.35. Hence we may write the energy stored by the capacitor as

$$\mathcal{E} = \frac{q^2(\epsilon_1 d_2 + \epsilon_2 d_1)}{2\epsilon_1\epsilon_2 A}. \tag{69}$$

Of course, since the gap between the electrodes is fixed, d_1 and d_2 must sum to d. Hence we may rewrite the energy as

$$\mathcal{E} = \frac{q^2\left[\epsilon_1(d - d_1) + \epsilon_2 d_1\right]}{2\epsilon_1\epsilon_2}. \tag{70}$$

The net interface force is given by $\partial\mathcal{E}/\partial d_1$:

$$F = \frac{q^2}{2A}\left[\frac{1}{\epsilon_2} - \frac{1}{\epsilon_1}\right]. \tag{71}$$

Equation (71) suggests that application of a voltage across the capacitor plates leads to a force at the interface between the two dielectrics. This force causes a movement of the interface. The exact amount of the movement is dependent upon the elastic compliances of the two media. The size of the interface force is largest if the two media have widely different permittivities. As in the tangential gradient case, this transduction mechanism could be used to impose a force at the interface which launches compressive waves in the media. Also, as in the tangential gradient case, the transduction mechanism is fundamentally quadratic.

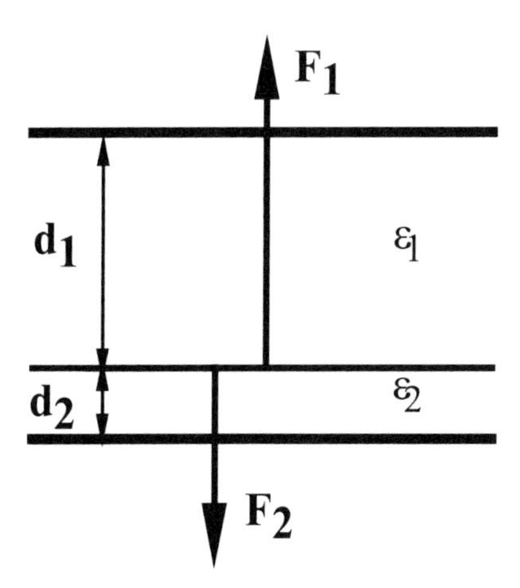

FIGURE 3.35. Interface stresses in a normal gradient transducer.

3.6 Electrostrictive Transduction

In the preceding sections of this chapter we have concentrated on the geometrical types of transduction mechanisms that result from changes in the energy stored in an electric field. There is another transduction mechanism that arises from electric field energy storage, but that is the result of a particular material property, namely electrostriction. Electrostrictive materials are those materials in which the electric permittivity is a function of the mass density. Many ceramics and crystals exhibit electrostrictive behavior. See, for example, Uchino [34] and Anderson et al. [35].

Clearly electrostrictive materials link the mechanical and electrical domains. Since they do so through storage of electric field potential energy, they are represented by a capacitance. Hence we will use a 2-port capacitance representation for electrostriction. It remains then to determine the constitutive relations for electrostriction, and to discuss how we could use the relations in transducers.

Let us imagine that we have a piece of electrostrictive material that is of dimensions $h \times w \times l$, and that there is an electric field in the direction in which the material has dimension h, as shown in Fig. 3.36. The capacitance of this material is still given by $C = \epsilon wl/h$, but now $\epsilon = \epsilon(\rho)$ where ρ is the mass density.

As in the previous discussions we express the energy stored by the capacitor and identify $\partial\mathcal{E}/\partial h$ as the force in the thickness direction. Taking the permittivity variation with dimensions into account, this yields

$$F = \left[\frac{q^2}{2wl\epsilon} - \frac{q^2 h}{2wl\epsilon^2}\frac{\partial\epsilon}{\partial h}\right]. \tag{72}$$

The first term in this force expression is the electrostatic force of attraction between the two capacitor plates. The second term, which appears for the first time in this equation, is the electrostrictive contribution.

Equation (72) can be further simplified by noting that although the density of the material may change, its mass remains constant. Hence

$$hlw\rho = \text{constant}. \tag{73}$$

Differentiating this equation permits us to relate changes in dimensions to changes in density:

$$hlw\,d\rho + \rho lw\,dh + \rho hw\,dl + \rho hl\,dw = 0. \tag{74}$$

Note that since h is the dimension in the direction of the field, dh is a direct result of electrostriction. The changes in the other dimensions cannot be related

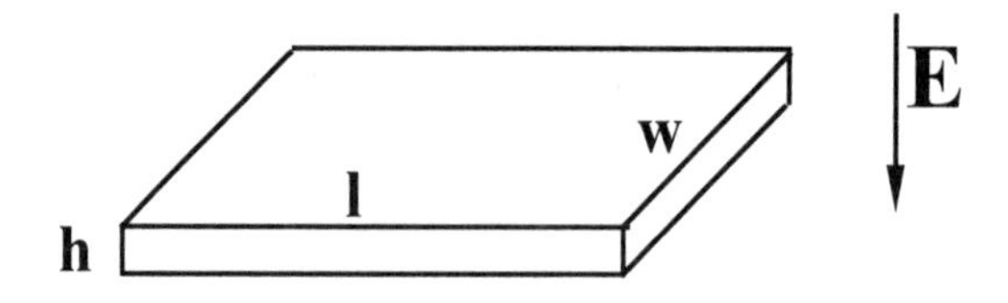

FIGURE 3.36. Electrostrictive material operation in transduction.

to electrostriction directly since there is no field in those directions. However, Hooke's law will generally guarantee changes in l and w due to the change in h. Assuming the material to be isotropic, we assume that $dl/l = dw/w = -\nu\, dh/h$, where ν is Poisson's ratio. Using these relations in (74) we have the following:

$$\frac{d\rho}{dh} = -\frac{\rho(1-2\nu)}{h}. \tag{75}$$

We use this relation in (72) so that we may eliminate $\partial\epsilon/\partial h$ in favor of $\partial\epsilon/\partial\rho$ which is better known for many electrostrictive materials. The result is

$$F = \frac{q^2}{2wl\epsilon}\left[1 + \frac{\rho(1-2\nu)}{\epsilon}\frac{\partial\epsilon}{\partial\rho}\right]. \tag{76}$$

For all known electrostrictive materials $\partial\epsilon/\partial\rho$ is positive. Hence, from (76), we see that the electrostrictive and electrostatic contributions to the force are in the same direction, namely attracting the two plates together regardless of the direction of the field. If an approximately linear transduction mechanism is desired in place of this fundamentally quadratic mechanism, we can use a bias charge and limit the operating range to small perturbations about the bias. The result is a transduction mechanism which greatly resembles another, fundamentally linear, mechanism referred to as piezoelectricity (to be discussed in Chapter 5). The major difference between a biased electrostrictive transducer and a piezoelectric transducer is that piezoelectric materials typically do not require application of a bias once they are poled.

The bulk of the recent work on electrostrictive materials has centered on lead magnesium niobate (PMN) based ceramics. Such composites have shown an ability to produce strains up to 0.1% at room temperature with no hysteresis. PMN also exhibits a high actuator sensitivity (ratio of displacement to applied voltage) compared to other ceramics. While electromechanical transducers using PMN composites are still rare, it is likely that this situation will change quickly in the next decade.

3.7 Summary

In this chapter we have introduced a number of transduction mechanisms which are based on changes in the energy stored in an electric field. Some of these result from geometrical variations; some from material property variations. All have been shown to be quadratic.

In the next chapter we will discuss the magnetic counterparts to the transduction mechanisms described here. As will be shown, transducers based on changes in the energy stored in a magnetic field are also fundamentally quadratic.

3.8 References

1. P. Lorrain and D. R. Corson, *Electromagnetic Fields and Waves* (W. H. Freeman, San Francisco, 1970).
2. E. M. Purcell, *Electricity and Magnetism* (McGraw-Hill, New York, 1965).
3. W. H. Hayt, Jr., *Engineering Electromagnetics* (McGraw-Hill, New York, 1981).
4. L. Eyges, *The Classical Electromagnetic Field* (Addison-Wesley, Reading, MA, 1972).
5. H. H. Woodson and J. R. Melcher, *Electromechanical Dynamics* (Wiley, New York, 1968).
6. G. M. Sessler and J. E. West, *Self-Biased Condenser Microphone with High Capacitance*, J. Acoust. Soc. Am. **34**, 1787 (1962). Reprinted in I. Groves (ed.), *Acoustic Transducers* (Hutchinson Ross, Stroudsburg, PA, 1981).
7. G. M. Sessler and J. E. West, *Foil Electret Microphones*, J. Acoust. Soc. Am. **40**, 1433 (1966).
8. G. M. Sessler (ed.), *Electrets*, 2nd ed. (Springer-Verlag, Berlin, 1987).
9. F. W. Fraim, P. V. Murphy, and R. J. Ferran, *Electrets In Miniature Microphones*, J. Acoust. Soc. Am. **53**, 1601 (1973).
10. Mechanical Technology Incorporated, Advanced Products Division, 968 Albany-Shaker Road, Latham, NY 12110, USA.
11. G. J. Philips and F. Hirschfeld, *Rotating Machinery Bearing Analysis*, Mechanical Engineering, **28** (1980).
12. M. R. Neuman and C. C. Liu, *Fabrication of Biomedical Sensors Using Thin and Thick Film Microelectronic Technology*, in C. Fung, P. W. Cheung, W. H. Ko, and D. G. Fleming (ed.), *Micromachining and Micropackaging of Transducers* (Elsevier, Amsterdam, 1985).
13. M. Despont, G. A. Racine, P. Renaud, and N. F. dr Rooij, *New Design of a Micromachined Capacitive Force Sensor*, J. Micromech. Microengng. **3**, 239 (1993).
14. Setra Systems, Inc., 159 Swanson Rd., Boxborough, MA 01719, USA.
15. S. D. Whitaker and D. B. Call, *A New Hand-Held Barometer/Altimeter Offers Portable Accuracy*, Sixth Symp. on Meterolog. Obs. and Instrum., Jan. 1987. New Orleans, LA, USA.
16. Atmospheric Instrumentation Research, Inc., 8401 Baseline Rd., Boulder, CO 80303, USA.
17. E. C. Wente, *The Sensitivity and Precision of the Electrostatic Transmitter for Measuring Sound Intensities*, Phys. Rev. **19**, 498 (1922). Reprinted in I. Groves (ed.), *Acoustic Transducers* (Hutchinson Ross, Stroudsburg, PA, 1981).
18. G. Rasmussen, P. V. Bruel, F. Skode, K. S. Hansen, and R. Frederiksen, *Measuring Microphones* (Bruel and Kjaer, Denmark, 1971).
19. M. Rossi, *Acoustics and Electroacoustics* (Artech, Norwood, MA, 1988).
20. J. J. Bernstein, *Micromachined Acoustic Sensors*, in M. D. McCollum, B. F. Hamonic, and O. B. Wilson (ed.), *Transducers for Sonics and Ultrasonics* (Technomic, Orlando, FL, 1992).

21. J. Fraden, *AIP Handbook of Modern Sensors: Physics, Designs, and Applications* (AIP Press, New York, 1993).

22. S. T. Cho and K. D. Wise, *A High Performance Microflowmeter with Built-In Self Test*, in *Digest of Technical Papers, Transducers '91* (IEEE, New York, 1991).

23. K. Sato and M. Shikida, *Electrostatic Film Actuator with a Large Vertical Displacement*, IEEE Micro Electro Mechanical Systems, Travemunde, Germany, February 1992.

24. T. W. Kenny, W. J. Kaiser, S. B. Waltman, and J. K. Reynolds, *Novel Infrared Detector Based on a Tunneling Displacement Transducer*, Appl. Phys. Lett. **59**, 1820 (1991).

25. K. Minami, S. Kawamura, and M. Esashi, *Fabrication of Distributed Electrostatic Actuator (DEMA)*, J. MEMS. **2**, 121 (1993).

26. GMC Instruments, Inc., 250 Telser Rd., Unit F, Lake Zurich, IL 60047.

27. L. S. Fan, R. M. White, and R. S. Muller, *A Mutual Capacitive Normal- and Shear-Sensitive Tactile Sensor*, Proc. IEEE Int. Electron. Devices Mtg., 1984, p. 220.

28. O. Jefimenko, *Electrostatic Motors* (Electret Scientific, Star City, WVA, 1973).

29. J. Brysek, K. Petersen, J. R. Mallon, Jr., L. Christel, and F. Pourahmadi, *Silicon Sensors and Microstructures* (NovaSensor, Fremont, CA, 1990).

30. M. J. Daneman, N. C. Tien, O. Solgaard, A. P. Pisano, K. Y. Lau, and R. S. Muller, *Linear Microvibromotor for Positioning Optical Components*, J. MEMS, **5**, 159 (1996).

31. M. G. Lim, J. C. Chang, D. P. Schultz, R. T. Howe, and R. M. White, *Polysilicon Microstructures to Characterize Static Friction, Proc. IEEE Micro Electro Mechanical Systems*, Napa Valley, CA, p. 82, 1990.

32. Turck, Inc., 3000 Campus Drive, Plymouth, MN 55441, USA.

33. Hy Cal Sensing Products, Division of Micro Switch, 9650 Telstar Avenue, El Monte, CA 91731, USA.

34. K. Uchino, *Electrostrictive Actuators: Materials and Applications*, Ceramic Bull. **65**, 647 (1986).

35. E. H. Anderson, D. M. Moore, J. L. Fanson, and M. A Ealey, *Development of an Active Truss Element for Control of Precision Structures*, Optical Engrg. **29**, 1333 (1990).

4

Transduction Based on Changes in the Energy Stored in a Magnetic Field

In Chapter 3 we discussed many transduction mechanisms associated with changes in the energy stored in an electric field. Similarly, we may build transducers in which the mechanism of energy conversion relies on changes in the energy stored in a magnetic field. These transducers are inductive devices, and are similar to the capacitive transducers described previously in many fundamental ways. The major difference between the two classes of transducers is that the magnetic field-based devices are often designed for higher power than their electric field counterparts. They are thus used quite frequently in actuation. Examples of inductive transducers include motors and moving coil disk drive heads.

In this chapter we will explore those transducers whose operation is based on changes in the energy stored in a magnetic field in a manner similar to that used to study transducers based on changes in an electric field. To emphasize the similarity, and to provide greater physical insight we will begin with an introduction to magnetic variables and magnetic circuits.

4.1 Magnetic Systems

In most modern approaches we choose to view magnetic elements from the electrical perspective, and model them as inductances. Although this is certainly acceptable, the more classical approach of developing magnetic circuits in a manner parallel to the development of electrical circuits provides greater insight into the behavior of magnetic elements. It is this process of developing magnetic analogs that we will pursue here. We begin with a brief overview of magnetism, based on material which can be found in many textbooks. For details and further descriptions, readers should consult, for example [1]–[4].

In 1819 Oersted observed that current-carrying wires cause a deflection of permanent magnetic dipoles in their neighborhood. Biot, Savart, and Ampére studied this phenomenon and determined the basic laws relating currents and forces. Of particular interest is Ampére's study of the force of one current-carrying wire on another. Suppose that we have two current-carrying wires as shown in Fig. 4.1. In these wires we identify infinitesimal segments as $\mathbf{dl}_i$, where the vector points in the direction of current flow. We identify $\mathbf{r}_{12}$ as the vector connecting the midpoints of the two wire segments. Stated mathematically, Ampére found that the force on each wire due to the presence of the other is given by

$$\mathbf{F} = I_1 I_2 \int_C \int_C \frac{(\mathbf{dl}_1 \cdot \mathbf{dl}_2)\mathbf{r}_{12}}{|\mathbf{r}_{12}|^3}, \tag{1}$$

where the line integrals are taken around the two loops. Note that this expression shows that the force decays as the inverse square of the distance between the wires, and that the force between any two line segments is along the vector connecting them.

As in the electric field case, it is clear that the force is dependent not only on the properties of a fixed object (current-carrying wire number 2 in this case), but also on the test object (current-carrying wire number 1) that we move near to the fixed object in order to measure the force. We define a field parameter that permits us to describe the physical properties of the magnetic environment independent of the test loop. This defining relation is known as the Biot–Savart law and is given by

$$\mathbf{dF} = I_2(\mathbf{dl}_2 \times \mathbf{B}). \tag{2}$$

where $\mathbf{B}$ is known as the magnetic induction field or magnetic flux density field, with typical units of Webers per meters squared (Wb/m^2) which are also known as Teslas (T). Note that it is clear from (2) that the magnetic field has units of force per current-length. Since current is dq/dt, we may alternatively think of the magnetic

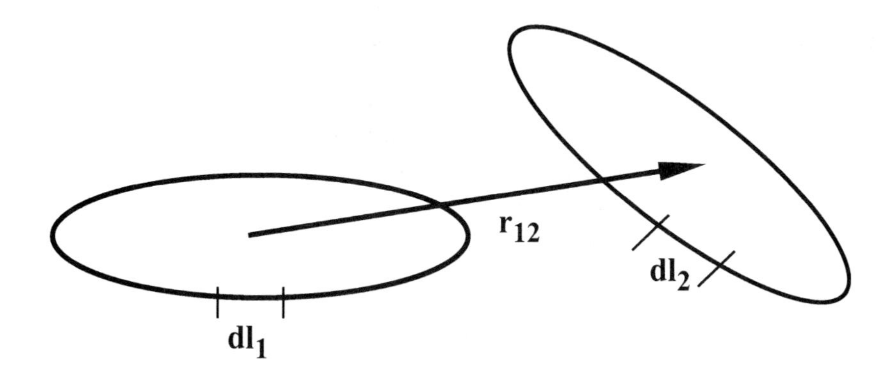

FIGURE 4.1. Interaction forces between two current-carrying wires.

flux density as having units of force per charge-velocity. This latter interpretation is more physically appealing because it emphasizes that in the absence of moving charge, there is no magnetic induction field established.

It is also convenient to define a magnetic flux ϕ, which has units of Webers, and which is related to **B** in a manner analogous to Gauss's law:

$$\phi = \int_S \mathbf{B} \cdot \mathbf{da}, \tag{3}$$

where **da** is an infinitesimal area with direction normal to the surface. It is because of this defining relation that **B** is often referred to as the magnetic flux density field. Note that by comparing this relationship to Gauss's law we can establish a logical analogy between electrical and magnetic variables: **D** must be analogous to **B**, and q to ϕ. Indeed, using the extended Maxwell system of analogies, ϕ is the magnetic displacement variable. The flow variable is the time rate of change of magnetic flux.

We can extend the electrical/magnetic analogy by requiring that there exist a second magnetic field variable analogous to **E**. This magnetic field parameter is **H** and is called the magnetic field. It is related to the magnetic induction through

$$\mathbf{H} = \mathbf{B} - 4\pi \mathbf{M}', \tag{4}$$

where $\mathbf{M}'$ is the magnetization, a field analogous to the electric polarization. The units of magnetic field are amp-turns per meter (A/m). For materials which are not stongly ferromagnetic and for weak fields, the magnetic induction field is proportional to the magnetic field:

$$\mathbf{B} = [\mu]\mathbf{H}. \tag{5}$$

Here $[\mu]$ is known as the magnetic permeability matrix. For isotropic materials, $[\mu]$ reduces to the scalar μ. Note that (5) shows that μ is directly analogous to ϵ.

At this point we have developed a partial description of magnetic environments and a partial analogy with electric variables. To complete the picture it is necessary to relate directly magnetic and electric behavior. Originally, magnetic and electric behavior were thought to be completely independent. This misperception was fostered by the fact that magnetic phenomena were discovered and studied significantly earlier than electric phenomena. In 1831, Faraday provided the first conclusive evidence that magnetic and electric phenomena are linked, by showing that a changing magnetic field can produce a voltage across a single loop of wire. Faraday's law can be expressed as

$$\int_C \mathbf{E} \cdot \mathbf{dl} = -\frac{d}{dt} \int_S \mathbf{B} \cdot \mathbf{da}. \tag{6}$$

Using (3) and the definition of the voltage, we can rewrite this equation as

$$e = -\frac{d\phi}{dt}, \tag{7}$$

which relates electric and magnetic variables. Note that as voltage is a relative measure, and there is ambiguity as to which terminal is defined as ground, the sign in this equation is somewhat arbitrary.

It is common for a conductor to be wound in a coil, in which case it is convenient to define a quantity called the flux linkage, λ, as the product of the number of turns in the coil and the magnetic flux through it:

$$\lambda = N\phi. \tag{8}$$

The terminal voltage then becomes

$$e = -N\frac{d\phi}{dt}, \tag{9}$$

which is the most common form in which Faraday's law is remembered.

Equation (9) shows that for a wound conductor there is a relation between the electric effort variable and the magnetic flow variable. Since we have allowed no loss or gain of energy in moving from one perspective to the other, this relation is described by a gyrator. Hence there must be another link between the electric and magnetic domains relating the electric flow variable to the magnetic effort variable, which we have yet to define. This relationship must be of the form

$$\text{magnetic effort} = Ni, \tag{10}$$

for energy to be conserved. The magnetic effort variable thus defined is referred to as the magnetomotive force (or MAGNETOMOTANCE), M, and the defining relation is a result of Ampére's law. Clearly, the units of MAGNETOMOTANCE are amp-turns.

In general form, Ampére's law may be stated as

$$\int_C \mathbf{B}\cdot\mathbf{dl} = \mu i. \tag{11}$$

This relation states that the line integral of **B** around any closed path is proportional to the total current passing through any surface bounded by that path. In the case of a coil of perfectly conducting wire, as shown in Fig. 4.2, we imagine the closed path to be through the coil center. Since **B** is along the path, the integral in (11) can be easily evaluated. Further, if there are N turns in the toroid, then the current passes a surface bounded by the path exactly N times. Hence, Ampére's law simplifies to

$$Bl = \mu i N. \tag{12}$$

Using the relation $\mathbf{B} = \mu\mathbf{H}$ we rewrite this equation as

$$iN = Hl = M. \tag{13}$$

Note that this equation is identical to (10) which was derived from the gyrator relations. It also shows that the magnetomotive force may be viewed as the magnetic field times the circumferential length of a conductor structure.

With the definition of the magnetomotive force as the magnetic effort variable, the magnetic system of variables is now completely defined. The displacement variable is the magnetic flux, ϕ. The flow variable is $d\phi/dt$. The effort variable is

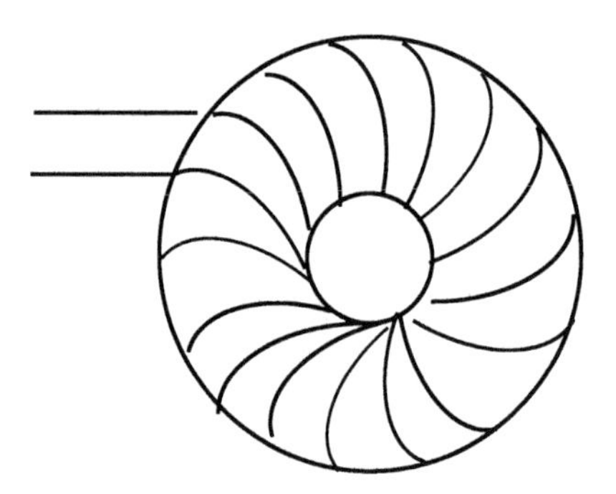

FIGURE 4.2. Toroidal geometry.

M. No momentum variable is defined. This is because there is nothing analogous to an electrical inductor in magnetic systems, i.e., there is no magnetic inertance element. Hence the momentum variable would never be useful as a state variable.

We can use the magnetic variables to demonstrate the additional physical insight they offer. Consider the toroidal coil discussed above. In the electrical domain it would be described by $\lambda = Li$. Viewed from the magnetic domain we note that

$$L = \frac{\lambda}{i} = \frac{N\phi}{M/N} = \frac{N^2 BA}{lH} = \frac{N^2 A\mu}{l}, \tag{14}$$

where A is the cross-sectional area of the toroidal core which we assume to contain all of the magnetic flux. Hence, by using magnetic variables and the electrical/magnetic gyrator relations provided by Faraday's law and Ampére's law, we can determine L precisely.

We may take this approach one step further by defining the permeance and reluctance of a magnetic element. The permeance, P, is defined as

$$P = \frac{\phi}{M}. \tag{15}$$

This ratio is used to define a magnetic capacitance. Since the toroidal coil stores energy, it must be represented as a magnetic capacitance or a magnetic inertance. From (14) we see that the toroid constitutive relation can be written in the form $e = g(q)$. Hence, the toroid is modeled as a magnetic capacitance. The value of the capacitance is given by the permeance:

$$P = \frac{\mu A}{l}. \tag{16}$$

The energy stored in this capacitance, by analogy to the electrical capacitance, is given by

$$\mathcal{E} = \frac{\phi^2}{2P}. \tag{17}$$

The reluctance, R, is the inverse of the permeance. We define it here because it is more commonly used than the permeance.

4.1.1 *Magnetic Materials*

One of the major differences between magnetic fields and electric fields lies in the material behaviors associated with each. All materials exhibit dielectric behavior which can be characterized by an electric field versus charge density field plot (E versus D). An ideal dielectric material appears as a straight line through the origin on such a plot, and many real materials approximate this model quite well. Typical deviations from this ideal electrical material behavior include small amounts of hysteresis, and some saturation effects at very high field strengths.

Magnetic field behavior of materials is similarly characterized by the magnetic field versus flux density field performance. Figure 4.3 shows what some BH material curves resemble. As Fig. 4.3 suggests, there are two classes of material behavior. The "ideal" materials show a straight line through the origin on the magnetic field versus magnetic flux density plot. Such ideal materials are not common, but materials which resemble the ideal behavior with a small amount of hysteresis are common. These materials are referred to as soft ferromagnetic materials.

Hard ferromagnetic materials exhibit a large amount of hysteresis in their BH curve. Two specific points on their BH curve are used to characterize this behavior: the remanence (or residual induction) and the coercivity. The remanence is the value of B when the magnetic field is zero. Note that the presence of significant magnetic flux density field intensity in the absense of an external magnetic field suggests the magnetic equivalent of electric polarization. This magnetic polarization leads, in the extreme, to a magnetic flux density which is relatively independent of the externally applied magnetic field. Such materials are referred to as permanent magnets, and may be modeled fairly well as ideal sources of magnetic flux.

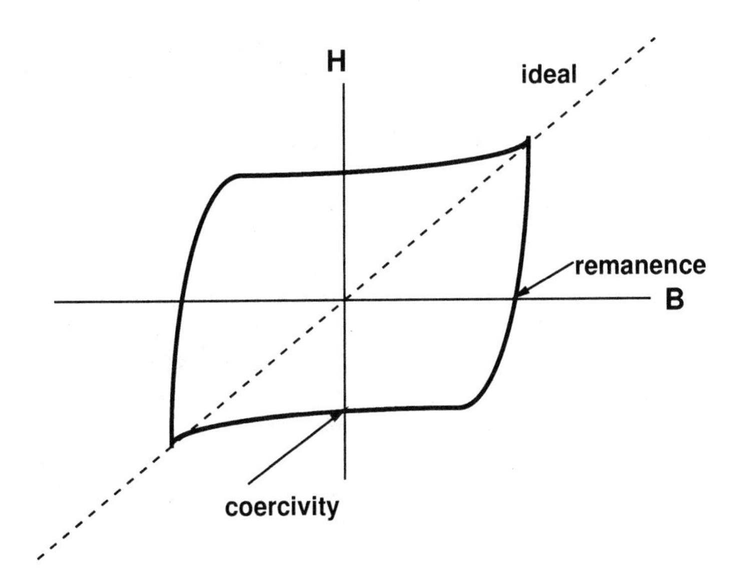

FIGURE 4.3. Magnetic field versus magnetic flux density field for typical magnetic materials.

TABLE 4.1. Remanence and coercivity of some common magnetic materials

Material	Remanence (Wb/m^2)	Coercivity ($\times$ 10^5 A/m)
Alnico 1, 2, 3, 4	0.55–0.75	0.33–0.57
Alnico 5, 6, 7	1.05–1.35	0.51–0.62
Alnico 8	0.70–0.92	1.19–1.51
Alnico 9	1.05	1.27
Copper–nickel–iron	0.55	0.42
Iron–chromium	1.35	0.48
Neodimium–iron–boron	1.15	8.67
Plastics	0.14	0.35–1.11
Rubbers	0.13–0.23	0.80–1.43
Samarium–cobalt	0.81	6.29

The coercivity of the material is the value of the magnetic field for which the flux density is zero. This is a measure of how high an external field is needed to overcome the magnetic flux fixed in the material. The strongest permanent magnetic materials are those which have a high remanence and high coercivity, thus ensuring that their permanent magnetic behavior is not easily overcome. An example of a strong permanent magnet is NdFeB, which has a remanence of 1.15 Wb/m^2 and a coercivity of 8.67×10^5 A/m. Table 4.1 provides the remanence and coercivity of some common magnetic materials.

4.1.2 Magnetic Circuits

Finally, with the definitions above we can model magnetic energy storage elements. The models require a gyrator for conversion from the electrical port variables to magnetic variables. The energy storage then is represented by capacitance elements, where the permeance gives the value of the capacitance. Energy loss due to leakage would be modeled using a magnetic resistance. A very simple model of the toroidal coil is shown in Fig. 4.4. Although the electrical model is simpler, it tends to obscure the physics presented in the magnetic model.

Using the relations described here it is possible to develop magnetic circuits in a manner directly analogous to electrical circuits. For example, consider the simple magnetic system shown in Fig. 4.5. In this system a soft ferromagnetic material is wrapped with wire at one place. Application of a current or a voltage at that electrical port induces a magnetic flux in the material. We model the system as shown in Fig. 4.6, where we have included a single lumped magnetic capacitance

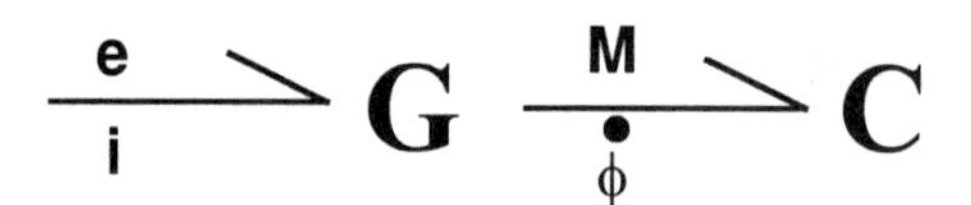

FIGURE 4.4. Bond graph model of a toroidal coil.

to represent each arm of the structure except for the arm containing an air gap. The 0-junctions in Fig. 4.6 serve to define the magnetomotive force at the structure corners. We have assumed that there is no loss in the circuit. Since the 0-junctions are 2-port junctions, the port variables associated with the input and output ports to each junction are identical. Hence they may be eliminated to produce Fig. 4.7. Note that in this figure there are four 1-junctions hooked together. Since all ports attached to a 1-junction have the same flow variable, this can be reduced to a single 1-junction as shown in Fig. 4.8. Finally, we can lump the capacitances to produce a single effective magnetic capacitance. We find the effective capacitance, C_{eff}, by noting that for capacitors in series, their inverses add. Hence,

$$C_{\text{eff}} = \frac{1}{\Sigma \frac{1}{C_i}}, \tag{18}$$

where each capacitance is given by the permeance of that segment of the structure. Since the magnetic permeability for typical soft ferromagnets is roughly three orders of magnitude greater than that for air, the air gap capacitance dominates in determining the effective capacitance. In other words, in typical examples of the structure shown in Fig. 4.5 most of the energy is stored in the air gap.

In the next sections, as we discuss transduction mechanisms which rely on changes in the energy stored in a magnetic field, we will introduce more complex magnetic circuits.

4.2 Variable Reluctance Transducers with Varying Gap

In Chapter 3 we discussed in detail transduction mechanisms in which the capacitance of the system is varied by intentionally varying a physical parameter on which the capacitance depends. This led us to consider variable gap, variable area, and variable permittivity capacitors. Similarly, we can build magnetic transducers by intentionally varying the parameters on which the magnetic capacitance depends: μ, A, and l. These transducers are generally grouped under the heading of variable reluctance transducers.

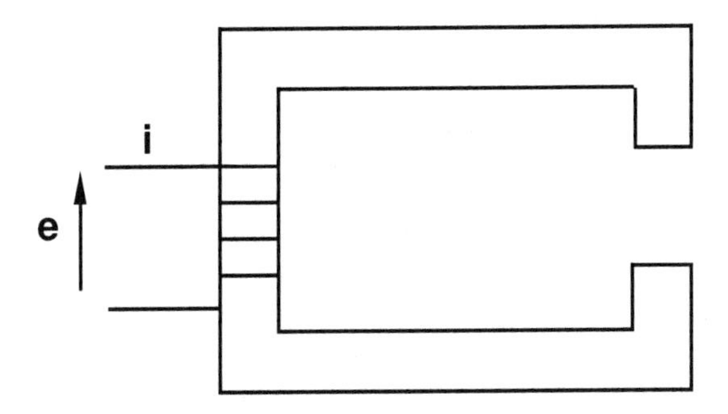

FIGURE 4.5. A simple magnetic system.

First consider the analogy to the variable gap electrical capacitor, i.e., the variable gap magnetic capacitor. Suppose that we have a magnetic circuit with some portion as indicated in Fig. 4.9. Now the energy stored in the air gap is clearly dependent on both the magnetic flux and the gap dimension. Hence the gap is now a 2-port capacitance as shown in Fig. 4.10. To determine the constitutive relations for this capacitance we express the energy stored, and differentiate to determine the force and Magnetomotance. This procedure is identical to that used for the capacitive systems presented in Chapter 3.

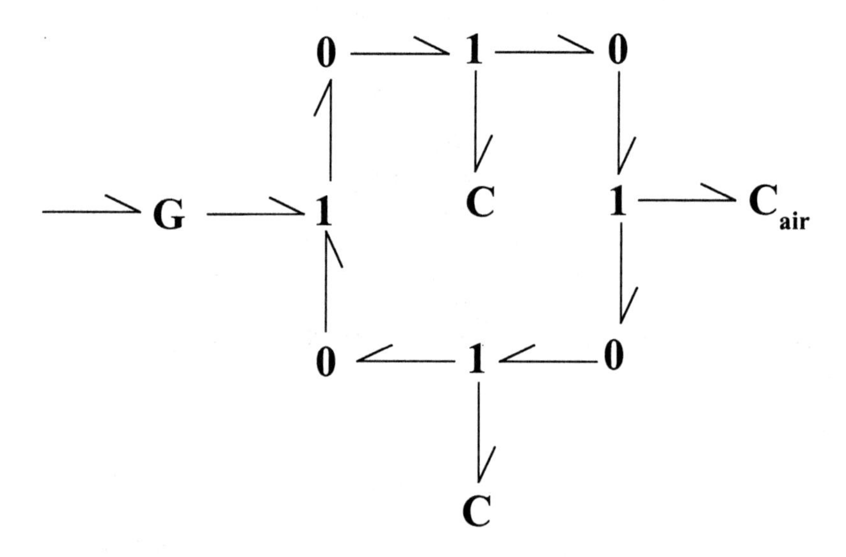

FIGURE 4.6. Model of a toroid including the magnetic capacitance.

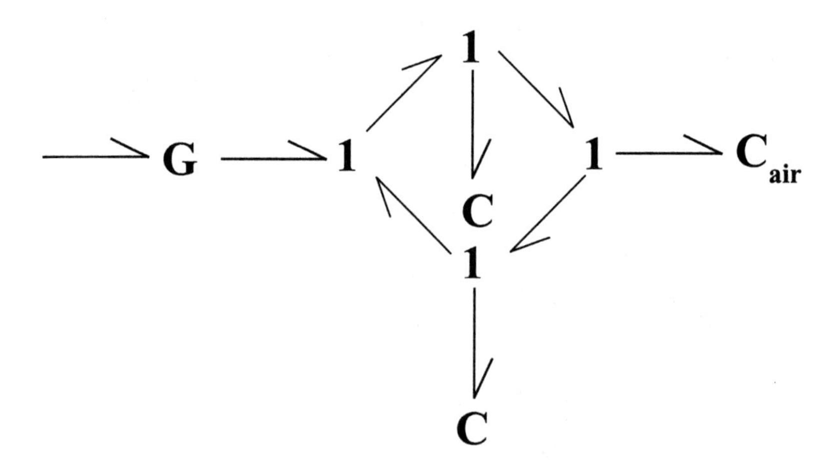

FIGURE 4.7. Reduced model of the toroid.

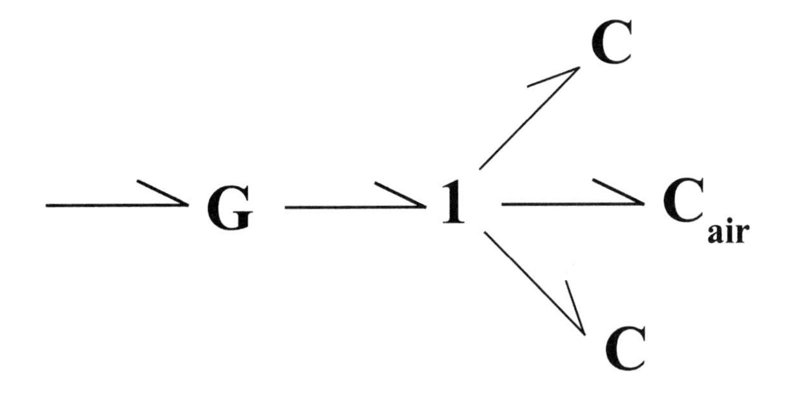

FIGURE 4.8. Further reduced model of the toroid.

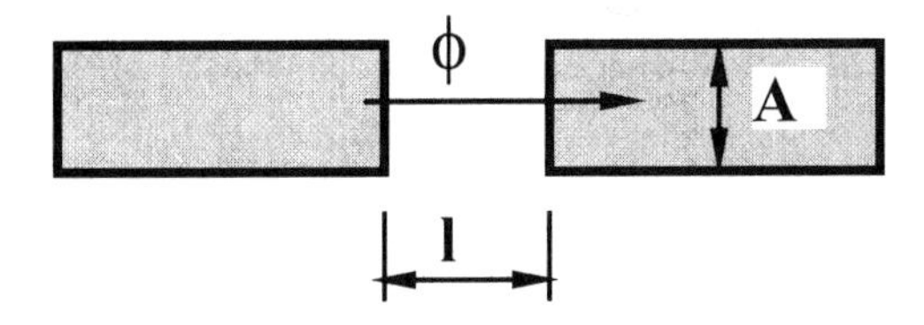

FIGURE 4.9. Magnetic element in which the gap dimension changes dynamically.

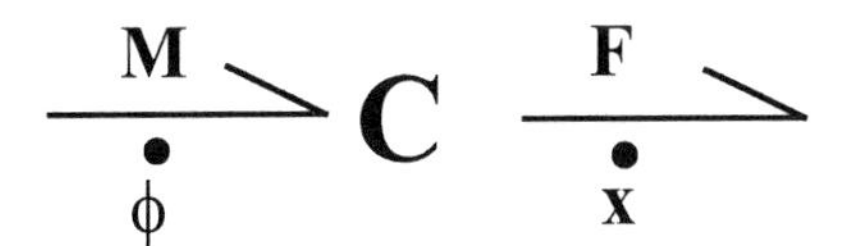

FIGURE 4.10. Model of magnetic element in which the gap dimension varies.

The energy stored in the air gap is given by

$$\mathcal{E} = \frac{\phi^2}{2C} = \frac{\phi^2}{2P} = \frac{\phi^2 l}{2\mu A}. \tag{19}$$

The power associated with the gap is the time rate of change of the energy stored:

$$\mathcal{P} = \frac{\partial \mathcal{E}}{\partial t} = \frac{\partial \mathcal{E}}{\partial \phi}\frac{d\phi}{dt} + \frac{\partial \mathcal{E}}{\partial l}\frac{dl}{dt}. \tag{20}$$

Since $d\phi/dt$ is the flow variable we identify $\partial\mathcal{E}/\partial\phi$ with M. Similarly, $\partial\mathcal{E}/\partial l$ is the magnetostatic force between the two pieces bordering the gap. Hence the constitutive equations are

$$M = \frac{\partial \mathcal{E}}{\partial \phi} = \frac{\phi l}{\mu A}, \tag{21}$$

$$F = \frac{\partial \mathcal{E}}{\partial l} = \frac{\phi^2}{2\mu A}. \tag{22}$$

Note that these equations are precisely analogous to those of a parallel plate capacitor with the appropriate parameter transformations from electric to magnetic domains (i.e., q to ϕ, ϵ to μ, d to l).

There are three means by which we can use the above constitutive relations in transduction associated with a variable gap: by relating l and M, by relating M and F, and by relating F and ϕ. Below we describe each of these transduction mechanisms, and present examples of transducers using them.

Knowing that there is a one-to-one correspondence between the magnetic and electric capacitance, we could immediately proceed to write down the various transduction relations for magnetic to mechanical coupling purely based on analogy. We do not take that approach here for two reasons. First, we choose to rederive most of the transduction relations in order to reinforce the methods which apply universally for derivation of constitutive relations from energy expressions. Second, there is an important practical distinction between the capacitive and inductive transducers which is obscured by consistently remaining in the electric domain. This distinction arises from the fact that there are two means to drive variable reluctance transducers: magnetically or electrically. Using a purely magnetic drive means applying a magnetic bias through use of a permanent magnet. Using an electric drive of variable reluctance transducers typically means using a voltage or current bias. The difference between the electric and magnetic drive is important because an electric drive cannot produce a dc magnetic flux. This is obvious from (9), which relates voltage to the *time rate of change* of magnetic flux. Throughout this chapter we will discuss both magnetically and electrically driven transducers.

4.2.1 Relating Displacement and MAGNETOMOTANCE

First consider transduction in which we relate l and M using (21). Note that this equation involves three variables: ϕ, l, and M. We wish to eliminate ϕ and relate l and M. There are two distinct ways to accomplish this goal: using a permanent magnet with the assumption that $\phi = \phi_0$, a constant, or using a solenoid driven at a specified frequency. Below we consider each of these approaches.

First imagine that we use a permanent magnet to establish the flux between the two poles whose separation distance is changing. As suggested by Fig. 4.3 a permanent magnet has a nearly constant value of ϕ over a large range of magnetic field, so let us assume that the flux is constant at a value ϕ_0. Then we have simply

$$M = Ni = \frac{\phi_0 l}{\mu A}, \tag{23}$$

which establishes a linear relationship between M or i and l. In theory, there is a dc current corresponding to the equilibrium gap dimension and a time-varying current corresponding to the time-varying gap dimension. In practice, it is nearly impossible to establish a dc current due to persistent conduction losses in real electronics. Thus, it is the time-varying current which is germane to the present discussion.

If the gap between the pole pieces is written as a nominal value, l_0, and a time-varying value, x, then we can use (23) to determine the sensitivity of the transducer. As a sensor, with x producing a signal current, $\hat{i}$, the sensitivity is given by

$$\psi = \frac{\hat{i}}{x} = \frac{\phi_0}{N\mu A}, \tag{24}$$

which suggests that the best sensitivities are obtained for strong permanent magnets, and small values of N and A. As an actuator, the sensitivity is given by the inverse of the expression in (24).

It should be noted that the results here were obtained without resorting to any sort of linearization, suggesting that this transduction mechanism is linear by its very nature. However, this conclusion would not be entirely justified, as the assumption that the permanent magnet produces a flux which does not vary with exposure to a magnetic field is not exactly correct. To the extent that a permanent magnet has a flux which is affected by the magnetic field produced in the solenoid, the change in the flux will distort the linear results presented here.

Now consider the same transduction mechanism driven with a solenoid providing the flux. Then we have

$$e = -N\frac{d\phi}{dt}, \tag{25}$$

so

$$\phi = -\frac{1}{N}\int e dt. \tag{26}$$

It is clear from (26) that we cannot produce a constant flux from a solenoid. Thus, it is necessary to resort to an alternative approach. The commonly chosen solution to this problem is to use an ac voltage at frequency ω to produce a magnetic flux at the same frequency. Thus

$$e = v_0 e^{j\omega t}, \tag{27}$$

and

$$\phi = \hat{\phi} = \frac{jv_0}{\omega N}e^{j\omega t}. \tag{28}$$

Substituting this expression for the flux into the expression relating M and l yields

$$\hat{M} = N\hat{i} = \frac{jv_0}{\omega N\mu A}e^{j\omega t}. \tag{29}$$

If we choose to rewrite the current using the following definition:

$$\hat{i} = \mathcal{I}e^{j\omega t}, \tag{30}$$

then we identify the current amplitude as

$$\mathcal{I} = \frac{j v_0 l_0}{\mu \omega N^2 A} + \frac{j v_0 x}{\mu \omega N^2 A} = \mathcal{I}_0 + \hat{\mathcal{I}}. \tag{31}$$

This expression shows that the current amplitude at the frequency at which the solenoid is driven has two terms. One is related to the nominal gap between the pole pieces. The other varies linearly with the gap dimension variation. Thus we can monitor this variation by looking at changes in the current amplitude. As a sensor then, the sensitivity is given by

$$\psi = \frac{|\hat{\mathcal{I}}|}{|x|} = \frac{v_0}{\omega N^2 \mu A}, \tag{32}$$

which is maximized for large driving voltages, low frequencies, and small values of N and A. As a practical matter, the key to making this approach work is to choose a value of ω which is outside the frequency range of interest in the sensor. This typically means choosing a higher driving frequency for the solenoid than the mechanical frequencies of interest.

Note that the transduction mechanism using a solenoid is linear provided that the voltage driving the magnetic flux is constrained to a set value. As a result there is no significant second harmonic distortion associated with this transduction approach.

An example of a transducer which relates the distance and magnetomotance (or current) in a variable gap variable reluctance transducer is a simple electromechanical relay such as shown in Fig. 4.11. In this device an arm of a magnetic circuit is attached to a moving object. As the arm rotates, the air gap in the magnetic circuit is modulated, thus producing a change in the magnetomotance which is detected as a current change. For small angles of rotation the motion of the relay arm is nearly translational. A spring is supplied to restore the relay to its equilibrium position. The relay is meant to serve as a simple on–off mechanism. Thus there is no attempt to linearize its operation by supplying a bias.

Figure 4.12 shows a bond graph model of the relay' neglecting any flux leakage or energy loss. The important aspect of this model from the point of view of transduction is the 2-port capacitance representation of the air gap. The constitutive relations of the air gap are those developed previously. We could include energy

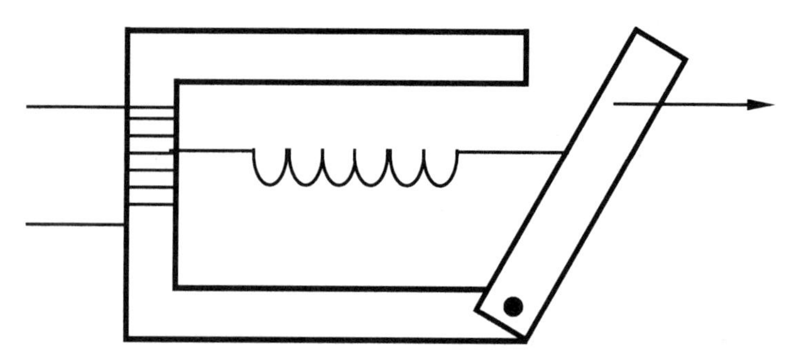

FIGURE 4.11. Magnetic relay.

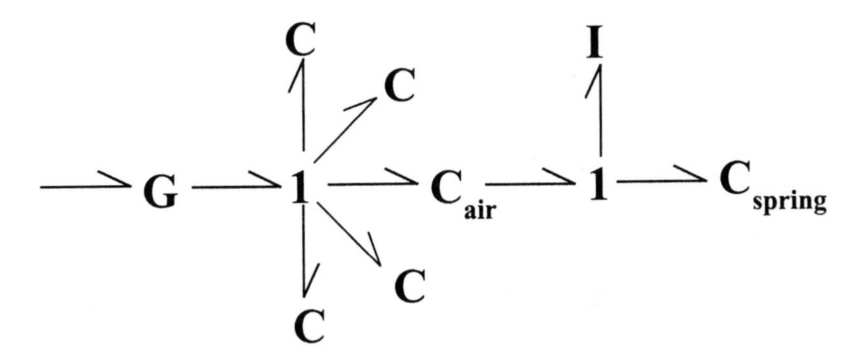

FIGURE 4.12. Model of a relay.

lost to the system by the addition of resistance along each magnetic arm. We could include internal leakage of flux by including a lumped leakage path along which the capacitance is given by C_{leak}. The new model would then be as given in Fig. 4.13 which assumes that the relay will be used as a sensor of displacement. A prescribed l, and thus given mechanical flow, produces an output current. As an actuator, in which a supplied current is used to produce a displacement of the relay arm, the system is as shown in Fig. 4.14. In either of these models, the multiple 1-port resistance and 1-port capacitance elements on the 1-junction, could be lumped into a single effective resistance (sum of resistances) and an effective capacitance (inverse of the sum of the inverse of the capacitances). The result for the actuator version, for instance, is shown in Fig. 4.15.

As modeled in Fig. 4.15, the relay is a fifth-order system. The leakage capacitance is a dependent energy storage element because the system is driven by a current which establishes the magnetomotive force in the relay body, independent of the existence of a leakage path. Using this model we can consider the low- and high-frequency limiting behavior of the actuated relay. At low frequencies, we

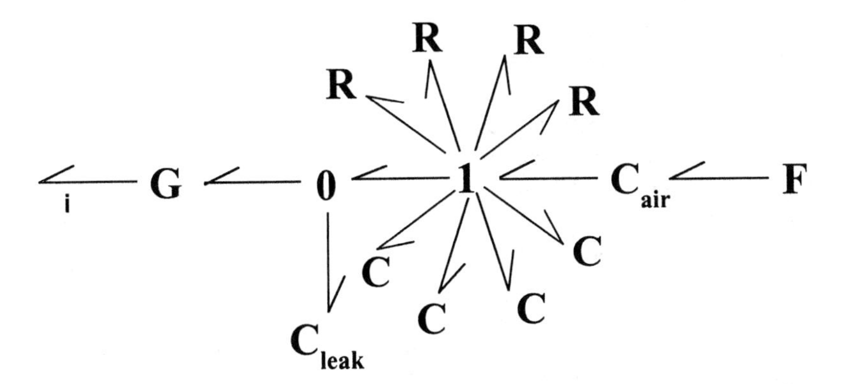

FIGURE 4.13. Model of a displacement sensing relay including leakage and flux loss.

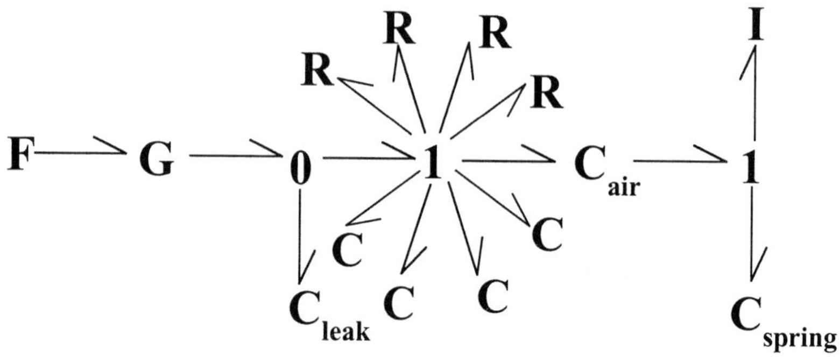

FIGURE 4.14. Model of an actuating relay including leakage and flux loss.

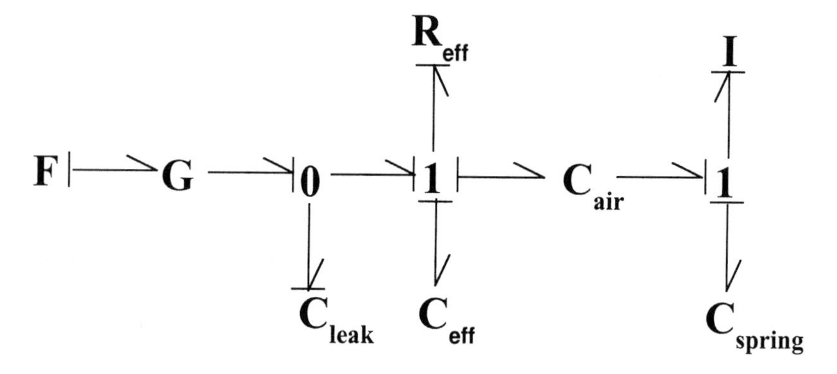

FIGURE 4.15. Simplified model of an actuating relay including leakage and flux loss.

consider the impedances at the 1-junction with the inertance and capacitance. The capacitance of the spring will dominate as the impedances add, and its impedance varies inversely with frequency while the mass's increases linearly with frequency. At the other 1-junction, the capacitance also dominates, since the impedance of the resistance is independent of frequency. Hence, a very much simplified model appropriate for low frequencies is shown in Fig. 4.16. Given only capacitive elements in the system, we expect the frequency response for the displacement versus magnetomotance (or current) to be flat. At high frequencies, the resistance and inertance previously ignored dominate over the capacitances previously included. For sufficiently high frequencies, the mass inertance dominates and we would anticipate the relay position versus magnetomotance to decay as $1/\omega^2$.

While the mechanical relay provided a good opportunity for us to examine a transducer whose operation exploits the magnetomotance/gap relation using previously developed modeling expertise, it is only one of many transducers which exploit this mechanism. For instance, Fig. 4.17 shows a pressure transducer made by CEC Vibration Products [5] in which pressure causes motion of a diaphragm relative to a core. A flux is induced in the core and the magnetic circuit closes

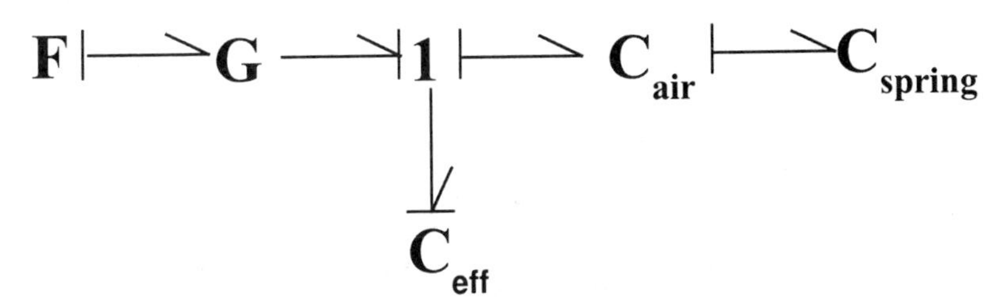

FIGURE 4.16. Simple low-frequency model of an actuating relay.

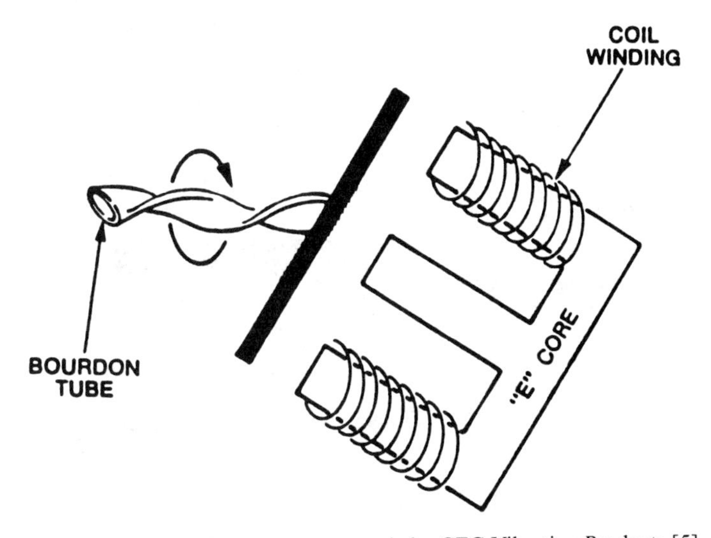

FIGURE 4.17. Pressure sensor made by CEC Vibration Products [5].

through the diaphragm. Thus, motion of the diaphragm directly affects the energy stored in the gap between it and the core.

Another sensor which exploits variable reluctance in which the gap is changing is shown in Fig. 4.18, taken from Cluley [6]. This transducer uses a solenoid in the center of an E-shaped core to establish a magnetic flux, and solenoids on the far arms of the E to sense the currents induced by angular rotation. The device uses two sensing solenoids so that the difference between the induced currents can be considered. Clearly this difference should be zero when the rocking armature is equidistant from the two core ends, and increases as the armature rotation increases.

One final example of a transducer using the relationship between magnetomotance and gap is shown in Fig. 4.19, taken from Trietley [7]. Here a permanent magnet establishes a flux. The magnetic circuit normally closes through the structure of the transducer. When a material with high magnetic permeability moves close to the transducer, the magnetic circuit changes its path, effectively changing the size of the gap. This induces a current in the coil. Thus, this device operates as a proximity sensor.

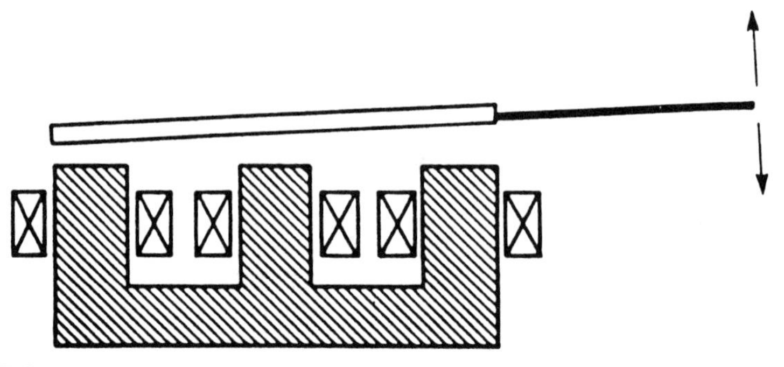

FIGURE 4.18. Rocking armature angular position sensor (taken from Cluley [6]).

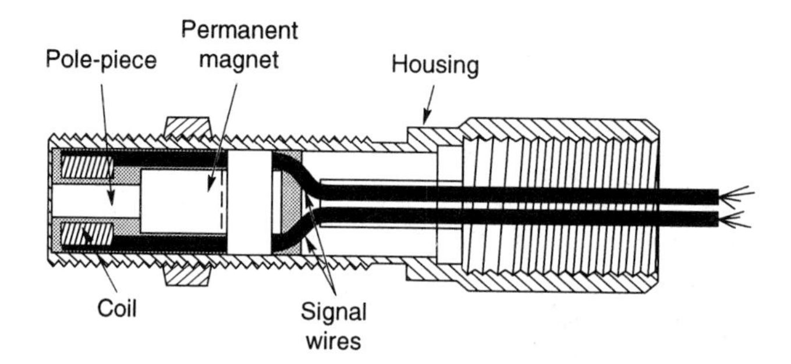

FIGURE 4.19. Magnetic proximity sensor taken from Trietley [7].

4.2.2 *Relating Force and Magnetomotance*

A second transduction process relates M and F in a manner similar to that used in relating e and F in the parallel plate capacitive transducer. Using (21) and (22) we see that

$$M^2 = \frac{2Fl^2}{\mu A}. \tag{33}$$

This relationship is not only nonlinear, but relates three variables rather than two. In that sense, it is quite analogous to the voltage/force relationship discussed in Chapter 3. While it is possible to analyze the conditions under which we could produce a linear transducer from this transduction relation, linearity is generally not an important issue in devices exploiting this mechanism. This is because (33) provides a relationship between current and force which can be used in actuation, and in actuation (in this case a current generating a force) we are often more concerned about generating large forces than about linearity. It is clear from (33) that one can maximize the force level by increasing the magnetomotance (or current), increasing A or μ, or decreasing l.

An example of a transducer using this mechanism in shown in Fig. 4.20, and was discussed by Dario et al. [8]. In this microvalve, current through a coil is used

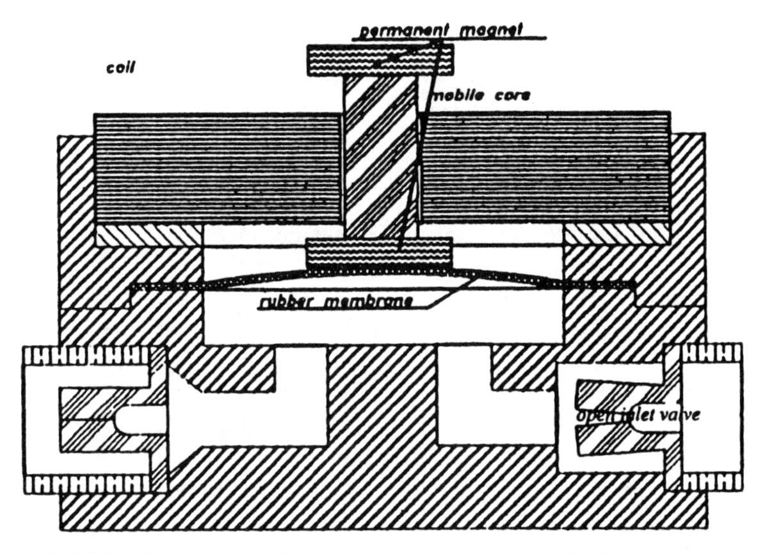

FIGURE 4.20. Magnetically driven microvalve proposed by Dario et al. [8].

to establish a force on a structure connected to a rubber membrane. The motion of this membrane controls the valves and fluid flow.

4.2.3 Relating Force and Flux

The third transduction mechanism relates F and ϕ in a system in which an air gap varies. For such a transducer we need only consider (22). This constitutive relation immediately indicates a quadratic nature. If the magnetic flux is established by a permanent magnet, then the resulting force is a static force described by

$$F = \frac{\phi_0^2}{2\mu A}. \tag{34}$$

We identify this as the magnetostatic force of attraction between the two pole pieces through which a flux of ϕ_0 flows.

Now consider an electrical bias, with

$$e = v_0 e^{j\omega t}. \tag{35}$$

This gives rise to a magnetic flux bias described by

$$\phi_{\text{bias}} = \frac{j v_0}{\omega N} e^{j\omega t}, \tag{36}$$

and a steady-state force described (using (22)) by

$$F_{\text{ss}} = \mathcal{F}_{\text{ss}} e^{2j\omega t} = \frac{v_0^2}{2\mu A \omega^2 N^2} e^{2j\omega t}. \tag{37}$$

If we assume that there is an additional component to the magnetic flux, which arises in response to a time-varying force at frequency Ω, or which is applied to create a force, then $\phi = \phi_{\text{bias}} + \hat{\phi}$ and $F = F_{ss} + \hat{F}$. Now we may make the appropriate substitutions into (22) and derive expressions for $\hat{F}$. The result is

$$\hat{F} = \hat{\mathcal{F}} e^{j(\omega+\Omega)t} = \frac{j v_0 \hat{\phi}}{\mu \omega A N} e^{j(\omega+\Omega)t}, \tag{38}$$

neglecting the second-order perturbation term.

Equations (37) and (38) show that this transduction mechanism is different from those expressed previously. The steady-state force component is a sinusoidal variation at twice the frequency of the electrical drive. The signal force–flux relationship shows that the force and magnetic flux do not have the same frequency. Indeed, as written above, the force perturbation is at a frequency which is the sum of that of the non-steady-state component of the magnetic flux and that of the electrical bias. Thus we effectively produce a frequency modulation by having a magnetic flux which varies at a frequency different from that used to drive the steady-state flux. At the sum frequency, the amplitude of the force allows us to determine the sensitivity. We could define this as the ratio of the amplitude of $\hat{F}$ to $\hat{\phi}$. However, we typically monitor the flux by looking at the voltage across a solenoid with n turns in it. Thus the sensitivity as a force sensor is

$$\psi = \frac{n^2 \Omega \omega \mu A}{v_0}, \tag{39}$$

which is maximized for large values of n and ω and small values of v_0. As an actuator, with flux inducing a force, the sensitivity is the inverse of that shown in (39).

A classic example of an actuator in which the force and magnetic flux are related is the solenoid. For instance, Fig. 4.21, taken from the literature of Ledex Products of Lucas Control Systems [9], shows how we might use a current in a coil to develop a magnetic flux in a solenoid core. This flux tends to draw closed a gap between the core and the stationary magnetic POLE PIECE, resulting in either a pulling force or a pushing force on the material connected to the core. Note that due to the nature of the force dependence on the square of the flux, the force is always directed so as to close the gap. Thus, the simple solenoids shown in Fig. 4.21 must be restored to their original position by means of some nonelectrical force. One possibility, a Ledex Product sold by Lucas Control Systems [9] is a solenoid with embedded restoring spring, as shown in Fig. 4.22.

Other interesting variations of solenoid actuators include models with noncylindrical cores. In these solenoids, the core shape is chosen to maximize some desired performance goal such as stroke length or force delivery or efficiency. In addition, rotary solenoids based on the approach shown in Fig. 4.21 have been developed. An example, marketed by Ledex Products, Lucas Control Systems [9] is shown in Fig. 4.23. In this solenoid actuator, the core rides on ball bearings in which the races deepen in one direction of rotation. As a result of a translational force tending to close a gap, the core rotates so as to move to the lowest position in the races. Again, this solenoid only moves in one direction unless a restoration force is provided by a spring.

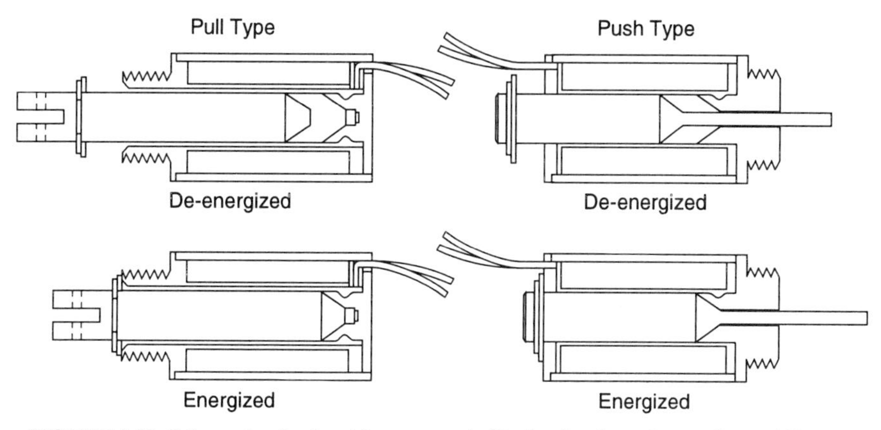

FIGURE 4.21. Schematic of solenoid actuators, (of Ledex Products, Lucas Control Systems [9]).

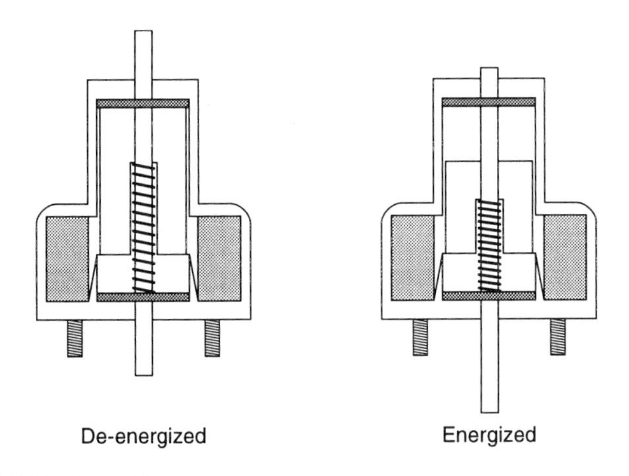

FIGURE 4.22. Solenoid actuators with a spring to provide restoring force (from Ledex Products, Lucas Control Systems [9]).

In all of the solenoids, the key aim is to open and close the gap quickly. Linearity is hardly ever a concern. Thus, the solenoids are driven by a dc voltage. In theory this leads to an infinite magnetic flux, but in practice, the flux increases rapidly until it saturates. This guarantees a rapid closing of the gap, although the relationships are far from linear.

Interesting examples of switch-type sensors based on the force and flux relations include devices using reed switches. A typical reed switch is shown in Fig. 4.24, taken from Omega Engineering, Inc. [10]. Reed switches are versions of electromagnetic relays in which thin ferromagnetic cantilevers (reeds) are normally separated by a small gap. However, when exposed to a permanent magnet, the flux through the reeds establishes a force drawing them together. This attractive force causes the reeds to make contact, thus closing the switch.

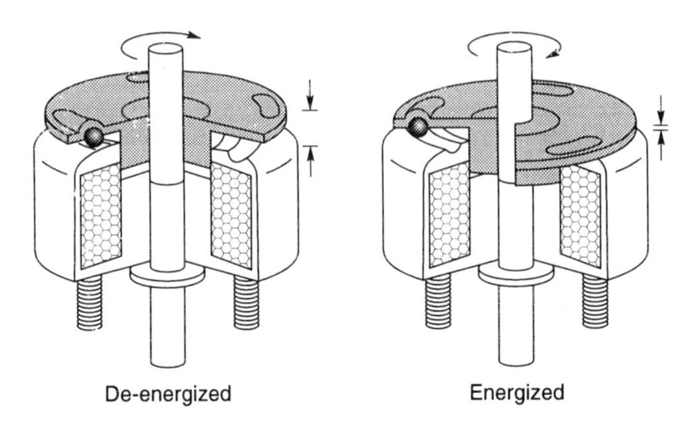

FIGURE 4.23. Rotary solenoids based on force and flux relations (from Ledex Products, Lucas Control Systems [9]).

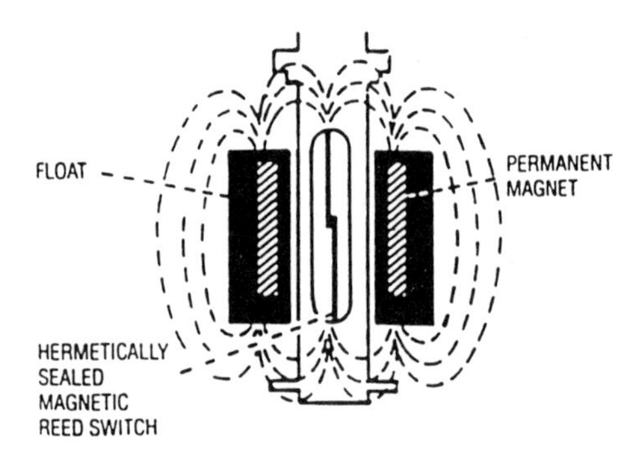

FIGURE 4.24. Reed switch Copyright 1995 Omega Engineering, Inc. All rights reserved. Reproduced with permission of Omega Engineering, Inc. Stamford, CT 06907 [10].

Figure 4.25 shows an example of a liquid level sensor using a reed switch. This device, sold by Malema Engineering Corporation [11], uses a reed switch sealed in a tube about which a float moves. The float contains permanent magnets which close the normally open switch when the liquid level float nears the reed switch position.

Another sort of sensor based on reed switch operation is a liquid flow sensor. Figure 4.26 shows a device marketed by Gems Sensors [12] in which the flow of a liquid displaces a spring-loaded shuttle which houses a reed switch. Permanent magnets in the housing induce a flux in the reed switch when the shuttle is raised sufficiently, thus changing the switch position. As Fig. 4.26 shows, the reed switch used in this particular device contains three reeds. The advantage of this over the conventional two-reed switch is that the three-reed switch can be either normally open or normally closed whereas the two-reed switch is normally open.

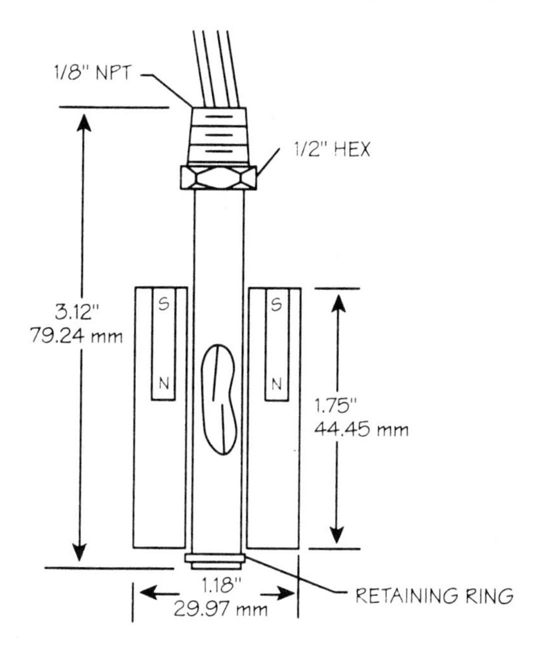

FIGURE 4.25. Liquid level sensor sold by Malema Engineering Corporation [11].

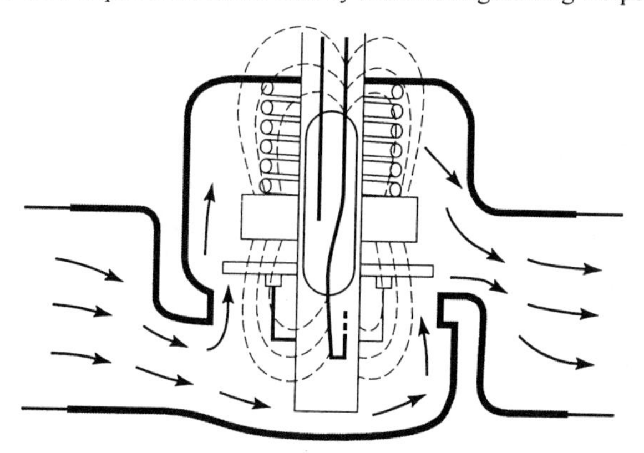

FIGURE 4.26. Liquid flow sensor made by Gems Sensors [12].

4.3 Variable Reluctance Transducers with Varying Permeability and Area

Now let us consider variable reluctance transducers in which a parameter other than the length of the magnetic path is varied. Given the reluctance as defined for air gap systems, there are two other parameters which could be modulated to produce an electrical signal, or stimulated to vary in response to electrical input, namely, the cross-sectional area of the magnetic path, and the magnetic permeability. We consider each below.

4.3.1 Variable Area

First consider variation of the cross-sectional area of the magnetic path. We could theoretically achieve this area variation in precisely the same manner as accomplished for the parallel plate capacitor transducer—by moving one of the pieces normal to the field direction, so that the effective area is given by the area of overlap of the metal parts. This situation is shown in Fig. 4.27. Now the magnetic capacitance is given by

$$C = \frac{\mu w y}{l}, \tag{40}$$

where w is the width of the metal parts into the page, and y is the length of the overlap. The energy stored by the magnetic capacitance is

$$\mathcal{E} = \frac{\phi^2 l}{2\mu w y}. \tag{41}$$

Note that this energy varies inversely with the position of the moving part.

We may use (41) to determine constitutive equations that are consistent with the energy expressions:

$$M = \frac{\partial \mathcal{E}}{\partial \phi} = \frac{\phi l}{\mu w y}, \tag{42}$$

$$F = \frac{\partial \mathcal{E}}{\partial y} = -\frac{\phi^2 l}{2\mu w y^2}. \tag{43}$$

Both of these equations are nonlinear relations in terms of the displacement y. They are identical to those derived for the analogous case of the area-varying variable capacitance parallel plate transducer with the appropriate parameter transformations. Indeed, these expressions could have been derived using these transformations rather than starting again with energy expressions.

Just as in the case of the parallel plate capacitive transducer, we could analyze the constitutive equations given in (42) and (43) and even seek to produce nominally linear versions of them. However, variable reluctance transduction based on dynamic changes in the cross-sectional area of a gap is essentially never exploited. This is because magnetic field lines bend quite easily. If the situation shown in Fig. 4.27 were to be created, then the field would simply bend between the pole pieces defining the air gap, thus eliminating the effective area change. In order to exploit the relations developed above, we would need to have alternative paths for

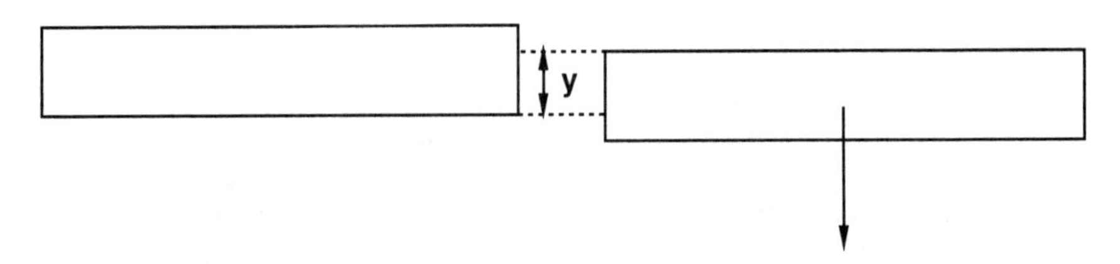

FIGURE 4.27. Variable area variable reluctance transducer.

the magnetic flux THAT ARE more energetically advantageous than simply bending to the moving pole piece. In other words, we would need some of the flux to switch its path to a nearby, fixed pole piece. This is difficult to establish in a well-controlled fashion, and thus this mechanism is not traditionally used.

4.3.2 *Variable Permeability*

Now consider transducers in which the air gap and cross-sectional area are fixed but the magnetic permeability varies. Unlike their electrical counterparts, this transduction mechanism is quite common, forming the principle upon which LVDTs (linear variable differential transformers) and RVDTs (rotational variable differential transformers) are based.

Suppose that we consider a strict analog of the parallel plate capacitor with time-varying permittivity. This magnetic analog is shown in Fig. 4.28. Fundamental to this system is the motion of a structure which has a discontinuity between two materials whose magnetic permeabilities differ. We now have a gap which has permeability μ_1 along a distance l_1, and which is some other material (permeability μ_2) along a distance l_2. The length of the gap and its total cross-sectional area are fixed. Also, since the coil is wound uniformly, the magnetic induction and magnetic flux are spatially uniform. Hence, since the time rate of change of magnetic flux is the flow variable in magnetic systems, the appropriate model for the system uses two capacitors in series. Further, the magnetic field and magnetomotance are different in the two regions defined by the two distinct permeabilities. We model the system, then, as shown in Fig. 4.29, and the total capacitance of the gap is given by

$$\begin{aligned} C_{\text{eff}} &= \frac{1}{1/C_1 + 1/C_2} \\ &= \frac{\mu_1 \mu_2 A}{(\mu_1 l_2 + \mu_2 l_1)}. \end{aligned} \tag{44}$$

Note that if one of the materials has a permeability much lower than that of the other, for example, $\mu_1 << \mu_2$, then the effective magnetic capacitance is simply that of the capacitor with low permeability, e.g., $C_{\text{eff}} = C_1$.

The energy stored in the gap is given by

$$\mathcal{E} = \frac{\phi^2}{2C_{\text{eff}}} = \frac{\phi^2(\mu_1 l_2 + \mu_2 l_1)}{2\mu_1 \mu_2 A}. \tag{45}$$

We may eliminate one of the variables by noting that $l_1 + l_2 = l$ is constant, so

$$\mathcal{E} = \frac{\phi^2[\mu_1 l + (\mu_2 - \mu_1) l_1]}{2\mu_1 \mu_2 A}. \tag{46}$$

We use this expression to determine the new constitutive relations

$$M = \frac{\phi[\mu_1 l + (\mu_2 - \mu_1) l_1]}{\mu_1 \mu_2 A}, \tag{47}$$

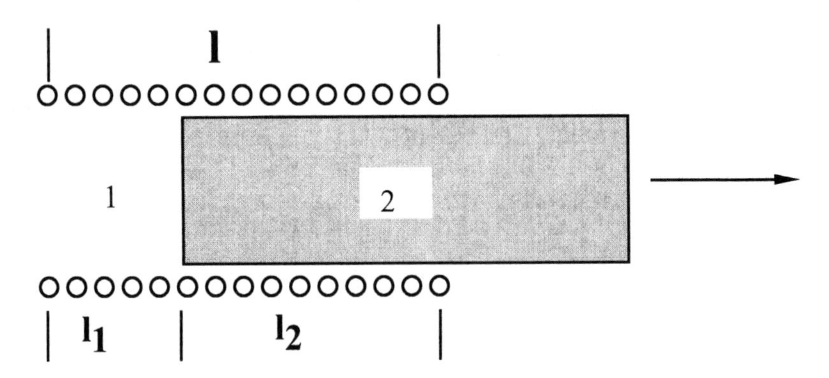

FIGURE 4.28. Transducer in which the magnetic permeability varies dynamically.

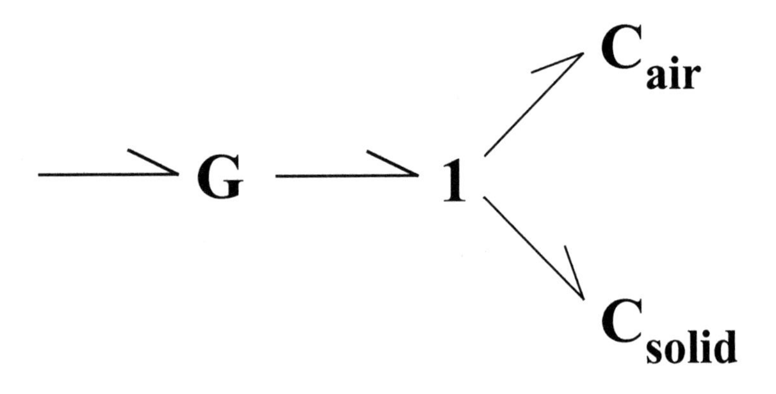

FIGURE 4.29. Model of the transducer in which the magnetic permeability varies dynamically.

$$F = \frac{\phi^2(\mu_2 - \mu_1)}{2\mu_1\mu_2 A}. \tag{48}$$

Here F is the force on the moving material in the direction of motion. Note that this force depends only on the difference in the permeabilities and the magnetic flux. It is not a function of the position of the material discontinuity in the coil.

Equations (47) and (48) are nonlinear constitutive relations linking the independent variables ϕ and l_1 to the dependent variables M and F. There are three means of producing a transducer using them: by relating M (or i) and l_1, by relating M (or i) and F, and by relating F and ϕ. Of these, the M and l_1 transduction is the most common because it directly relates the motion of the intentionally moving object to the magnetomotance which is sensed, using Ampére's law, as a current. It is this transduction mechanism which is presented in detail below. The other two relations are left for the reader to consider.

To establish a relationship between M and l_1 we exploit (47). Note that this equation involves three variables: ϕ, l_1, and M. Thus we seek to establish a known ϕ in order to eliminate it as a variable. As with most of the magnetic systems, there are two ways to do this: with a permanent magnet, and with a solenoid.

First consider a permanent magnet establishing a magnetic flux ϕ_0. Then (47) can be rewritten as follows:

$$M = \frac{\phi_0 l}{\mu_2 A} + \frac{\phi_0(\mu_2 - \mu_1)l_1}{\mu_1 \mu_2 A} = M_0 + \hat{M}. \tag{49}$$

Note that there is a static magnetomotance term and a time-varying term which corresponds to the signal in which we are interested. This time-varying term depends on the difference between the magnetic permeabilities of the two media, and produces a time-varying current in a solenoid with $\hat{M} = N\hat{i}$. As a position sensor, the sensitivity, then is given by

$$\psi = \frac{\hat{i}}{l_1} = \frac{\phi_0(\mu_2 - \mu_1)}{\mu_1 \mu_2 A N}. \tag{50}$$

If we assume that $\mu_2 >> \mu_1$, then the sensitivity is approximately given by

$$\psi \approx \frac{\phi_0}{\mu_1 A N}, \tag{51}$$

which is maximized for a strong magnet, and small values of A and N.

Note that the transduction relation given in (49) is linear. To the extent that there is distortion, it is the result of the permanent magnet flux not being a constant. Thus, a real advantage of this approach to transduction is that it is quite linear. However, the geometry can be difficult and the cost and weight of permanent magnets can be prohibitive in many situations.

Now consider a flux bias which is produced by a solenoid so that $\phi = \phi' e^{j\omega t}$. Then (47) becomes

$$M = \frac{\phi' l}{\mu_2 A} e^{j\omega t} + \frac{(\mu_2 - \mu_1) l_1 \phi'}{\mu_1 \mu_2 A} e^{j\omega t} = (\mathcal{M}_0 + \hat{\mathcal{M}}) e^{j\omega t}. \tag{52}$$

Again, it is the non-steady-state terms which produce the signal of interest in transduction.

From (52) we may find the sensitivity of the transducer used in position sensing. We define this as the signal current amplitude divided by l_1:

$$\psi = \left| \frac{\hat{\mathcal{I}}}{l_1} \right| = \left| \frac{(\mu_2 - \mu_1)\phi'}{\mu_1 \mu_2 A N} \right|. \tag{53}$$

Note that in the limit that $\mu_2 >> \mu_1$, this reduces to the same sensitivity as the permanent magnet approach, with the substitution that $\phi' = \phi_0$.

The most common examples of variable magnetic permeability transducers are LVDTs and RVDTs. These devices are based on the transduction principles described above but tend to be somewhat more complicated simply because of the

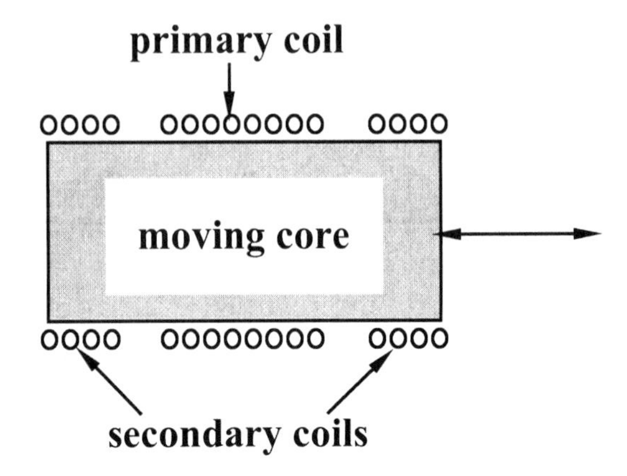

FIGURE 4.30. Model of the transducer in which the magnetic permeability varies dynamically.

introduction of multiple electrical ports. Figure 4.30 shows a typical LVDT structure. In this device a primary coil is used as a source of magnetic flux through the core which moves. Two secondary coils are used as sensors.

To understand the operation of the LVDT, first consider the case in which the core is centered between the three coils. Then the gap between both of the secondary coils is occupied entirely by the core material. If the structure is precisely symmetric, the current induced in each secondary coil (due to the magnetomotance) is the same. Hence, if these currents are added out of phase, the output signal will be null.

Now suppose that the core moves slightly to the right. Then the secondary coil on the right still is occupied entirely by the core material, but the secondary coil on the left is only partly occupied by the core. Using (47) we can determine the difference in the magnetomotance at the two sensing coils from

$$\Delta M = \frac{\phi l}{\mu_2 A} - \frac{\phi[\mu_1 l + (\mu_2 - \mu_1) l_1]}{\mu_1 \mu_2 A} = \frac{\phi(\mu_1 - \mu_2) l_1}{\mu_1 \mu_2 A}. \tag{54}$$

Assuming that the primary coil supplies an ac bias which produces flux $\phi' e^{j\omega t}$, we may approximate (54) as

$$\Delta M = \frac{\phi'(\mu_1 - \mu_2) l_1}{\mu_1 \mu_2 A} e^{j\omega t}. \tag{55}$$

Note that this linear relation between M and l_1 is valid as long as the magnetic flux amplitude is approximately a constant.

If the core moves to the left of center rather than to the right, then (54) and (55) are correct except for a sign change. Hence, by using two sensing coils, we can

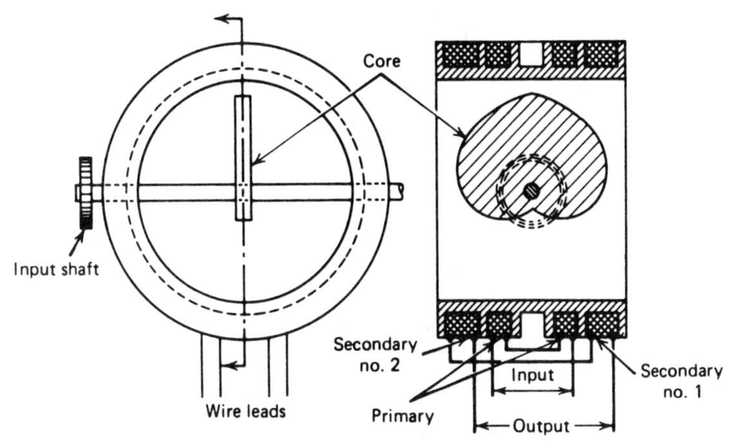

FIGURE 4.31. Example RVDT sold by Schaevitz Sensors, Lucas Automation and Control Engineering, Inc. [17].

produce a transducer in which the sign of the output signal shows the direction, and the magnitude is almost linearly related to the core displacement. The signal which is sensed is a current, which is simply the net magnetomotance divided by the number of turns in the secondary coils. RVDTs are based on the same principles but use rotational motion of a noncircular core rather than translation motion. Both LVDTs and RVDTs are widely used commercially. They are described more fully in References [13]–[16].

There are many manufacturers of LVDTs and RVDTs, and a large variety of types of these transducers are available commercially. Included in the inventory are dc versions, which operate using the standard ac LVDT or RVDT and conditioning processors to modulate and demodulate at the input and output ends, respectively. Figure 4.31 shows an example of an RVDT commercially available from Lucas Automation and Control Engineering, Inc.[17]. Note that the core shape is intentionally chosen to be asymmetric so as to facilitate measurement of the angular position of a shaft while providing a linear operating regime.

In addition to conventional position sensing, LVDT and RVDT methods are common for other types of sensing. For instance, Fig. 4.32 (taken from Trietley [7]) shows a load cell made by placing an LVDT into an elastic structure which deforms under load. While this shows a proving ring structure, many elastic structures can be used.

A very different example of a device which operates by taking advantage of the variable permittivity magnetomechanical interaction is shown in Fig. 4.33, which shows a pressure sensor manufactured by Dwyer Instruments, Inc. [17]. In this sensor, pressure causes motion of the free end of a cantilevered leaf spring and the magnet attached to that end. A high permeability helical-shaped material is between the magnet at the free end of the spring. Motion of the spring thus results in a change in the fraction of the gap which is the high-permeability material. This produces a torque on the helix which tends to keep it aligned with the magnet. The

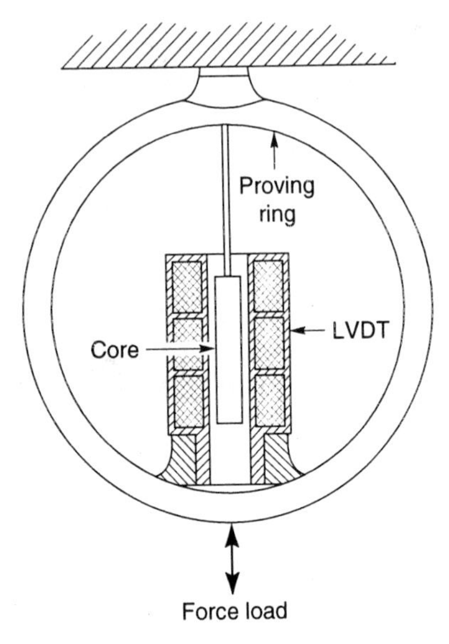

FIGURE 4.32. Load cell based on LVDT operation, taken from Trietley [7].

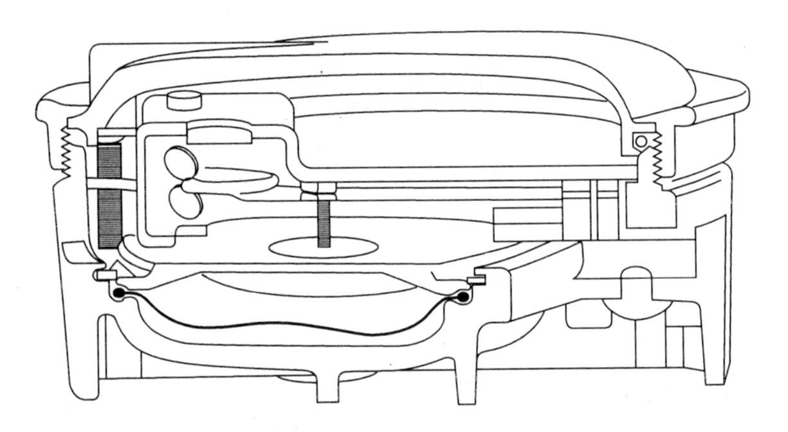

FIGURE 4.33. Magnehelic® operating principle for a pressure gauge. This product is sold by Dwyer Instruments, Inc. [17].

pointer for the sensor's indicator is attached to the helix and thus moves to indicate the pressure on a meter.

Still another example of a variable permeability, a variable reluctance sensor, is shown in Fig. 4.34, taken from Sydenham [18]. Here a pair of sensors look at the eccentricity of a tube by monitoring the fraction of set gaps which the tube wall occupies. The signals are added differentially, so that the electrical output signal is zero for wall thickness of equal size.

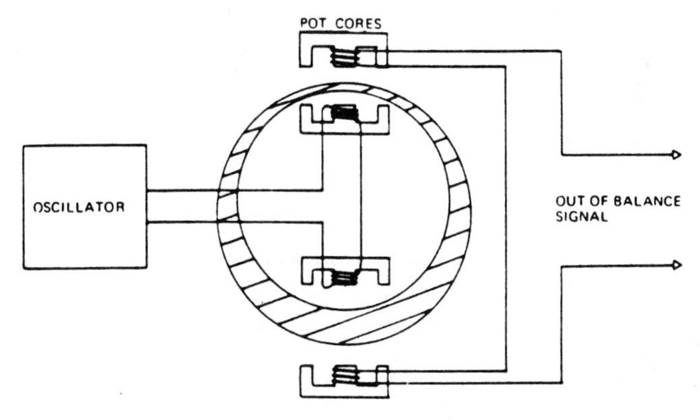

FIGURE 4.34. Tube eccentricity sensor (taken from Sydenham [18]).

4.4 Gradient Transduction with Two Ferromagnetic Materials

In the previous chapter we showed that using two solid dielectrics between the plates of a capacitor gives rise to interfacial stresses which can be used in transduction. Similarly, if we place two solid ferromagnets in a magnetic structure gap, interfacial stresses arise between them, and this can be used in transduction. There are two cases of interest, the normal gradient and the tangential gradient. These are precisely analogous to the normal and tangential gradient capacitive transducers. Further, like their capacitive cousins, the magnetic gradient transduction mechanisms are not often exploited commercially, as there tend to be superior ways to achieve the same results. Thus, gradient transduction mechanisms are presented quickly here.

Figure 4.35 shows two ferromagnets in a gap in a magnetic structure. The materials are oriented so that their interface is normal to a line connecting the two gap endpoints. This arrangement is referred to as the normal gradient case. In the structure, the magnetic flux through the two materials is the same. Thus the magnetic induction is the same in the two materials. This necessitates a step discontinuity in the magnetic field.

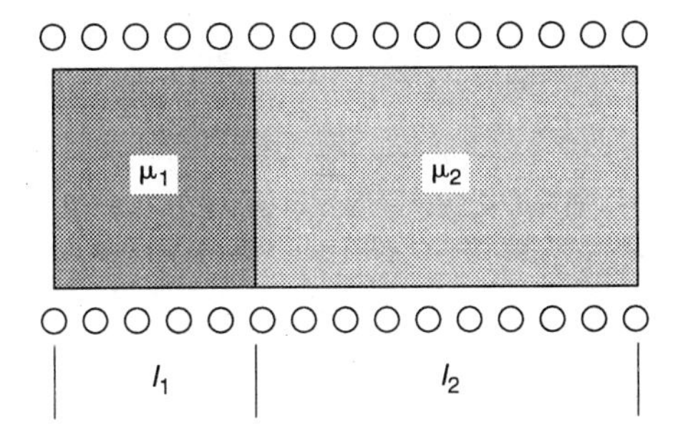

FIGURE 4.35. Normal gradient magnetic transducer.

Clearly the normal gradient case is identical to the case just described for LVDTs. We can model the magnetic gap as two magnetic capacitances with common flow and the force at the interface of the materials is described by (48), which we rewrite here in the form

$$F_{\text{net}} = \frac{\phi^2}{2A}\left[\frac{1}{\mu_2} - \frac{1}{\mu_1}\right]. \tag{56}$$

We may put this in terms of a field parameter by noting that $\phi = BA$.

Equation (56) shows that large interface forces are generated for normal gradient devices in which the permeabilities of the two materials are quite different. This equation also shows that the constitutive relation is quadratic, as are virtually all of the relations considered so far. A fairly linear transducer could be produced by using the standard technique of applying a large static magnetic flux so that the signal is a small perturbation.

Now consider the tangential gradient case, in which two ferromagnets are placed in a magnetic gap so that the material interface is tangent to the magnetic flux flow. In this case the flux will divide in the gap, but the magnetomotance is the same in the two materials. Although we could still determine the transducer relations using the standard computation of the energy and determination of the constitutive relations from partial derivatives, it is more convenient to write the energy in terms of M, the parameter that is common. We can do this by resorting to the co-energy expression rather than the energy expression. For electrical capacitors the co-energy expression is $\mathcal{E}^* = Ce^2/2$. By analogy, the co-energy expression for magnetic capacitance is given by $\mathcal{E} = CM^2/2$. Using this we can write

$$\mathcal{E}^* = \frac{M^2(\mu_1 A_1 + \mu_2 A_2)}{2l}, \tag{57}$$

where A_1 is the area of the gap associated with medium 1 and A_2 is the area associated with medium 2. If we assume that the area is cylindrical, with total area $A = \pi r^2$ and $A_1 = \pi r_1^2$, then the energy can be rewritten as

$$\mathcal{E} = \frac{\pi M^2[\mu_1 r_1^2 + \mu_2(r^2 - r_1^2)]}{2l}. \tag{58}$$

From this we calculate the interface forces using $f = \partial\mathcal{E}/\partial r_1$. The result is given by

$$f_{\text{net}} = \frac{\pi M^2 r_1(\mu_1 - \mu_2)}{l}. \tag{59}$$

By noting that $M = Ni$, we can rewrite (59) in a manner which relates the net interface force to the current. In this form it is clear that the tangential gradient mechanism can be used in transduction in two fundamental ways: either an applied current will develop an interface force which results in motion of the interface, or a force applied at the interface can be measured by monitoring the output current. In either case, the force/current relationship is quadratic and it is largest when the two permeabilities are very different. The relationship can be rendered nominally linear through the application of a bias current.

4.5 Magnetostrictive Transduction

Up to this point of the chapter we have discussed transduction mechanisms which rely on specific geometries but will work in general for a whole host of ferromagnetic materials. In addition to these transduction mechanisms there is a transduction mechanism which relies on changes in the energy stored in a magnetic field and which takes advantage of a particular material property referred to as magnetostriction. Magnetostrictive materials are those materials in which the magnetic permeability is a function of the mass density of the medium. Although all materials are magnetostrictive to some extent, the materials which exhibit a strong magnetostrictive effect include iron, cobalt, nickel, alloys of these materials, and a new rare earth compound known as Terfenol.

Magnetostriction is a well-known effect that has been exhaustively studied for over a century. It was first discovered in nickel by Villari and studied by Joule in the 1840s. Magnetostriction in nickel, cobalt, and iron became quite important in transduction in the 1930s and 1940s, when magnetostrictive transducers were routinely used in sonar applications. They fell out of favor, beginning in the 1960s, as advances in piezoelectricity offered transducers with greater efficiencies and more linear behavior. However, work by Clark and others [20]–[22] has led to the recent development of a rare earth compound which is magnetostrictive and offers about a factor of three improvement in efficiency over piezoelectric devices. Hence, there is once again great interest in magnetostriction.

Magnetostriction is analogous to electrostriction, but there is a fundamental difference. Electrostrictive materials tend to have a high electric polarization. This makes electrostrictive materials appropriate for nearly linear transducers. By contrast, magnetostrictive materials generally have a low magnetic polarization (magnetization). Thus, to use magnetostrictive materials in development of a nearly linear transducer, we must apply a bias. This bias can be a magnetic bias, applied using either a permanent magnet or an ac voltage, or a mechanical bias, applied by prestressing the material. In all other fundamental ways, magnetostriction is analogous to electrostriction. Hence, we could immediately write the constitutive relations simply drawing on the analogy to electrostriction. Rather than take that approach here, we will derive the magnetostrictive relations from scratch, thus reinforcing the derivations of both electrostrictive and magnetostrictive relations.

Let us imagine that we have a magnetostrictive material that has dimensions $h \times w \times l$, and that there is a magnetic field oriented in the direction in which the material dimension is l, as shown in Fig. 4.36. The magnetic capacitance of this material is given by $C = \mu wh/l$, but now $\mu = \mu(\rho)$.

We determine the force on the material, as usual, from $F = \partial \mathcal{E}/\partial l$. This yields

$$F = \frac{\phi^2}{2\mu wh} - \frac{\phi^2 l}{2\mu^2 wh}\frac{\partial \mu}{\partial l}. \tag{60}$$

The first term in (60) is the magnetostatic force, and the second is the magnetostrictive contribution.

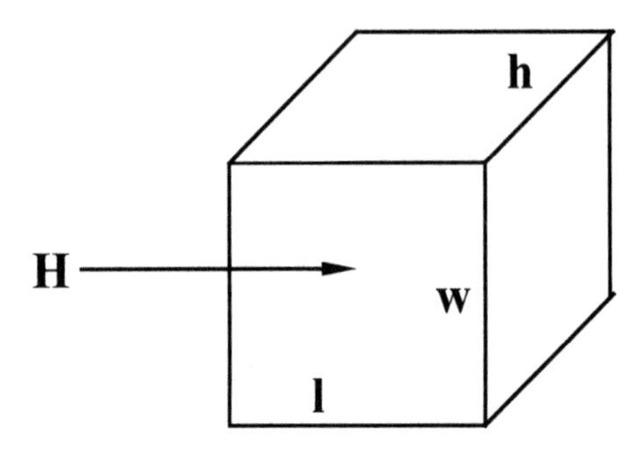

FIGURE 4.36. Magnetostrictive material.

We can simplify the force expression by invoking conservation of mass:

$$hwl\rho = \text{constant}. \tag{61}$$

We assume that the material is isotropic and obeys Hooke's law so that a change in the dimension l causes a corresponding change in dimensions h and w, with Poisson's ratio scaling the h and w direction strains. Then by differentiating (61) we have

$$\frac{\partial \rho}{\partial l} = -\frac{\rho(1-2\nu)}{l}. \tag{62}$$

Using this expression in (60) we can write

$$F = \frac{\phi^2}{2wh\mu}\left[1 + \frac{\rho(1-2\nu)}{\mu}\frac{\partial \mu}{\partial \rho}\right]. \tag{63}$$

The advantage of this expression is that $\partial\mu/\partial\rho$ is well known for magnetostrictive materials. For a given material, $\partial\mu/\partial\rho$ is always positive (inducing a compressive force) or always negative (inducing a tensile force). This means that the magnetostrictive force developed in a specified material has a sign which is independent of the sign of the magnetic flux. In order to use this as an approximately linear transduction mechanism, we can use a bias flux and limit the operating range to small perturbations about the bias, or apply a mechanical prestress in the direction of the magnetostrictive effect.

With the advent of Terfenol, there has been a dramatic increase in the number of magnetostrictive transducers under study. This is because Terfenol ($Tb_0.3Dy_0.7Fe_1.93$) exhibits coupling efficiencies of 50–55% and can produce strains up to 2×10^{-3}. It is sold commercially by Etrema [23]. An example of a three-dimensional positioner is shown in Fig. 4.37 and described fully in Wang and Busch-Vishniac [24]. Here Terfenol-D bars are used to provide high-speed positioning with better than 0.1 μm accuracy over a workspace of $1 \times 1 \times 0.5$ cm. Terfenol offers advantages in this application because of its speed of response and load-carrying ability.

Figure 4.38 shows a typical linear actuator built using Terfenol. In this actuator described by Goodfriend et al. [25], a rod attached to Terfenol is moved as the

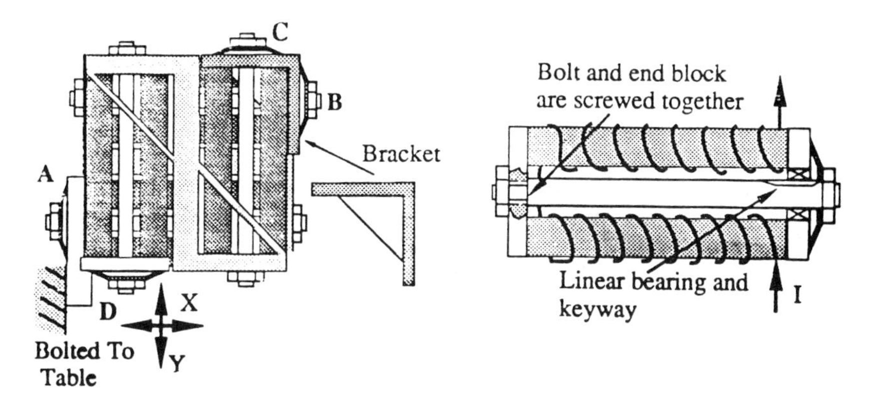

FIGURE 4.37. Magnetostrictive micropositioner. Reprinted with premission from W. Wang and I. Busch-Vishniac, *A High Precision Micropositioner Based on Magnetostriction Principle*, Rev. Sci. Instrum. **63**, 249 (1992). Copyright 1992 American Institute of Physics.

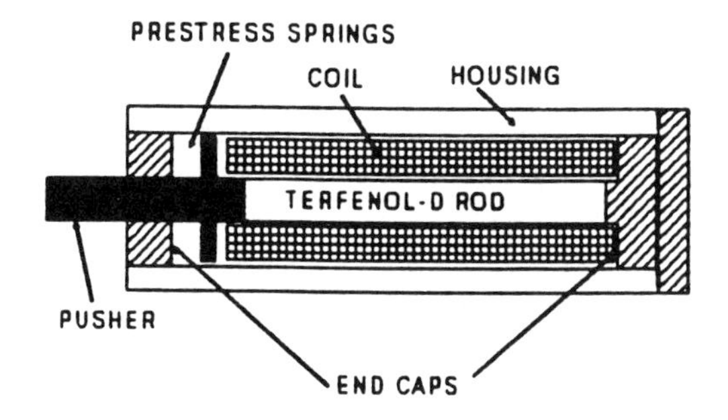

FIGURE 4.38. Terfenol linear actuator as described by Goodfriend et al. [25].

magnetostrictive material expands or contracts in response to the current running through a solenoid. It is a relatively simple matter to extend this design to the construction of a shaker, and this transduction approach has been described by Flatau et al. [26]. Further, by incorporating multiple rods, we could build a rotational motor using the same sort of approach.

In addition to application in position actuation, magnetostrictive transducers are finding their way into sound sources. Figure 4.39, taken from Moffett et al. [27], shows a flextensional sound source exploiting Terfenol. In this device, current in a solenoid causes expansion and contraction of a Terfenol rod. This expansion and contraction couples to a nominally curved surface roughly parallel to the rod axis. Thus expansion straightens the surface, and contraction makes it more curved. The net effect is to pump far more of the surrounding medium than the end of the rod could on its own. Thus the curved surface is simply a mechanical amplifier designed to produce greater sound intensity. This geometry (which did not originate with Moffett) is refered to as a flextensional device.

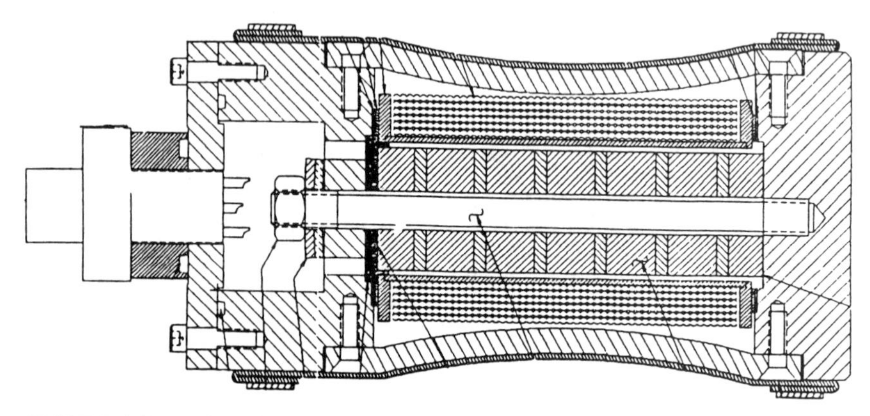

FIGURE 4.39. Terfenol-based, flextensional sound source as described by Moffett et al. [27].

4.6 Eddy Current Transducers

Every transduction mechanism discussed in this chapter so far has an analogous transduction mechanism associated with changes in the energy stored in an electric field. In this section we discuss a mechanism with no such analog, namely, transduction based on eddy currents induced in metals.

When nonferromagnetic metals are exposed to a high-frequency magnetic field, currents are induced in the metals. The current direction is such that the magnetic field it induces is in the opposite direction to the applied magnetic field. The net result is a reduction in the magnetic field strength. The amount of reduction depends on the target material, with aluminum, copper, and brass generally considered to be the best targets. By detecting the magnetic field strength, we can determine the location of the target.

To better understand eddy currents, we start with Faraday's law. In our discussion of Faraday's Law the sign of the relationship between voltage and time rate of change of magnetic flux was said to be somewhat arbitrary. In actuality, Lenz's Law gaurentees that the voltage induced by a time varying magnetic flux tends to oppose the magnetic field. Thus, we must have

$$e = -N\frac{d\phi}{dt}, \tag{64}$$

where the minus sign is similar to that which appears in the constitutive equations for linear springs and dampers, and indicates forces or fields in opposition to those driving the reaction.

Now let us suppose that we have a time-varying magnetic flux. If we expose a conductor to this magnetic field, we induce a voltage across it according to Faraday's law, (64). This voltage gives rise to a current in the conductor which we call an eddy current. Note that this description suggests that eddy currents can be generated in all materials. While this is, strictly speaking, true, it is important to note that the current generated in response to the induced voltage varies inversely with the material electrical resistance, i.ė., $e/R = i$. Thus eddy currents are far greater for materials with low resistance (conductors) than for any other materials. As a result, we tend to think of eddy currents as being present only in conductors.

Eddy currents induced in a conductor are not the same as the currents through a coil. In particular, they differ in that eddy currents exist primarily at the surface of a conductor. They are therefore a skin effect. The classical model of eddy currents is that their intensity decreases exponentially with depth into a conductor:

$$i_{\text{eddy}} = i_{\text{surface}} e^{-z\sqrt{c\omega\mu\sigma}}, \tag{65}$$

where i_{surface} is the current at the surface, z is the depth coordinate, c is a constant, and σ is the conductivity of the material. If we define the skin depth to be that depth at which the intensity has decayed to e^{-1} of the surface value (or to 36.8% of the surface value), then the depth is given by

$$\delta = \frac{1}{\sqrt{c\omega\mu\sigma}}. \tag{66}$$

Note that this relation indicates that for a given material, the current is closer to the surface for higher frequencies than low. At microwave frequencies, the eddy currents penetrate only a few molecular spacings. To provide perspective on these numbers, we provide the skin depths of a few common metals at 1 MHz: for copper—0.066 mm, for aluminum—0.082 mm, and for mild steel—0.159 mm.

The fact that eddy currents reside near the surface of a conductor has a significant effect on the apparent resistance of the conductor. In general the electrical resistance depends on the conductivity of the material and the area through which current passes. The apparent resistance to eddy currents varies inversely with the product of skin depth and conductivity. Thus

$$R \propto \sqrt{\frac{c\omega\mu}{\sigma}}, \tag{67}$$

which indicates that the size of the current induced by the time-varying flux decreases with frequency for a given material, and increases with the ratio of conductivity to permeability for a set frequency.

Eddy current sensors take advantage of the eddy currents induced in conductors in response to time-varying magnetic fields. Generally, they use two coils, one to set up the time-varying flux, and the other to sense the induced current. This is shown in Fig. 4.40. If no conductor is present, then the reference coil establishes a time-varying magnetic flux which induces a current in the sensing coil. When a conductor is moved into the magnetic field, an eddy current is induced in the conductor which tends to resist the magnetic field. Thus, the sensing coil responds to a lower intensity magnetic field, and the induced current is lower. It is this reduction in current in the sensing coil that is used in typical eddy current proximity sensing, i.e., for sensing how close a conductor is located to the source of the magnetic field. A typical frequency range of operation for an eddy current sensor is 50 kHz to 10 MHz.

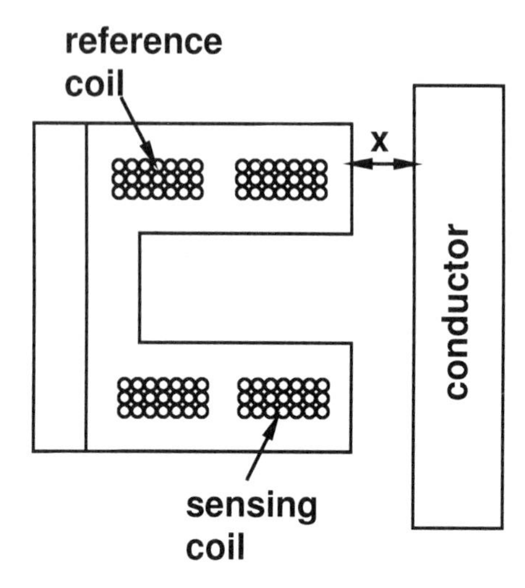

FIGURE 4.40. Typical eddy current sensor.

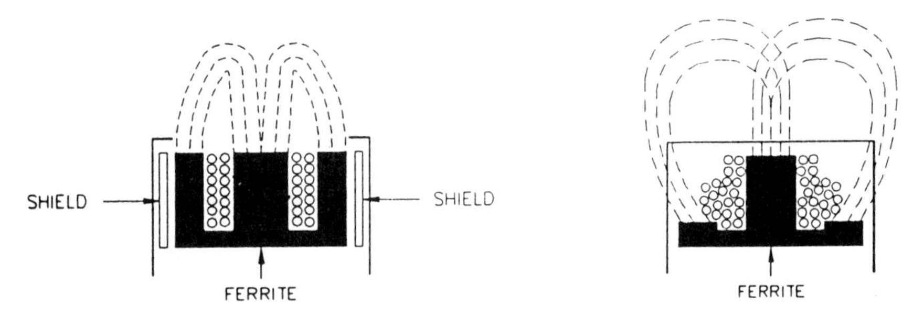

FIGURE 4.41. Turck, Inc. [28] eddy current sensors: a). shielded and b). nonshielded.

Eddy current sensors can be used in a number of applications. The most common usage is as a proximity sensor. Figure 4.41 shows two examples of eddy current sensors manufactured by Turck, Inc. [28]. While these devices can be operated as analog sensors, they are often used as proximity switches. In the proximity switch mode, a field amplitude threshhold is set, below which a solid-state switch changes the signal output from high to low or vice versa. In this manner, a digital signal is produced which indicates whether a metal object has approached closer than a prescribed value.

Another use of eddy current sensors is in material identification. This can be seen in consideration of (67). If a conductor is fixed at a given position, and exposed to a time-varying field of known frequency, then the apparent resistance and the eddy current intensity are dependent on the material conductivity to permeability ratio. By sensing the eddy current intensity, we can thus determine the material which is present. The differences in the common materials is summarized in Fig. 4.42, taken from Dally et al. [29].

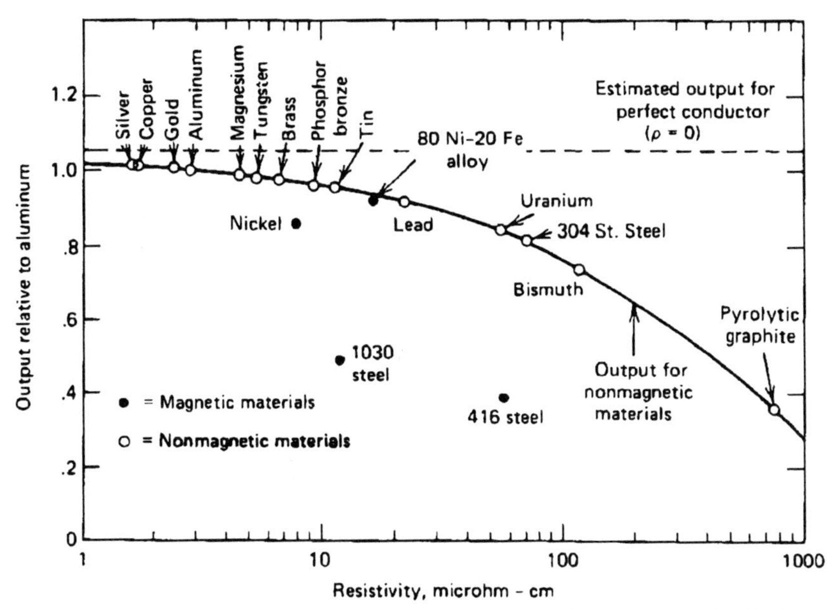

FIGURE 4.42. Eddy current sensor response as a function of material described in Instrumentation for Engineering Measurements, 2nd ed., F. W. Dally, W. F. Riley, and K. G. McConnell, 1993 [29]. Figure reprinted by permission of John Wiley and Sons, Inc.

Similarly, if the material composition is known, as well as its location and the drive frequency, then the sensor can be used to determine the temperature of the conductor, as both conductivity and permeability are functions of temperature. Used in this manner, the eddy current sensor is a form of resistance thermometer. Such resistance devices are discussed in more detail in Chapter 7.

Still another use of eddy current sensors is in flaw detection. For instance, consider a metal tube moving horizontally as shown in Fig. 4.43. The solenoid tends to establish a circumferential eddy current which affects the current in a sensing coil. The existence of a flaw in the tube can be sensed as a change in the current induced in the sensing coil. This approach, which is fast, reliable, and easily tuned to pass flaws of small size while catching flaws of large size, is gaining in popularity as a process control technique. For instance, it is routinely used to check for flaws in the skins of aircraft.

4.7 Summary

In this chapter we have introduced a number of transduction mechanisms based on changes in the energy stored in a magnetic field. These mechanisms parallel those based on changes in the energy stored in an electric field. All are nonlinear but can be rendered nominally linear using bias fields.

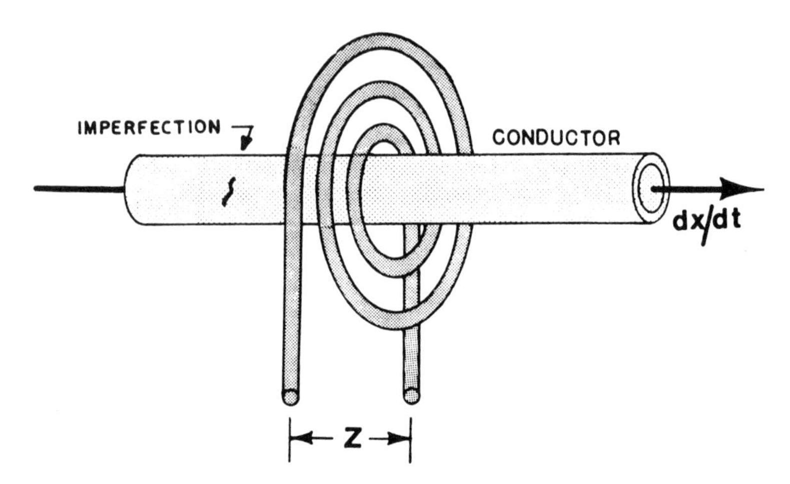

FIGURE 4.43. Eddy current sensor for flaw detection (from Burke[30]).

In the next two chapters we discuss inherently linear transduction mechanisms based on electric and magnetic field variations. As will be discussed, these mechanisms are very common in sensors and actuators of all sorts.

4.8 References

1. P. Lorrain and D. R. Corson, *Electromagnetic Fields and Waves* (W. H. Freeman, San Francisco, 1970).
2. E. M. Purcell, *Electricity and Magnetism* (McGraw-Hill, New York, 1965).
3. W. H. Hayt, Jr., *Engineering Electromagnetics* (McGraw-Hill, New York, 1981).
4. L. Eyges, *The Classical Electromagnetic Field* (Addison-Wesley, Reading, MA, 1972).
5. *CEC Pressure Tranducer Handbook*, CEC Vibration Products, Division of Sensortronics, 746 Arrow Grand Circle, Covina, CA 91722, USA.
6. J. C. Cluley, *Transducers for Microprocessor Systems* (Springer-Verlag, New York, 1985).
7. H. L. Trietley, *Transducers in Mechanical and Electronic Design* (Marcel Dekker, New York, 1986).
8. P. Dario, N. Croce, M. C. Carrozza, and G. Varallo, *A Fluid Handling System for a Chemical Microanalyzer*, J. Micromech. Microengrg., **6**, 95 (1996).
9. Ledex Products, Lucas Control Systems, 801 Scholz Drive, PO Box 427, Vidalia, OH 45377-0427, USA.
10. Omega Engineering Inc., P.O. Box 4047, Stamford, CT 06907, USA.
11. Malema Engineering Corp., 1101 S. Rogers Circle, Unit 15, Boca Raton, FL 33487, USA.

12. Gems Sensors, Imo Industries Inc., 1 Cowles Road, Plainville, CT 06062-1198, USA.

13. H. H. Woodson and J. R. Melcher, *Electromechanical Dynamics* (Wiley, New York, 1968).

14. E. O. Doebelin, *Measurement Systems, Application and Design* (McGraw-Hill, New York, 1983).

15. C. L. Nachtigal (ed.), *Instrumentation and Control, Fundamentals and Applications* (Wiley, New York, 1990).

16. E. E. Herceg, *Handbook of Instrumentation and Control*, (Schaevitz Engineering, Pensauken, NJ, 1986).

17. Dwyer Instruments Inc., P.O. Box 373, Michigan City, IN 46360, USA.

18. P. H. Sydenham, *Transducers in Measurement and Control* (Instrument Society of America, Research Triangle Park, 1980).

19. A. E. Clark and H. S. Belson, *Giant Room-Temperature Magnetostrictions in* $TbFe_2$ *and* $DyFe_2$, Phys. Rev. B, **5**, 3642 (1972).

20. H. T. Savage, A. E. Clark, and J. M. Powers, *Magnetomechanical Coupling and* ΔE *Effect in Highly Magnetostrictive Rare Earth–*Fe_2 *Compounds*, IEEE Trans. Magn. Mag., **11**, 1355 (1975).

21. A. E. Clark, in E.P. Wohlfarth, *Ferromagnetic Materials* (North-Holland, Amsterdam, 1980).

22. Etrema Products Inc., 2500 North Loop Dr., Ames, IA, USA.

23. W. Wang and I. Busch-Vishniac, *A High Precision Micropositioner Based on Magnetostriction Principle*, Rev. Sci. Instrum., **63**, 249 (1992).

24. M. J. Goodfriend, K. M. Shoop, and O. D. McMasters, *Characteristics of the Magnetostrictive Alloy Terfenol-D Produced for the Manufacture of Devices*, in Proc. Conf. on Recent Advances in Adaptive and Sensory Materials and Their Applications, Blacksburg, VA, April 1992.

25. A. B. Flatau, D. L. Hall, and J. M. Schlesselman, *Magnetostrictive Active Vibration Control Systems (A Preliminary Investigation)*, in Proc. Conf. on Recent Advances in Adaptive and Sensory Materials and Their Applications, Blacksburg, VA, April 1992.

26. M. B. Moffett, J. F. Lindberg, E. A. McLaughlin, and J. M. Powers, *An Equivalent Circuit Model for Barrel-Stave Flextensional Transducers*, in M. D. McCollum, B. F. Hamonic, and O. B. Wilson (ed.), *Transducers for Sonics and Ultrasonics* (Technomic, Lancaster, PA, 1992).

27. Turck Inc., 3000 Campus Dr., Plymouth, MN 55441, USA.

28. F. W. Dally, W. F. Riley, and K. G. McConnell, *Instrumentation for Engineering Measurements*, 2nd ed., (Wiley, New York, 1993).

29. H. E. Burke, *Handbook of Magnetic Phenomena* (Van Nostrand Reinhold, New York, 1986).

5

Piezoelectricity and Pyroelectricity

In the previous chapters on transduction we have concentrated on those mechanisms which depend on changes in the energy stored in magnetic and electric fields. Virtually all of the constitutive relations linking mechanical and electrical or magnetic variables have been nonlinear (generally quadratic). In addition to these mechanisms there are linear transduction processes, some of which are capacitive and some of which are inductive. Of these processes, the most common and well studied is piezoelectricity, a phenomenon exhibited by some materials in which application of a strain causes the establishment of an electric field and vice versa. In this chapter we focus on piezoelectricity and piezoelectric transducers. The related phenomenon of pyroelectricity is also presented.

Strictly speaking, piezoelectricity belongs under the heading of those transducers based on changes in the energy stored in an electric field. One view of piezoelectricity is that it is a material property which is a special form of electrostriction in which there is a sizeable electric polarization in the material. This renders the normally quadratic electrostrictive relations nearly linear. The truly piezoelectric materials (i.e., those which are piezoelectric in their natural form) are all crystals, although several ceramic materials and some polymers can be made to behave like piezoelectric materials once they are poled. For our purposes, the distinction between natural piezoelectric materials and those which act as piezoelectric materials is moot. We will, therefore, simply refer to all such materials as piezoelectric. Common piezoelectric materials include quartz, barium titanate ($BaTiO_3$), lead zirconate titanate (PZT), and polyvinylidene fluoride (PVDF). Although the materials themselves can be expensive when compared to other transducer materials, piezoelectric transducers are one of the most common transducer types on the market. They are relatively inexpensive for high-quality devices, easy to manufacture, rugged, and able to work without an external power supply. As a result, piezoelectric transducers have been used to measure almost anything: strain, pressure, force, acceleration, light modulation, etc.

Jacques and Pierre Curie are generally credited with being the first to discover piezoelectricity in 1880 [1]. The term *piezoelectric* comes from the Greek *piezein* which means to press and, for a time, the phenomenon was referred to as *pressure electricity*. The Curies concentrated their efforts on quartz, which was a leading piezoelectric material for many years subsequent to their work. In 1918, Langevin suggested the design for the first piezoelectric transducer based on the Curies's studies [2]. Although there was significant research in this field, it was not until 1947, when Gray discovered a new piezoelectric material, that piezoelectric transducers really came of age [3]. This new material was barium titanate, which was found to behave as though it is piezoelectric when it is poled with a high electric field. The discovery of such behavior in barium titanate was an important breakthrough in piezoelectricity for two reasons. First, the piezoelectric effect in poled barium titanate is substantially greater in magnitude than for the previously known materials. Hence, there was suddenly a material available that could provide significantly more sensitive transducers. Second, prior to the discovery of piezoelectric behavior in poled barium titantate the only piezoelectric materials known were crystals such as quartz and Rochelle salt. In these crystals the lattice structure of the material supplies the electric polarization. In barium titanate and other ceramics, the naturally occurring polarization is small but exposure to a high electric field while the ceramic is being formed tends to align the molecules in a manner analogous to a crystal lattice, thus introducing a sizable electric polarization.

In the early 1950s continued research with ceramic materials led to the discovery of a strong piezoelectric effect in lead zirconate titanate (PZT) and this material has remained the preferred piezoelectric material since its discovery by Jaffe [4]. In recent years work has focused on piezoelectric polymers, polymer composites, and thin films. These materials are of intense interest primarily because their flexibility and size permits piezoelectricity to become important in new arenas. The first polymer found to be significantly piezoelectric was polyvinylidene fluoride (PVDF) which was discussed by Kawai in 1969 [5]. This semicrystalline material is not piezoelectric in its natural form, but can be made to exhibit piezoelectric behavior through poling and stretching procedures. In addition there has been substantial work by Newnham et al. [6], Halliyal et al. [7], Safari et al. [8], and others on polymer composites using small pieces of ceramic piezoelectric materials suspended in a polymer matrix. The advantage of this approach is that the material can be designed so that its mechanical properties are primarily determined by the polymer matrix, while the transduction behavior is dominated by the material suspended in the matrix. In this manner, researchers have been able to produce a new material in which the coupling normally seen between orthogonal field directions is significantly reduced.

Most recently, Jiang and Cross have shown that we can make thin films of PZT (less than 1 μm) and sputter them directly onto semiconductor substrates [9]. They have shown that the properties of these thin films are almost the same as bulk PZT material. Although this newest breakthrough has not yet led to new transducers, it is clear that it will in the near future. Of particular interest is the possibility of mass production of very small and sensitive transducers.

5.1 Piezoelectric Relations

In order to understand the operation of piezoelectric transducers it is necessary to derive the constitutive equations which are used to describe the mechanical/electrical coupling. We derive these relations here based on the treatment in Berlincourt et al. [10].

To derive the piezoelectric relations we begin with the Gibbs free energy of a material which is defined as

$$G = U - S_i T_i - E_m D_m - H_m B_m - \sigma\theta, \tag{1}$$

where U is the internal energy of the material, S the mechanical strain, T the mechanical stress, σ the entropy, and θ the temperature. Note that these labels are standard in the treatment of piezoelectricity and are used here even though different symbols have been introduced for some of these variables in previous chapters. The Einstein convention is used in (1), meaning that a sum is performed over repeated indices in any term. The index i goes from 1 to 6, and the index m from 1 to 3. While the meaning of the index m is clear, the interpretation of index i is less obvious. Indeed, stress and strain are usually regarded as tensors, with two indices used to designate specific components. Normal strain and stress are described by components with indices kk, while shear components have indices lk with $l \neq k$. In this standard notation, the stress and strain tensors have nine elements. However, only six of them are independent, because $F_{lk} = F_{kl}$ where F is either stress or strain. Thus there are only six independent elements of the stress and strain tensors. It is these six independent elements which are represented as a vector in the standard treatment of piezoelectricity. The notation assigns $i = 1, 2$, and 3 to the normal stress/strain components in the three principal directions. Indices $i = 4, 5$ and 6 indicate shear stress/strain about the 1 (23 and 32 components), 2 (31 and 13 components), and 3 (12 and 21 components) directions, respectively. With these assignments, the tensor notation for mechanical field variables is reduced to a six-component vector. As will be shown shortly, this compact representation is advantageous because it permits a two-dimensional representation of electromechanical coupling coefficients, rather than a three- or four-dimensional representation.

We manipulate (1) using the first and second laws of thermodynamics:

$$dQ = \theta\, d\sigma, \tag{2}$$

where Q is the heat, and

$$dU = dQ + T_i\, dS_i + E_m d\, D_m + H_m\, dB_m. \tag{3}$$

Taking the total derivative of Eq. (1) and using the two thermodynamic laws we can write

$$dG = -S_i\,dT_i - D_m\,dE_m - B_m\,dH_m - \sigma\,d\theta. \tag{4}$$

Equation (4) defines the independent thermodynamic variables as T_i, E_m, H_m, and θ. Therefore we can define the dependent variables as follows:

$$S_i = -\left(\frac{\partial G}{\partial T_i}\right)_{E,H,\theta}, \tag{5}$$

$$B_m = -\left(\frac{\partial G}{\partial H_m}\right)_{E,T,\theta}, \tag{6}$$

$$D_m = -\left(\frac{\partial G}{\partial E_m}\right)_{T,H,\theta}, \tag{7}$$

$$\sigma = -\left(\frac{\partial G}{\partial \theta}\right)_{T,E,H}. \tag{8}$$

Here the notation indicates which variables are held constant in the process of differentiation. Of course we can also define each of these variables in functional form in terms of the independent thermodynamic variables:

$$S_i = S_i(T_i, E_m, H_m, \theta), \tag{9}$$

$$D_m = D_m(T_i, E_m, H_m, \theta), \tag{10}$$

$$B_m = B_m(T_i, E_m, H_m, \theta), \tag{11}$$

$$\sigma = \sigma(T_i, E_m, H_m, \theta). \tag{12}$$

Using these functional relations we define the total differentials of the dependent variables:

$$\begin{aligned} dS_i = {} & \left(\frac{\partial S_i}{\partial T_j}\right)_{E,H,\theta} dT_j + \left(\frac{\partial S_i}{\partial E_m}\right)_{T,H,\theta} dE_m \\ & + \left(\frac{\partial S_i}{\partial H_m}\right)_{T,E,\theta} dH_m + \left(\frac{\partial S_i}{\partial \theta}\right)_{T,E,H} d\theta, \end{aligned} \tag{13}$$

$$\begin{aligned} dD_m = {} & \left(\frac{\partial D_m}{\partial T_j}\right)_{E,H,\theta} dT_j + \left(\frac{\partial D_m}{\partial E_m}\right)_{T,H,\theta} dE_m \\ & + \left(\frac{\partial D_m}{\partial H_m}\right)_{T,E,\theta} dH_m + \left(\frac{\partial D_m}{\partial \theta}\right)_{T,E,H} d\theta, \end{aligned} \tag{14}$$

$$\begin{aligned} dB_m = {} & \left(\frac{\partial B_m}{\partial T_j}\right)_{E,H,\theta} dT_j + \left(\frac{\partial B_m}{\partial E_m}\right)_{T,H,\theta} dE_m \\ & + \left(\frac{\partial B_m}{\partial H_m}\right)_{T,E,\theta} dH_m + \left(\frac{\partial B_m}{\partial \theta}\right)_{T,E,H} d\theta,, \end{aligned} \tag{15}$$

$$d\sigma = \left(\frac{\partial \sigma}{\partial T_j}\right)_{E,H,\theta} dT_j + \left(\frac{\partial \sigma}{\partial E_m}\right)_{T,H,\theta} dE_m$$

$$+\left(\frac{\partial\sigma}{\partial H_m}\right)_{T,E,\theta} dH_m + \left(\frac{\partial\sigma}{\partial\theta}\right)_{T,E,H} d\theta. \tag{16}$$

Now we rewrite the total differentials above using the definitions for the dependent variables in terms of the Gibbs free energy, (5)–(8). This results in

$$dS_i = \frac{\partial}{\partial T_j}\left[-\frac{\partial G}{\partial T_i}\right] dT_j + \frac{\partial}{\partial E_m}\left[-\frac{\partial G}{\partial T_i}\right] dE_m + \frac{\partial}{\partial H_m}\left[-\frac{\partial G}{\partial T_i}\right] dH_m + \frac{\partial}{\partial\theta}\left[-\frac{\partial G}{\partial T_i}\right] d\theta, \tag{17}$$

$$dD_m = \frac{\partial}{\partial T_j}\left[-\frac{\partial G}{\partial E_m}\right] dT_j + \frac{\partial}{\partial E_m}\left[-\frac{\partial G}{\partial E_m}\right] dE_m + \frac{\partial}{\partial H_m}\left[-\frac{\partial G}{\partial E_m}\right] dH_m + \frac{\partial}{\partial\theta}\left[-\frac{\partial G}{\partial E_m}\right] d\theta, \tag{18}$$

$$dB_m = \frac{\partial}{\partial T_j}\left[-\frac{\partial G}{\partial H_m}\right] dT_j + \frac{\partial}{\partial E_m}\left[-\frac{\partial G}{\partial H_m}\right] dE_m + \frac{\partial}{\partial H_m}\left[-\frac{\partial G}{\partial H_m}\right] dH_m + \frac{\partial}{\partial\theta}\left[-\frac{\partial G}{\partial H_m}\right] d\theta, \tag{19}$$

$$d\sigma = \frac{\partial}{\partial T_j}\left[-\frac{\partial G}{\partial\theta}\right] dT_j + \frac{\partial}{\partial E_m}\left[-\frac{\partial G}{\partial\theta}\right] dE_m + \frac{\partial}{\partial H_m}\left[-\frac{\partial G}{\partial\theta}\right] dH_m + \frac{\partial}{\partial\theta}\left[-\frac{\partial G}{\partial\theta}\right] d\theta. \tag{20}$$

Finally, we note that since differentiation is a linear process, the order of differentiation is interchangable. This permits us to equate various partial derivatives, typically linking variables in different energy domains. In addition to noting these equalities, we typically define the physical descriptors of a material in terms of these partial derivatives. Hence we have the following list:

$$\text{piezoelectric constants}: \frac{\partial S_i}{\partial E_m} = \frac{\partial D_m}{\partial T_i} = d_{mi}^{H,\theta},$$

$$\text{piezomagnetic constants}: \frac{\partial B_m}{\partial T_i} = \frac{\partial S_i}{\partial H_m} = d_{mi}^{E,\theta},$$

$$\text{pyroelectric constants}: \frac{\partial D_m}{\partial\theta} = \frac{\partial\sigma}{\partial E_m} = p_m^{T,H},$$

$$\text{pyromagnetic constants}: \frac{\partial B_m}{\partial\theta} = \frac{\partial\sigma}{\partial H_m} = i_m^{T,E},$$

$$\text{magnetodielectric constants}: \frac{\partial D_m}{\partial H_k} = \frac{\partial B_k}{\partial E_m} = m_{mk}^{T,\theta},$$

$$\text{thermal expansion coefficients}: \frac{\partial S_j}{\partial\theta} = \frac{\partial\sigma}{\partial T_j} = \alpha_j^{E,H},$$

$$\text{elastic compliance}: \frac{\partial S_i}{\partial T_j} = s_{ij}^{E,H,\theta},$$

$$\begin{aligned}
\text{electric permittivity}: \quad & \frac{\partial D_m}{\partial E_k} = \epsilon_{mk}^{T,H,\theta}, \\
\text{magnetic permeability}: \quad & \frac{\partial B_m}{\partial H_k} = \mu_{mk}^{T,E,\theta}, \\
\text{heat capacity}: \quad & \frac{\partial \sigma}{\partial \theta} = \frac{\rho c^{E,H,T}}{\theta}.
\end{aligned} \tag{21}$$

This list defines 112 parameters in all, each of which is generally referred to as a particular type of material constant. It should be noted, however, that the derivation above does not depend on any of these so-called constants being truly constant as various properties of the system are changed.

Using (21) we now can write the functional form of the relations defined in Eqs. (5)–(8):

$$\begin{aligned}
S_i &= s_{ij}^{E,H,\theta} T_j + d_{mi}^{H,\theta} E_m + d_{mi}^{E,\theta} H_m + \alpha_i^{E,H} d\theta, \\
D_m &= d_{mi}^{H,\theta} T_i + \epsilon_{mk}^{T,H,\theta} E_k + m_{mk}^{T,\theta} H_k + p_m^{T,H} d\theta, \\
B_m &= d_{mi}^{E,\theta} T_i + m_{km}^{T,\theta} E_k + \mu_{mk}^{T,E,\theta} H_k + i_m^{T,E} d\theta, \\
d\sigma &= \alpha_i^{E,H} T_i + p_m^{T,H} E_m + i_m^{T,E} H_m + \frac{\rho c^{E,H,T}}{\theta} d\theta.
\end{aligned} \tag{22}$$

Note that in this equation, the differentials, with the exception of temperature and entropy, have been replaced by the variable itself. This is justified by noting that the nominal values of all of the variables except for entropy and temperature are zero. Thus we define the differential as the comparison to this zero-value state.

Equation (22) is valid for any material and shows not only electromechanical coupling terms but also magnetomechanical, electromagnetic, thermoelectric, thermomagnetic, and thermomechanical coupling. For standard piezoelectric materials, it is safe to assume that the magnetic effects are negligible. We also will assume temporarily that thermal effects may be neglected. This is not a particularly good assumption, as virtually all piezoelectric materials are also significantly pyroelectric, i.e., a thermal change establishes an electric field and vice versa. However, with these assumptions (22) reduces to a set of two equations

$$\begin{aligned}
S_i &= s_{ij}^{E} T_j + d_{mi} E_m, \\
D_m &= d_{mi} T_i + \epsilon_{mk}^{T} E_k,
\end{aligned} \tag{23}$$

which can be written in matrix form as

$$\begin{aligned}
\mathbf{S} &= [s^E]\mathbf{T} + [d]^T \mathbf{E}, \\
\mathbf{D} &= [d]\mathbf{T} + [\epsilon^T]\mathbf{E}.
\end{aligned} \tag{24}$$

In these equations, the superscripts on the piezoelectric constants have been dropped for simplicity. The *direct* piezoelectric effect is described in the second equation of each of these pairs, and shows that a charge density field results from not only an electric field, but also a stress in the material. The first equation in each pair is called the *converse* piezoelectric effect. It shows that a strain will result from an electric field in the material as well as a stress.

Equations (24) are one form of the constitutive relations associated with piezoelectricity. It is a simple matter to rearrange the piezoelectric equations by manipulation of (24). This produces three alternate forms:

$$\begin{aligned}\mathbf{S} &= [s^D]\mathbf{T} + [g]^T\mathbf{D},\\ \mathbf{E} &= -[g]\mathbf{T} + [\beta^T]^T\mathbf{D},\end{aligned} \tag{25}$$

$$\begin{aligned}\mathbf{T} &= [c^E]\mathbf{S} - [e]^T\mathbf{E},\\ \mathbf{D} &= [e]\mathbf{S} + [\epsilon^S]\mathbf{E},\end{aligned} \tag{26}$$

$$\begin{aligned}\mathbf{T} &= [c^D]\mathbf{S} - [h]^T\mathbf{D},\\ \mathbf{E} &= -[h]\mathbf{S} + [\beta^S]\mathbf{D}.\end{aligned} \tag{27}$$

In these equations $[s^D]$ is the compliance matrix measured with fixed charge density field, $[\beta^T]$ is the matrix of elements each of which is the inverse of a permittivity measured with fixed stress, $[\beta^S]$ is the same measured with fixed strain, $[c^E]$ is the stiffness measured with fixed electric field, and $[c^D]$ is the stiffness measured with fixed charge density field. (There is a certain irony to the choice of s for a compliance, and c for a stiffness, but as this choice is standard it is kept here.) The remaining unfamiliar matrices are referred to as piezoelectric constants. They are related by

$$d_{mi} = \epsilon^T_{nm} g_{ni} = s^E_{ji} e_{mj} = \epsilon^S_{nm} s^E_{ji} h_{nj}. \tag{28}$$

None of the constitutive equations given in Eqs. (24) – (27) are in the standard form we have used to this point, in which the effort variables (e and F) are given in terms of the displacement variables (q and x). We may rewrite the equations in the standard form by relating the strain and displacement, electric displacement and charge, electric field and voltage, and stress and force, according to the following:

$$\begin{aligned}\mathbf{x} &= [l]\mathbf{S},\\ \mathbf{F} &= [A]\mathbf{T},\\ \mathbf{q} &= [A^*]\mathbf{D},\\ \mathbf{e} &= [l^*]\mathbf{E}.\end{aligned} \tag{29}$$

Here $[l]$ is the 6×6 matrix whose diagonal components are the equilibrium lengths that scale the strain, $[A]$ is the 6×6 matrix whose diagonal components are the areas that scale the stress, $[A^*]$ is the 3×3 matrix whose diagonal components are the areas of the electrodes normal to each principle direction, and $[l^*]$ is the 3×3 matrix whose diagonal components are the lengths of the electrodes normal to each principle direction. The off-diagonal elements of all of these matrices are zero. It is (27) which permits us to write the constitutive relations in the standard form, with the displacement variables as independent and the effort variables as dependent. Thus, we use (29) in (27) to get

$$\begin{aligned}\mathbf{F} &= [A][c^D][l]^{-1}\mathbf{x} - [A][h]^T[A^*]^{-1}\mathbf{q},\\ \mathbf{e} &= -[l^*][h][l]^{-1}\mathbf{x} + [l^*][\beta^S][A^*]^{-1}\mathbf{q}.\end{aligned} \tag{30}$$

This set of equations is now in the standard form we have used for capacitive transducers, with the exception that it is in vector rather than scalar form. If we focus on some particular application of a piezoelectric transducer in which we need consider only one component of the force and voltage vectors, then this set of constitutive relations reduces to the identical form used previously. In general, we have nine independent displacement variables, and nine dependent effort variables linked in the constitutive equations. This general, three-dimensional piezoelectric case can thus be modeled as a 9-port capacitance. Six of the ports are associated with mechanical energy, and three with electrical. As a practical matter, we are hardly ever concerned with piezoelectric transducers with more than one direction of electric field application or sensing. Further, the matrix of piezoelectric constants tends to be sparse. Thus, it is quite normal to simplify the model of the piezoelectric transducer material to a multiport capacitance with far fewer than nine ports. Of course, the minimum number of ports is still two, as there is at least one nonzero piezoelectric constant for the material to be useful.

From (30) it is clear that there are three transduction relations that could be exploited in piezoelectric sensors and actuators. One could relate force and charge, voltage and displacement, or force and voltage. All three approaches are commonly used. For example, force transducers, load cells, and accelerometers relate force and charge. Position sensors and positioners (such as used in scanning tunneling microscopy) relate voltage and displacement. Piezoelectric pressure sensors often relate force and voltage. Because (30) are linear, it is a simple matter to consider the various transduction relations in detail, and we do not belabor the point here.

Clearly, (24)–(27) show that the magnitude of the piezoelectric effect is described by the piezoelectric constants. Table 5.1 shows typical maximum values for these constants for some common transducer materials. Also shown in the table is k, the fraction of electrical energy converted to mechanical energy or vice versa in the piezoelectric process. Note the big discrepancy between quartz and the more modern piezoelectric materials. This will be discussed further in the next section, which focuses on the material properties of various piezoelectric materials used in transduction.

5.2 Piezoelectric Materials

Before considering the specific piezoelectric behavior of various materials, it is appropriate to consider a qualitative model of the phenomenon of piezoelectricity

Material	$d_{max}(pC/N)$	k
Quartz	5	0.1
$BaTiO_3$	242	0.15–0.2
PZT	496	0.5–0.7
PVDF	23	0.1

TABLE 5.1. Piezoelectric properties of some common transduction materials.

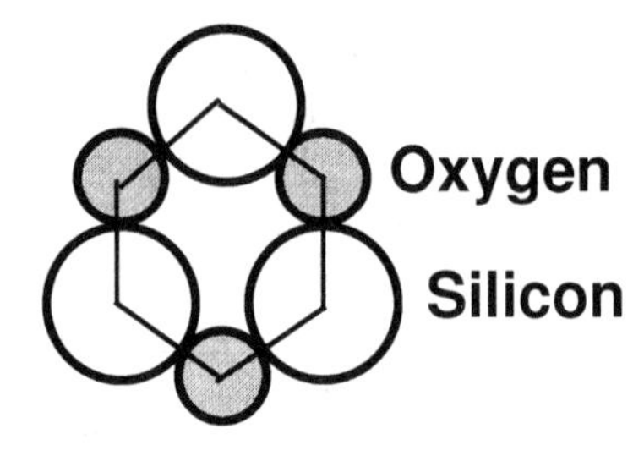

FIGURE 5.1. Quartz crystal in its unstressed state.

FIGURE 5.2. Quartz crystal in a stressed state.

in materials. Suppose that we have a crystal which exhibits piezoelectric activity. For instance, let us consider quartz. Such a crystal in its unstressed state looks as shown in Fig. 5.1. While the crystal has no net charge, the oxygen atoms each have a negative charge, and the silicon atoms in the lattice each have a positive charge. If the crystal is now stressed longitudinally, it assumes a shape shown exaggerated in Fig. 5.2. The displacement of the atoms causes a net charge to be seen by the electrode surfaces, which induces a charge in them. It is this charging of the electrodes which causes the electric fields observed in response to mechanical strain, the phenomenon we call piezoelectricity. Piezoelectricity thus may be thought of as the disturbance of electric dipoles from their equilibrium state.

With the constitutive relations given in the previous section we have now defined four 3×6 matrices of piezoelectric constants, i.e., 72 piezoelectric constants. Fortunately, most materials of interest have only a few significantly nonzero entries in each matrix. In order to quickly illustrate the important properties of a piezoelectric material, a standard graphical convention has been adopted. In this approach, the significant nonzero elements of the $[s^E]$, $[d]$, and $[\epsilon^T]$ matrices are shown as circles. Lines connect values of the same magnitude with filled circles indicating the same sign sense, and an open circle indicating the opposite sign sense. An "X" indicates that the value is $2(z_{11} - z_{12})$ where z is any of the properties, and an open circle on top of an "X" indicates multiplication by a factor of -2. The result is a picture that permits quick determination of the important physical couplings and properties, without giving specific values for any parameter.

Figure 5.3 shows an example of this graphical representation used to show the piezoelectric behavior of quartz. From this figure we can draw two important conclusions. First, quartz is symmetric in two directions but different in the third

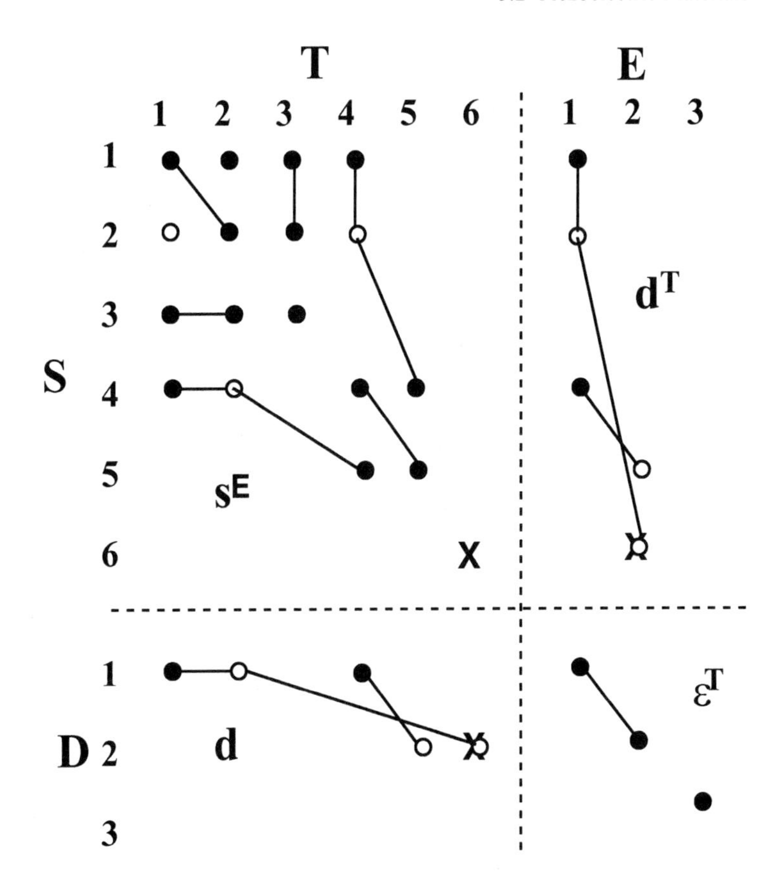

FIGURE 5.3. Physical properties of quartz.

direction. By convention the coordinate direction in which physical properties are most unlike the other principal directions is called the 3-direction. Second, we see that quartz has significant piezoelectric coupling in shear. While it is not obvious from the figure, since magnitudes are not provided, the shear electromechanical coupling in quartz is greater than the compressive electromechanical coupling. This is clear in Table 5.2, which contains data primarily from Berlincourt et al. [10]. In particular, a shear stress about the 1-axis will produce an electric displacement in the 1-direction. Similarly, a shear stress about the 2-axis will produce an electric displacement in the 2-direction. These two effects are the same size but opposite sign.

Figure 5.4 shows how quartz might be used in a transducer based on the information supplied in Fig. 5.3. Vertical motion of the top plate shears the quartz

TABLE 5.2. Properties of quartz at 25°C (data from Berlincourt et al. [10]).

Parameter	Dielectric	Piezoelectric ($\times 10^{-12}$ C/N)	Compliance ($\times 10^{-12}$ m^2/N)	Density (kg/m^3)
$\epsilon_{11}^T/\epsilon_0$	4.52			
$\epsilon_{33}^T/\epsilon_0$	4.68			
d_{11}		2.31		
d_{14}		0.727		
d_{26}		−4.62		
s_{11}^E			12.77	
s_{12}^E			−1.79	
s_{13}^E			−1.22	
s_{33}^E			9.60	
s_{44}^E			20.04	
s_{66}^E			29.12	
s_{14}^E			4.50	
ρ				2648.50

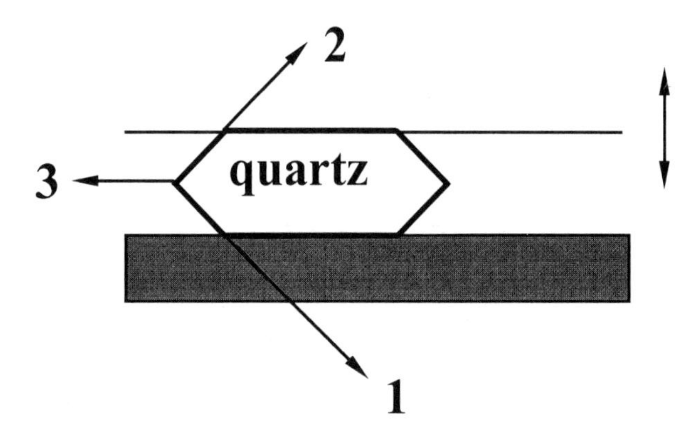

FIGURE 5.4. Example of how quartz is used in a transducer.

positively in the 1-direction and negatively in the 2-direction. Hence the two effects can be added in phase to produce a doubled signal. This approach is used regularly in shear accelerometers.

Figure 5.5 shows the graphical representation of the electromechanical properties of barium titanate and lead zirconate titanate. Again we observe that the materials exhibit a two-dimensional symmetry. However, Tables 5.3 and 5.2, based on data from Wilson [11], show that barium titanate and lead zirconate titanate offer greater coupling in compressive operation than in shear. In these tables d_h is the piezoelectric constant associated with hydrostatic pressure. Figure 5.5 shows one of the inconveniences associated with using these ceramics in the compressive modes, namely, that an electric field in the 3-direction will produce strains in the

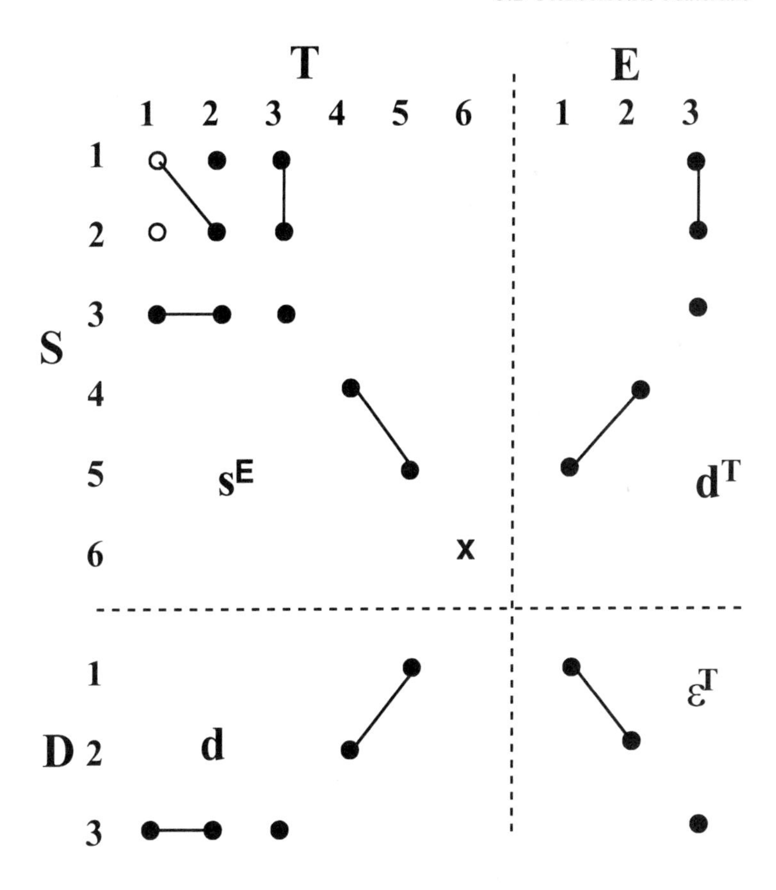

FIGURE 5.5. Barium titanate physical properties.

1-, 2-, and 3-directions. In spite of this coupling of principal dimensions, barium titanate and lead zirconate titanate are usually used in the compressive modes.

As Tables 5.3 and 5.2 also show, the piezoelectric effect in lead zirconate titanate is greater in magnitude than that for barium titanate. However, in terms of the parameters shown in the tables, the two materials are remarkably similar. A difference not shown in the tables is that the Curie temperature of barium titanate is about 115°C while for PZT it varies from about 300–365°C.

Figure 5.6 shows the graphical representation of the electromechanical properties of poled PVDF and Table 5.2, with data taken from Rossi [12], shows the numerical values. While the strength of the piezoelectric activity in PVDF is substantially below that of PZT or $BaTiO_3$, it offers advantages associated with being a polymer. In particular, PVDF is very flexible, and can be manufactured in thin,

TABLE 5.3. Properties of barium titanate at 25°C (data from Wilson [11]).

Parameter	Dielectric	Piezoelectric ($\times 10^{-12}$ C/N)	Compliance ($\times 10^{-12}$ m^2/N)	Density (kg/m^3)
$\epsilon_{11}^T/\epsilon_0$	1300			
$\epsilon_{33}^T/\epsilon_0$	1200			
d_{31}		−58		
d_{33}		149		
d_{15}		242		
d_h		33		
s_{11}^E			8.6	
s_{12}^E			−2.6	
s_{13}^E			−2.7	
s_{33}^E			9.1	
s_{44}^E			22.2	
s_{66}^E			22.4	
ρ				5550

TABLE 5.4. Properties of lead zirconante titanate (PZT 4) (data from Wilson [11]).

Parameter	Dielectric	Piezoelectric ($\times 10^{-12}$ C/N)	Compliance ($\times 10^{-12}$ m^2/N)	Density (kg/m^3)
$\epsilon_{11}^T/\epsilon_0$	1475			
$\epsilon_{33}^T/\epsilon_0$	1300			
d_{31}		−123		
d_{33}		289		
d_{15}		496		
d_h		43		
s_{11}^E			12.3	
s_{12}^E			-4.05	
s_{13}^E			-5.31	
s_{33}^E			15.5	
s_{44}^E			39.0	
s_{66}^E			32.7	
ρ				7500

flat sheets. It also has an acoustical impedance close to that of salt water, making it very attractive as an underwater transducer. Unfortunately, the manufacture of PVDF is a complicated and thus expensive process. In addition to requiring poling, PVDF generally requires stretching in one direction.

Let us consider one of the more common piezoelectric materials, PZT, in use in a compressive mode as a thickness expander. Then we can use the dot matrix of Fig. 5.5 to produce expressions for the strain and electric displacement field in the

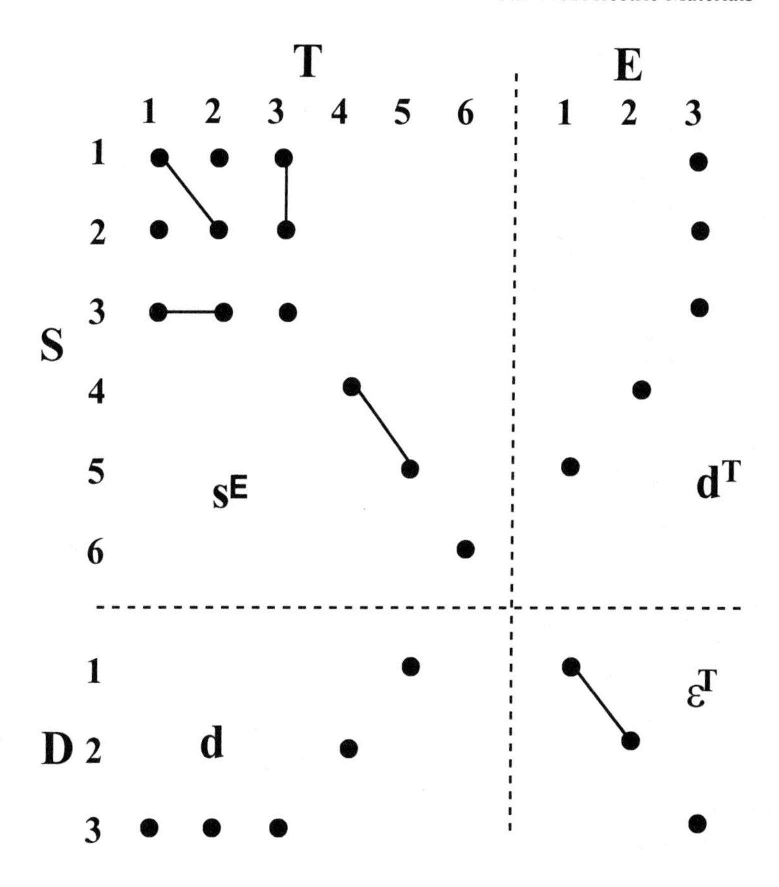

FIGURE 5.6. PVDF physical properties.

thickness directions:

$$
\begin{aligned}
S_3 &= s_{31}^E(T_1 + T_2) + s_{33}^E T_3 + d_{33} E_3, \\
D_3 &= d_{31}(T_1 + T_2) + d_{33} T_3 + \epsilon_{33}^T E_3.
\end{aligned}
\tag{31}
$$

Note that these relations show significant electromechanical cross-talk between the thickness direction and the two other principle directions. This is generally viewed as undesirable because it makes it difficult to distinguish direct causes for signals. For instance, D_3 could clearly arise from a stress in any of the three principle directions. Because of this problem, there has been significant research into materials which decouple the piezoelectric activity in the principle directions. Figure 5.7, taken from Smith [13], shows a schematic of the composite materials that have resulted from this research. These materials are referred to as 1–3 com-

TABLE 5.5. Properties of PVDF (data from Rossi [12]).

Parameter	Dielectric	Piezoelectric ($\times 10^{-12}$ C/N)	Compliance ($\times 10^{-12}$ m^2/N)	Density (kg/m^3)
$\epsilon_{11}^T/\epsilon_0$				
$\epsilon_{33}^T/\epsilon_0$	12			
d_{31}		23		
d_{32}		3		
d_{33}		21		
s_{11}^E			300–600	
s_{12}^E			400–1000	
s_{13}^E				
s_{33}^E				
s_{44}^E				
s_{66}^E				
ρ				1780

posites. The piezoelectric rods in the material are connected in one dimension, while the polymer matrix in which the rods are embedded is connected in three dimensions. Given this geometry, the effect of d_{31} and d_{32} is effectively nulled, producing a new expression for the electric displacement field in the 3-direction:

$$D_3 = d_{33}T_3 + \epsilon_{33}^T E_3, \tag{32}$$

which completely decouples the thickness direction from the other principle directions. Commerical versions of these materials are now available. For instance, Materials Systems Inc. [14] sells injection-molded PZT-polymer composites for use in hydrophones and sound generation underwater.

While the materials described here are some of the more common piezoelectrics, there are certainly other options. In particular, the desire to build transducers using solid-state fabrication techniques has produced a new set of piezoelectric materials which are being examined and exploited for miniaturized sensors. These materials include zinc oxide, aluminum nitride, lead lanthanum zirconate titanate, and even gallium arsenide. Interested readers are referred to Rosen et al. [15] for information on other piezoelectric materials.

5.3 Piezoelectric Structures in Transducers

Because many of the piezoelectric materials are ceramics, and they are formed from powder, they can take almost an arbitrary finished shape. As a practical matter, the common shapes for piezoelectric materials in transducers are disks, donuts, cylindrical shells, and rectangular constructions. Even within these rather simple geometries, there are a number of ways we can use the materials, depending on which of the nonzero piezoelectric coefficients we wish to exploit.

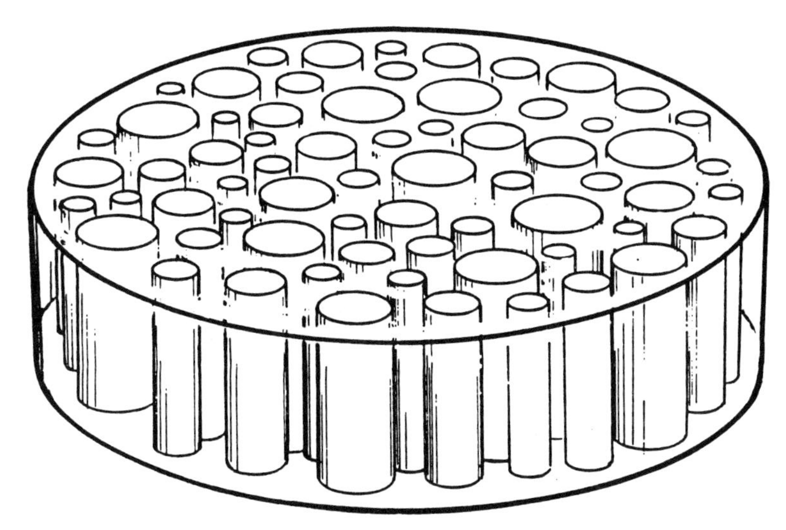

FIGURE 5.7. Schematic of a 1–3 piezoelectric composite with piezoelectric rods embedded in a polymer matrix.

Some of the various means of using piezoelectric materials to obtain electromechanical coupling are shown graphically in Figs. 5.8–5.14. Figure 5.8 shows the most common approach, in which there is a coupling between the stress and strain in a principal direction and the electric field in the same direction (i.e., the d_{ii} piezoelectric coefficient is being exploited). This is commonly referred to as a longitudinal expansion mode. Since there are three principal directions in each material there are clearly three axes which could be chosen. However, the 3-direction is by far the most common choice. Often the expansion/compression is in the principal direction in which the material is small relative to the other dimensions. Transducers of this sort are referred to as thickness expansion transducers. They are used in compression accelerometers, hydrophones, and pressure sensors.

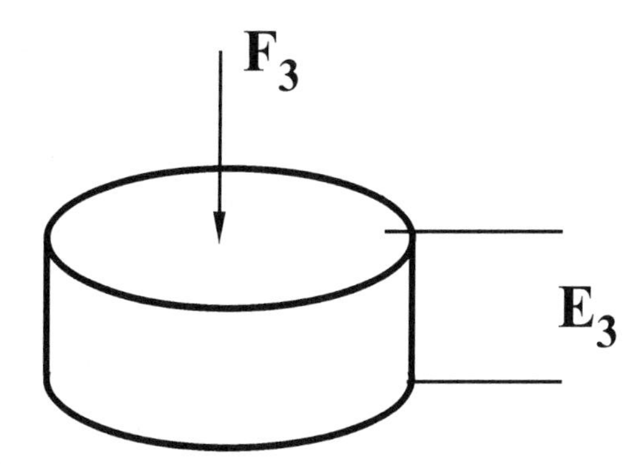

FIGURE 5.8. Thickness expansion mode of a piezoelectric transducer.

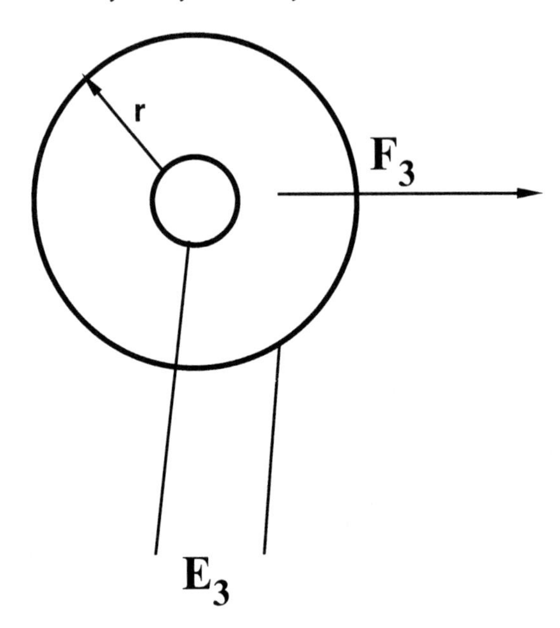

FIGURE 5.9. Thickness expansion using radial geometry in piezoelectric material.

An alternative geometry which also relies on the d_{ii} coefficient primarily is the radial mode expander shown in Fig. 5.9. This approach is particularly useful for coupling between the piezoelectric solid medium and a surrounding fluid medium, such as might be required in sound generation. Again, it is typical that the 3-direction is the expansion direction.

Figure 5.10 shows an example of a transducer geometry which takes advantage of a shear mode in barium titanate. A thin plate is clamped at one end and electroded along the clamped and free edges. Application of an electric potential results in a vertical force (normal to the electric field and the surface of the plate), which bends the thin piezoelectric. Alternatively, application of a force induces a voltage. Because this device operates in bending, it is referred to as a bender. If a single layer of piezoelectric material is used, it is a bender monomorph. If two layers are used and oriented so as to effectively double the sensitivity of the device, it is called a bender bimorph. Bender monomorphs and bimorphs are very common transducer elements. They may be found, for example, in some phonograph stylus cartridges. There, motion of the needle in the grooves on the album is translated into bending of four bender bars. Addition and subtraction of the induced voltage signals produces the two channels of signals.

Sometimes, segments of piezoelectric tubes, such as shown in Fig. 5.11, are used for transducers. The advantage of a segment of a cylinder over a rectangular strip is added structural rigidity due to the curvature. At times, multiple segments are combined in a single transducer. For example, the structure shown in Fig. 5.12 is used by Burleigh Instruments, Inc. [16] for an aligner/translator. Exciting the

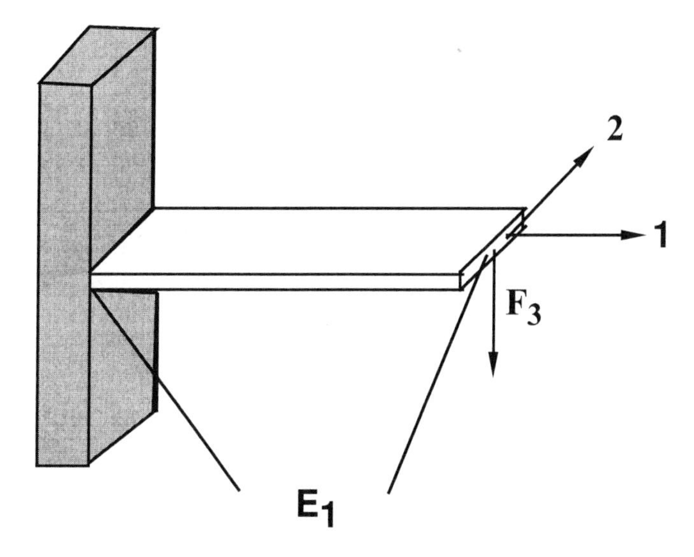

FIGURE 5.10. Thickness shear transducer used in bender bimorphs and monomorphs using barium titanate.

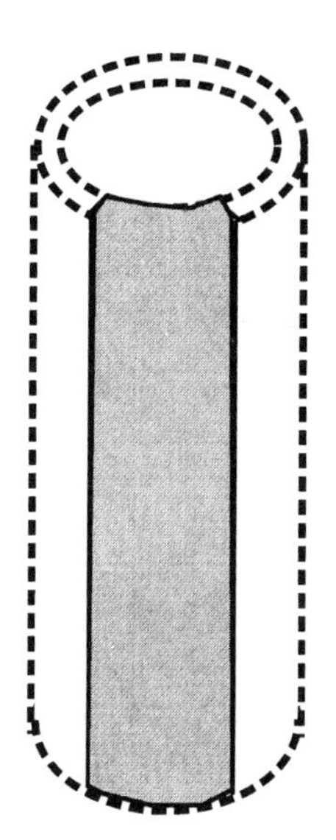

FIGURE 5.11. Example of a piezoelectric structure made from a segment of a cylindrical tube.

three segments simultaneously with the same input translates the top ring, while exciting the segments with different signals tilts the top ring.

The examples so far assume that a single pair of electrodes is used for each piezoelectric material in the transducer. It is also possible to put multiple electrodes

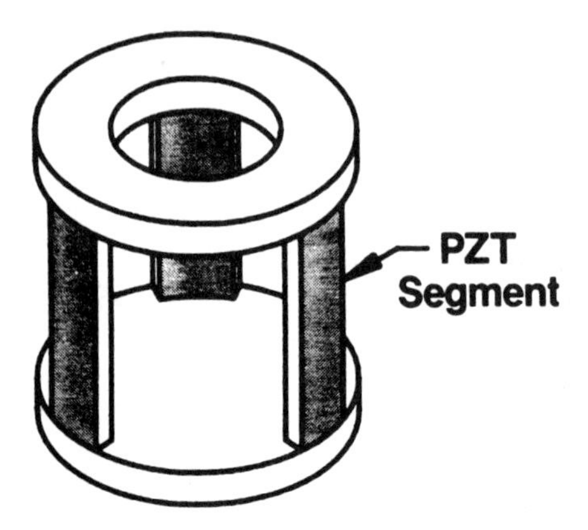

FIGURE 5.12. Aligner/translater made by Burleigh Instruments, Inc. [16] using three cylindrical segments.

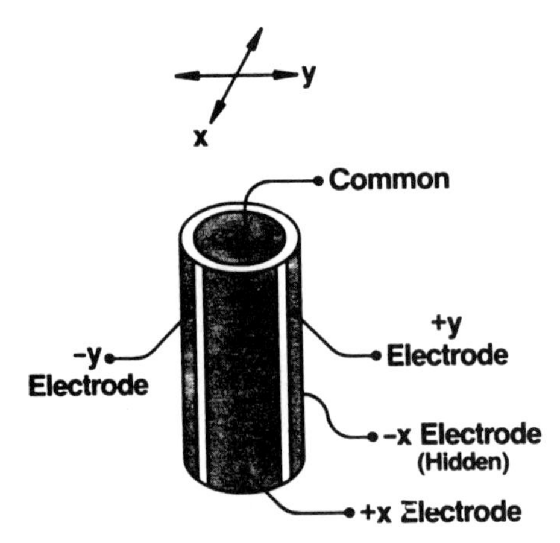

FIGURE 5.13. Piezoelectric actuator used in the scanning tunneling microscope of Binnig and Rohrer [17] and made by Burleigh Instruments, Inc. [16].

on the material in order to produce or sense motion in more than one direction. For example, the scanning tunneling microscope described by Binnig and Rohrer [17] used the Burleigh Instruments, Inc. [16] piezoelectric cylinder design shown in Fig. 5.13. Application of different voltages to the four electrodes bends the cylinder, while x- or y-motion can be obtained or sensed by grounding the other pair of electrodes.

It is also common to use multiple piezoelectric elements in a stack as shown in Fig. 5.14. The primary advantage of this approach over use of a single element of material with the same total dimensions is an increase in the generated charge by a factor equal to the number of elements in the stack. We can see this clearly from consideration of the constitutive relation for the electric displacement. Suppose that we consider thickness expansion, and that we neglect all but the d_{33} piezoelectric coupling term. Then we have (32) which relates D_3 to T_3 and E_3 for each element of

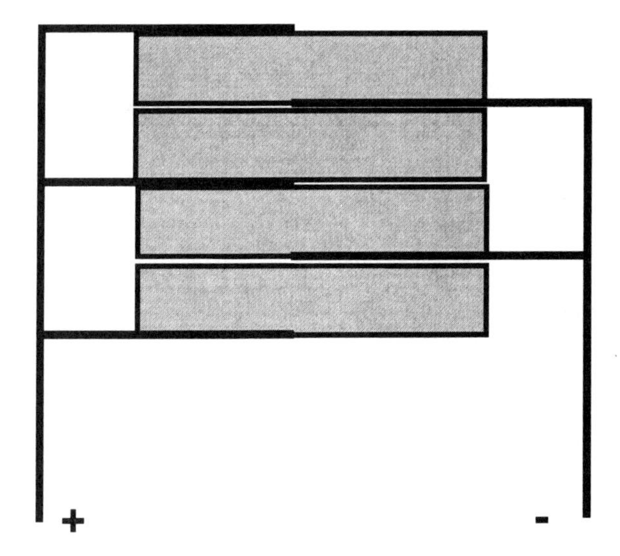

FIGURE 5.14. Piezoelectric stack as used in many sensors.

the stack. Since D_3 is proportional to the charge that would be developed on plates orthogonal to the 3-direction, we have a linear relation between q_3 and T_3 and E_3. The net charge developed will be the sum of the charges from each element in the stack. Assuming that the elements mechanically behave as stiffnesses, the stress is the same on each element, and the electrode structure ensures that the voltage is the same. The net voltage is the net charge divided by the net capacitance. Given the multiple electrodes, the net capacitance is the capacitance of each element times n. Thus the net voltage is unchanged by stacking elements. Hence, if a *voltage* amplifier were to be used with a sensor stack, there would be no voltage enhancement over that of a single element of the same net size!

Before proceeding to discuss some typical examples of piezoelectric transducers, we will discuss models of piezoelectricity. This will permit us to develop logical models of some commercially available piezoelectric transducers which are suitable for use in their analysis.

5.4 Models of Piezoelectricity

Let us consider a piezoelectric transducer in which the material dimensions are large in two of the principal directions and small in the third, thickness direction. We assume the thickness direction to be the direction of orientation of the electric field and mechanical strain, and that the piezoelectric material has planar perpendicular surfaces normal to the thickness direction. These surfaces are coated with a metal electrode.

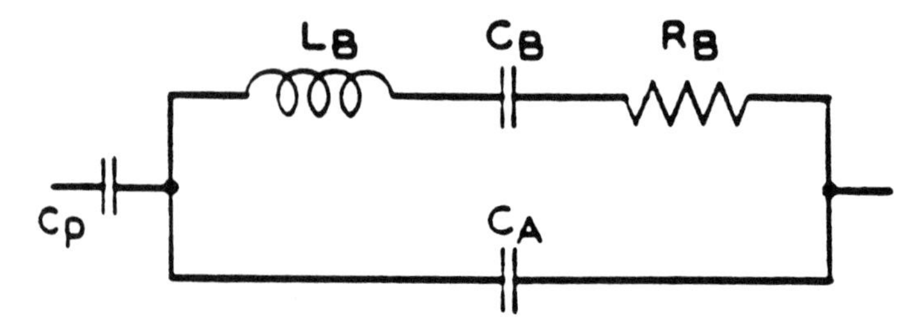

FIGURE 5.15. Cady's model of a piezoelectric material [18].

One of the earliest models of a piezoelectric crystal of this type was developed by W.G. Cady [18]. Figure 5.15 shows his pure circuit model, developed using the impedance analogy. Mason extended this model to include a description of the crystal used to couple to an external system, i.e., as a transducer. This model is shown in Fig. 5.16 [19]. Here C_0 is the static electrical capacitance of the crystal, C_E its elastic mechanical compliance, L_E half the mass of the crystal, R the internal friction and support friction, and C_M the so-called mutual capacitance-compliance (the ratio of the charge applied to the crystal to the force required to keep the crystal from moving). Note that C_M enters the model in a very nonintuitive manner. In particular, there are two negative capacitances which would tend to indicate that the elastic–electric interaction is destabilizing. This interpretation is clearly nonphysical and results from the insistence on using 1-port energic elements only.

Subsequent to development of the model shown in Fig. 5.16, Mason produced more sophisticated circuit models of piezoelectric transducers using ideal lossless 2-ports to separate the mechanical and electrical components [20] and [21]. One of these models that is commonly used for low-frequency regimes is shown in Fig. 5.17, where m is the transformer modulus. The equivalent bond graph model is shown in Fig. 5.18. Using impedance techniques we can derive the expression

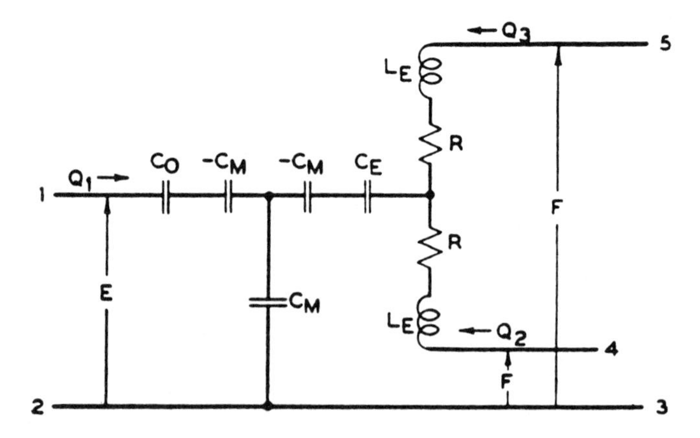

FIGURE 5.16. An early Mason model of a piezoelectric transducer [19].

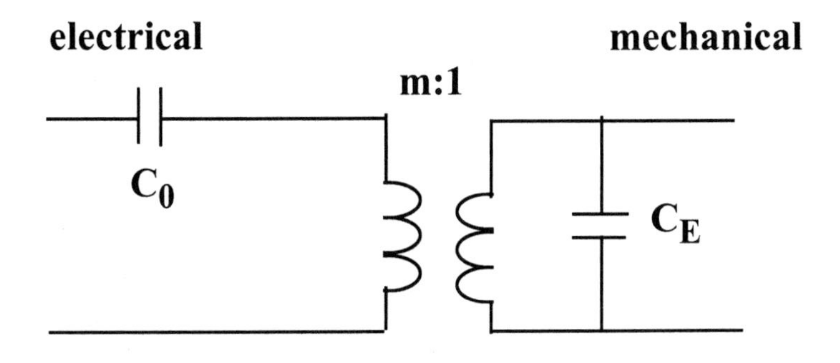

FIGURE 5.17. A Mason model of a piezoelectric transducer suitable for low-frequency use.

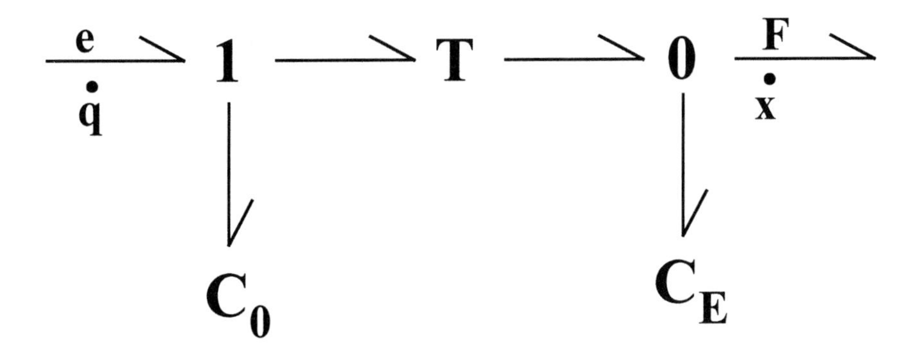

FIGURE 5.18. Equivalent bond graph model of the Mason model.

for the voltage across the capacitor:

$$e = \frac{q(m^2 C_E + C_0 + x/m^2 F)}{C_0(m^2 C_E + x/m^2 F)}. \tag{33}$$

As described by Busch-Vishniac and Paynter [22], the voltage expression given in (33) indicates two serious problems with the model. First, so long as we use only 1-port capacitances, we find that the voltage expression cannot be decoupled from the force. This is the reason that most transducer designers insist that we can only discuss the piezoelectric behavior once a load is assumed. Typically transducer designers take the voltage expression to the limit of an open circuit (mechanical side) or a shorted circuit. However, the behavior of the constituent material itself should possess a characteristic transduction property which can be examined in isolation. It should not depend upon the load since the external load cannot change the internal material properties. If we were to solve this problem by assuming that the terms involving F in (33) could be neglected, then the voltage would have no dependence on the strain—clearly a nonphysical result.

Second, we note that in the limit of no motion of the piezoelectric ($x = 0$) (33), reduces to the following:

$$e = q\left[\frac{1}{C_0} + \frac{1}{m^2 C_E}\right]. \tag{34}$$

However, if the material is not permitted to strain, the mechanical compliance cannot be relevant. Hence the term must disappear in the zero strain limit. Since the compliance is a constant for the material, this requires that the transformer modulus must be a function of x. This nonphysical result is directly caused by the attempt to model the energetic coupling between the elastic and electric fields using separate 1-port elements for each.

Although pure circuit models of piezoelectric activity are still the norm, we maintain that these models introduce more confusion than they dispell. The alternative to pure circuit models using 1-port energic elements is to model piezoelectricity using a 2-port capacitor, much as we did for other capacitive transduction processes. We begin by noting that the voltage across the piezoelectric is given by

$$e = \frac{\partial \mathcal{E}}{\partial q} = \frac{q(d_0 + x)}{\epsilon A}. \tag{35}$$

We perform the partial integration to determine the energy stored:

$$\mathcal{E} = \frac{q^2(d_0 + x)}{2\epsilon A} + E_p + \phi(x), \tag{36}$$

where E_p is the polarization energy and $\phi(x)$ is some function of x that is undetermined by the partial integration. We identify this additional energy term as that associated with the spring-like behavior of the piezoelectric material:

$$\phi(x) = \frac{x^2}{2C_E}. \tag{37}$$

Note that now we have included two terms in the energy stored in the material that were not included in previous discussions of capacitive transduction. The first of these is the polarization energy, which was assumed to be small in nonpiezoelectric materials. It is the absence of a strong polarization that necessitates an external power supply, the energy of which is included in the q^2 term for the nonpiezoelectric, capacitive transducers. The second new term is the material stiffness. This term was neglected in previously discussed capacitive transducers because the medium between the electrodes was assumed to be air. Although the assumption that air stiffness can be neglected is reasonable when the gap between the electrodes is large, air stiffness can be important for small electrode gaps. For piezoelectric transducers, the pertinent stiffness is that of the piezoelectric material, which typically is many orders of magnitude higher than that of air of the same size.

We may combine (36) and (37) to determine the force between the electrodes on the piezoelectric. The result is

$$F = \frac{\partial \mathcal{E}}{\partial x} = \frac{q^2}{2\epsilon A} + \frac{x}{C_E}, \tag{38}$$

where we have assumed that the polarization energy is independent of strain. This equation and (35) are the two energetically consistent constitutive relations that define the piezoelectric material. They may be easily interpreted despite the presence of coupling and nonlinearities.

Compared to the results for the pure circuit model, there are two other important differences. First, we note that the energy, force, and voltage have each been found to be functions of the two displacement variables only. This means that the piezoelectric material can indeed be discussed independent of the load. In the case of $\mathcal{E}$ and e, the expressions explicitly couple the energy variables, indicating that it is not possible to obtain these results using any combination of ideal nonenergic 2-ports and 1-port energic elements. Indeed, (35), (36), and (38) identify the material as an essential 2-port capacitance. Second, consider the limiting behavior as no strain is permitted. In this limit, (35) reduces to the correct expression for a fixed gap parallel plate capacitor. Similarly (38) yields the electrostatic force between plates in a parallel plate capacitor.

From the arguments above, we conclude that the pure circuit model of a piezoelectric transducer simply is not capable of providing an accurate picture of the transduction process. In contrast, a bond graph model is simple, energetically consistent, and easily interpreted. Recent work by Moon and Busch-Vishniac [23]–[27] uses a bond graph modeling approach to include the thermal storage of energy as well as the mechanical and electrical, and to develop a modeling procedure for piezoelectric transducers that marries well with finite-element analysis. This will permit the logical design of piezoelectric transducers which minimize the influence of thermal gradients or of transducers which sense thermal changes and minimize the effect of the piezoelectric coupling.

5.5 Examples of Piezoelectric Transducers

Up to this point we have mentioned the fundamental principles upon which piezoelectricity depends, discussed the general means in which piezoelectric materials are used in transduction, and developed a model of piezoelectric materials which is suitable for analysis. In this section we present some examples of common piezoelectric transducers. Our aim is to show a broad range of applications and geometries, and to reinforce some of the physical concepts through modeling of a couple of the transducers. We begin with transducers which use a single piezoelectric element or stack of parallel elements to obtain a broadband response. Next we discuss devices in which a single conventional piezoelectric element is used in a resonant mode. We then introduce a transducer which relies on use of multiple piezoelectric elements (not oriented parallel to one another). Finally we present a few examples of unusual piezoelectric transducer applications and structures.

5.5.1 *Examples of Broadband Transducers Using One Piezoelectric Element or a Stack of Parallel Piezoelectric Elements*

First consider a standard compression accelerometer such as shown in Fig. 5.19. In this transducer a mass is supported by leaf springs and attached to a stack of piezoelectrics. The other end of the springs and of the stack is attached to the accelerometer case. The piezoelectric elements add stiffness to the system as well as converting mechanical motion of the inertial mass to an electrical output. A simple model of the transducer is shown in Fig. 5.20. Analysis of this model shows that at low frequencies (below resonance), the inertial mass dynamic behavior dominates over the stiffness characteristics and the output voltage is proportional to the acceleration of the case. There are many manufacturers of compression accelerometers of this type. A couple of examples made by Dytran Instruments, Inc. [28] and Kistler Instrument Corp. [29] are shown in Fig. 5.21.

A slightly different piezoelectric accelerometer is shown in Fig. 5.22, which pictures shear accelerometers made by PCB Piezotronics [30] and Kistler Instrument Corp. [29]. In this accelerometer a ring or block of piezoelectric material is cemented to a central support and a cylindrical or rectangular inertial mass hung on the opposite side. The piezoelectric material is thus operated in the shear mode rather than the longitudinal mode operation of a compression accelerometer.

Yet another example of a piezoelectric accelerometer is given in Fig. 5.23. In this device, marketed by PCB Piezotronics [30], a beam-shaped element is supported at a point and responds to vibration by flexing. It thus operates very much like the bender element shown previously. The advantages of this geometry are low profile, low weight, and good thermal stability.

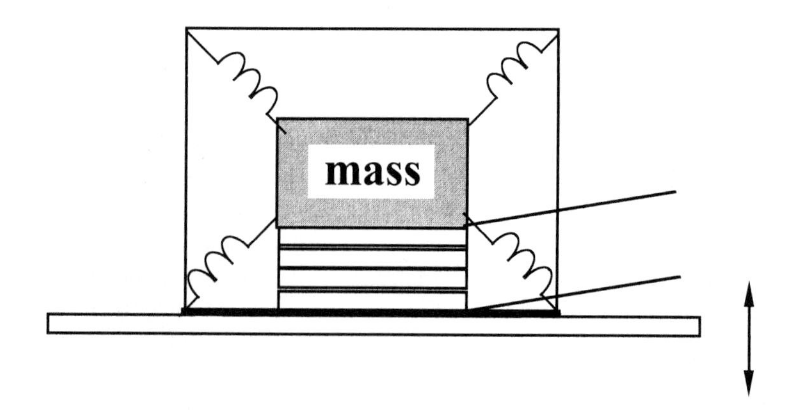

FIGURE 5.19. A piezoelectric compression accelerometer.

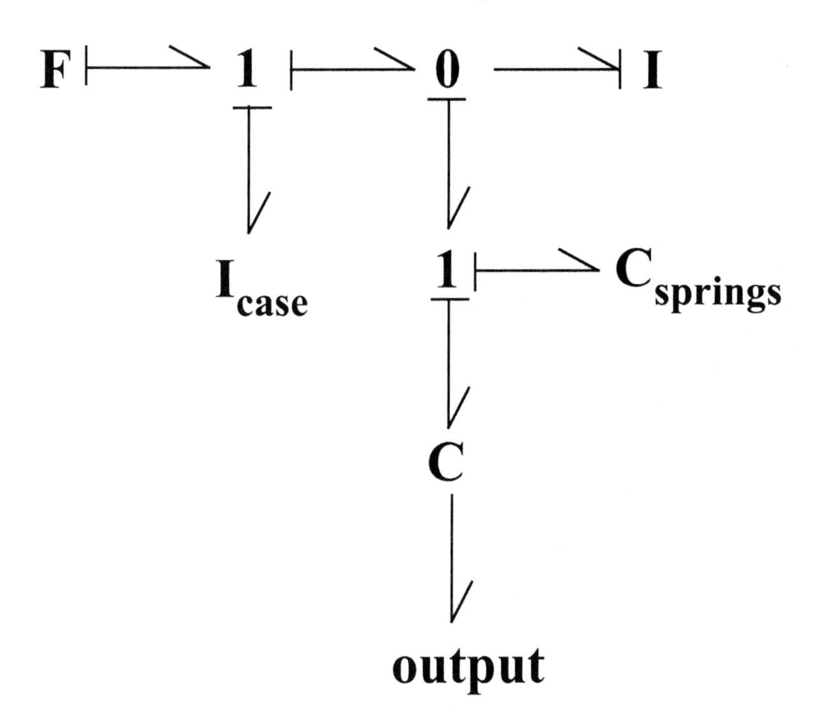

FIGURE 5.20. Model of a piezoelectric compression accelerometer.

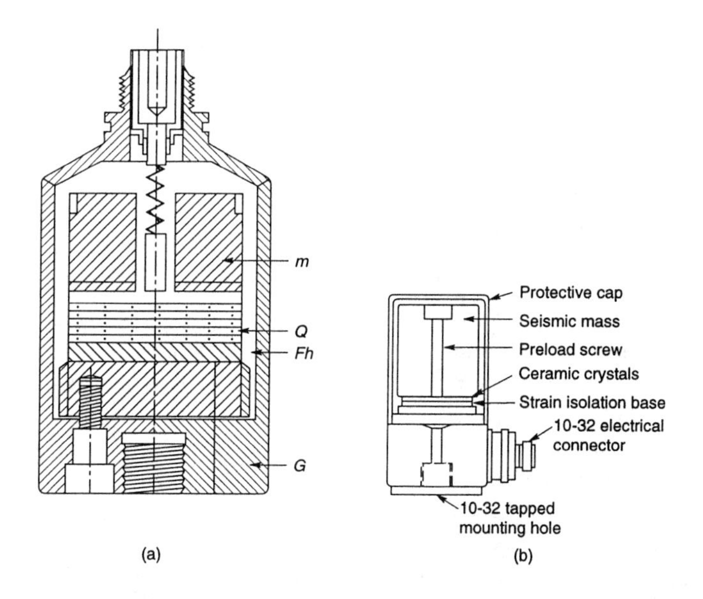

FIGURE 5.21. Piezoelectric compression accelerometers from (a) Kistler Instrument Corp. [29] and (b) Dytran Instruments, Inc. [28].

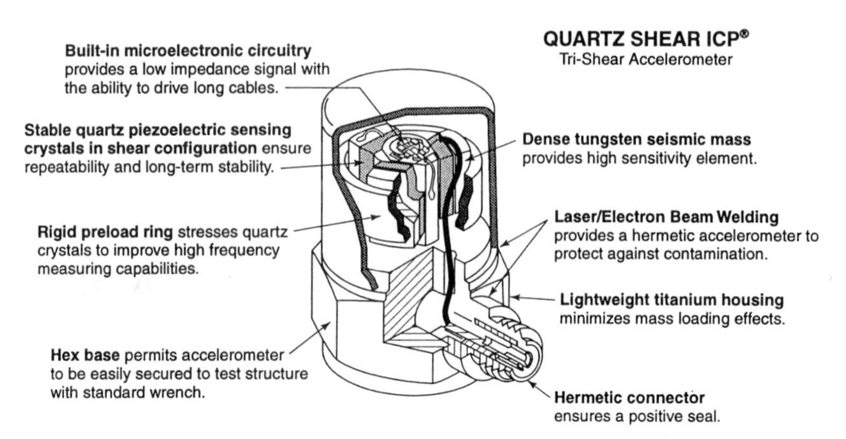

FIGURE 5.22. Shear accelerometers made by a). PCB Piezotronics [30], and b). Kistler Instrument Corp. [29].

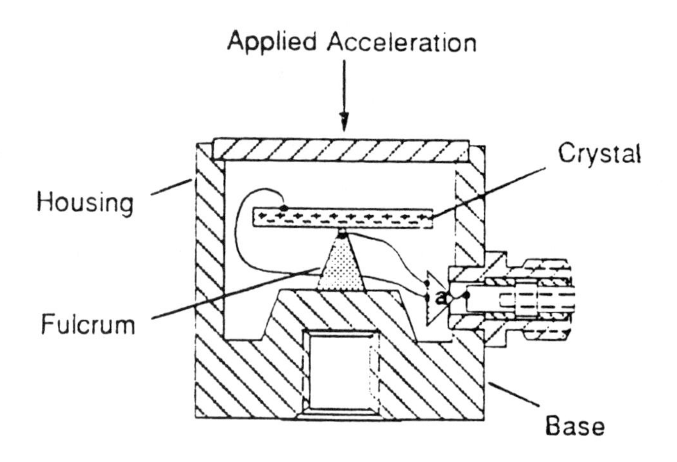

FIGURE 5.23. A flexural accelerometer made by PCB Piezotronics [30].

The low-frequency limit of these piezoelectric accelerometers is determined by the system time constant, i.e., the product of the piezoelectric resistance and capacitance. The high-frequency response limit is dictated by the resonance of the mass–spring system. Commercially available accelerometers vary from 1.5 g to about 1 kg and offer flat frequency response from about 20 Hz to up to 15 kHz. The advantages of piezoelectric accelerometers are that they operate with high sensitivity, over a fairly broad frequency range, and have no need for an external power supply. However, they require that amplifiers be closely located (because they are very high capacitance devices) and they often show undesirable crosstalk characteristics. The crosstalk, or production of a signal due to motion orthogonal

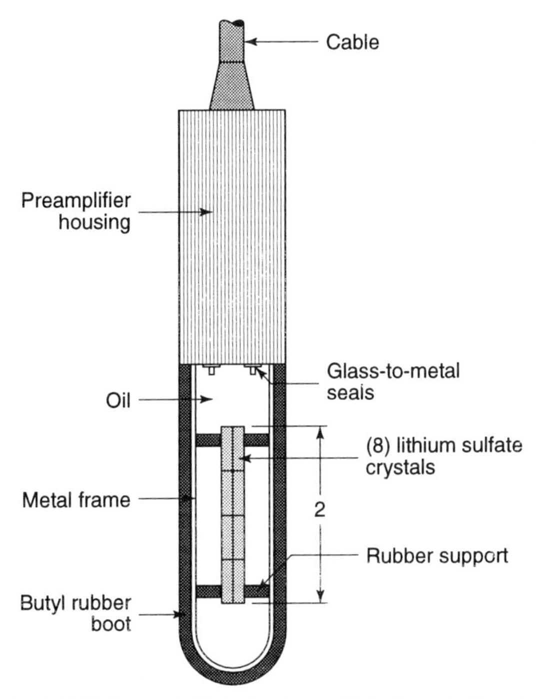

FIGURE 5.24. USRD-type H23 hydrophone. This figure is from Bobber [31].

to the direction of maximum acceleration sensitivity, results from the mechanical and electromechanical coupling between the orthogonal coordinates. In most piezoelectric accelerometers, the sensitivity to acceleration which is orthogonal to the main axis of sensitivity is roughly 0.1–1% of the primary axis sensitivity.

Figure 5.24, taken from Bobber [31] shows an example of a piezoelectric hydrophone, used for detection of sound underwater. In this device a set of lithium sulfate crystals are used as the piezoelectric transduction element. Incident sound pressure causes compression and expansion of the crystals which results in a small output voltage. This voltage is amplified using a preamplifier which is located close to the transducer. Oil in the transducer cavity permits the use of the device at substantial depths in the ocean without collapse of the support structure due to increases in hydrostatic pressure. As shown in Fig. 5.25, also from Bobber [31], the frequency response of this hydrophone is flat over a range from about 400 Hz to 20 kHz.

Figure 5.26 shows a piezoelectric pressure sensor which closely resembles the compression accelerometers shown in Fig. 5.21. This pressure sensor, made by Dytran Instruments, Inc. [28], uses a thin stainless steel diaphragm to convert the pressure to a force on the end piece which pushes against a stack of piezoelectric crystals. A seismic mass and additional piezoelectric crystal with opposite polarity from the stack are scaled to form an accelerometer which exactly cancels the inertial effect of the end piece and diaphragm mass.

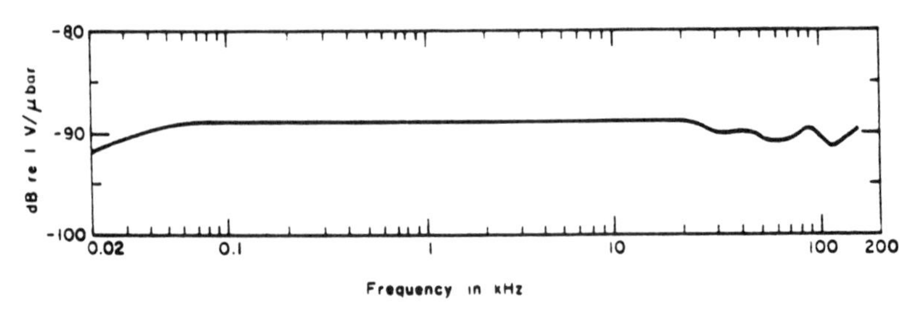

FIGURE 5.25. Free-field voltage sensitivity of the H23 piezoelectric hydrophone (from Bobber [31]).

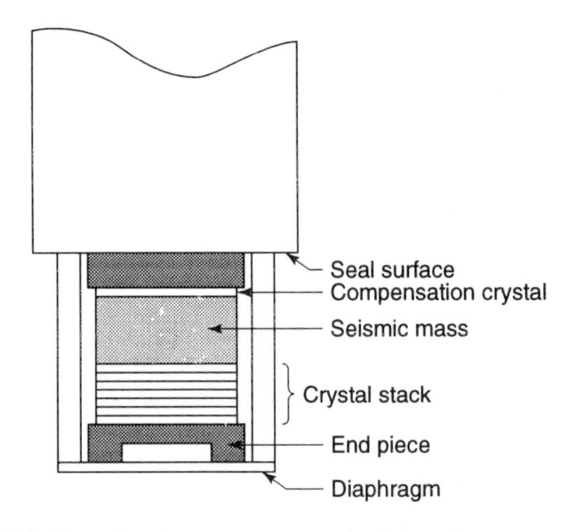

FIGURE 5.26. Piezoelectric pressure sensor by Dytran Instruments, Inc. [28].

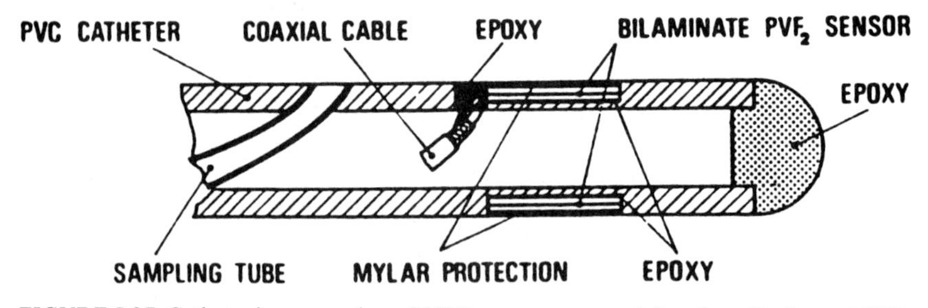

FIGURE 5.27. Catheter incorporating a PVDF pressure sensor, taken from Dario et al. [32].

Another sort of pressure sensor is shown in Fig. 5.27, taken from Dario et al. [32]. This figure shows a catheter which uses a bimorph made of PVDF to monitor the blood pressure of a patient. PVDF is ideal for this application because of its flexibility, and because its impedance is much more similar to that of blood than the ceramic or crystalline piezoelectric materials.

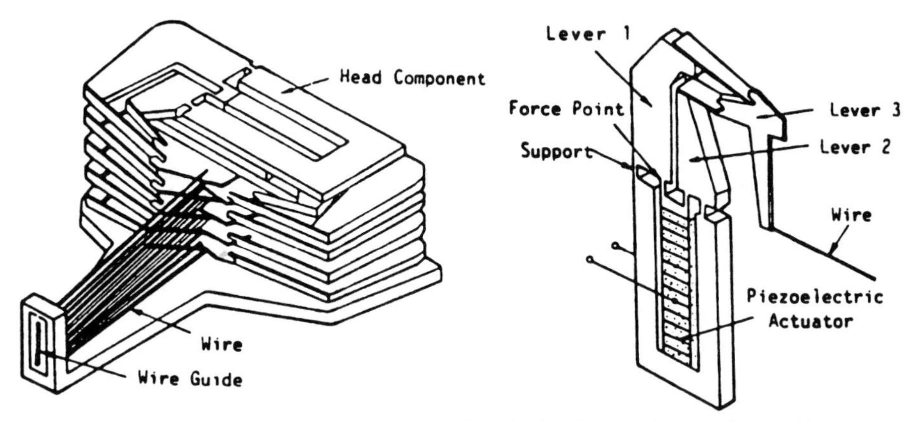

FIGURE 5.28. Piezoelectric dot-matrix print head, taken from Uchino [33].

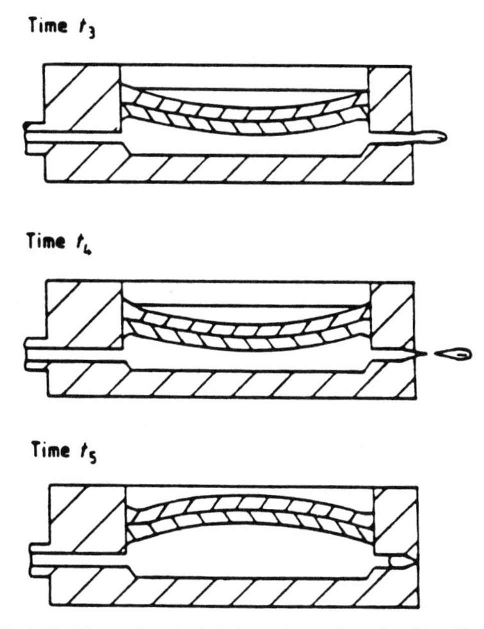

FIGURE 5.29. Piezoelectric ink jet printer described by Keeling [34].

Figure 5.28, taken from Uchino [33] shows a piezoelectric dot-matrix print head. In this device a piezoelectric stack is used to develop a stress in a mechanical structure attached to a wire. This moves the wire, which is one of a set, each corresponding to a potential print dot. Figure 5.29, taken from Keeling [34] shows another sort of printer, also using piezoelectric actuation. In this ink jet printer a bimorph element is used to draw ink into a cavity and then force it out to the paper. Not shown in this figure are the valves that are required for this mechanism to work.

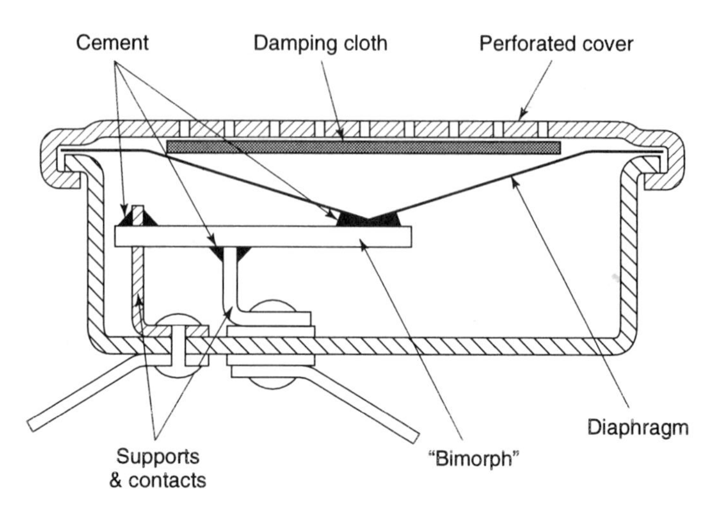

FIGURE 5.30. Piezoelectric microphone sold by Vernitron [35].

As a final example of a broadband, conventional piezoelectric transducer, Fig. 5.30 shows a piezoelectric microphone marketed by Vernitron [35]. The piezoelectric element in this transducer is a bimorph element which is excited by a force at its end. This force is developed from sound pressure on a diaphragm attached to the bimorph at one point. A perforated cover and damping cloth protect the diaphragm from moisture and from large transient pressures.

5.5.2 *Examples of Resonant Transducers Using a Single Piezoelectric Element or Element Stack*

Many piezoelectric transducers are intentionally operated in a resonant mode in order to enhance performance while limiting sensitivity to a narrow frequency band. In general, the motion produced in piezoelectric material is quite small. While this may be acceptable for a sensor, it is undesirable in an actuator. Operation in a resonant mode provides a means of ensuring that the actuation level is as high as practically obtainable for a given material. In this subsection, we show examples of transducers of this sort.

Figure 5.31 shows a piezoelectric device used for ultrasonic cleaning made by Vernitron [35]. In this transducer, two PZT disks are held between a light metal support and a reaction mass. The system is intentionally driven at its fundamental resonance frequency in order to get high levels of motion. This causes violent agitation of the fluid in the tank. The purpose of the clamp and reaction mass are to ensure that virtually all of the mechanical energy is expended in moving the support up and down. PZT is particularly well suited for operation in a resonant mode because it has sharp resonances with little damping. Q-factors of up to 500 are obtainable with commercially available material.

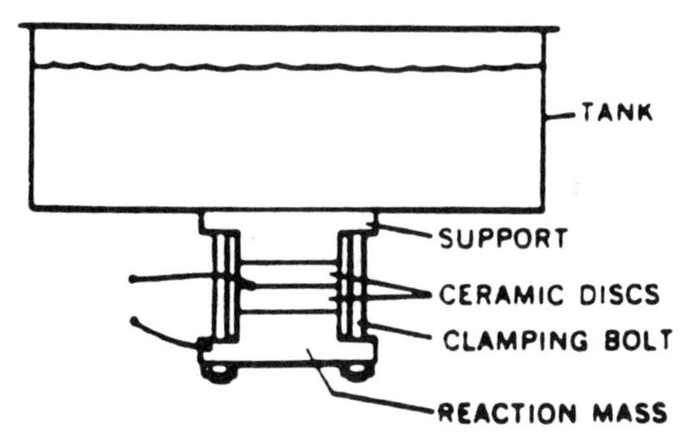

FIGURE 5.31. Ultrasonic cleaning device.

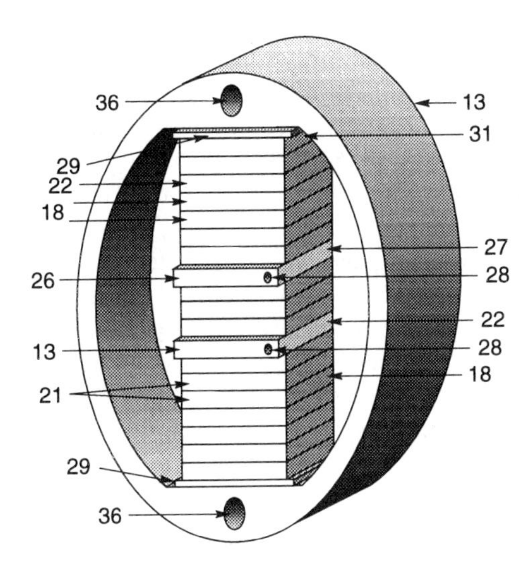

FIGURE 5.32. Flextensional sound source described by Toulis [36].

A second example of a transducer which is operated in a resonant mode is the simple flextensional sound source shown in Fig. 5.32, which is taken from the Toulis patent which first describes flextensional devices [36]. The heart of this sound actuator is a ceramic element which is operated so that it expands in the axial direction. As Fig 5.32 shows, this results in motion of an oval shell surrounding the piezoelectric material. The purpose of the shell is to amplify the sound-generating capabilities by producing much greater compression and rarefaction of the medium than would be produced simply by exposing the ceramic itself to the environment. In other words, the shell acts to increase the effective area over which pressure fluctuations are generated.

Yet another example of a piezoelectric transducer operated in resonant mode is shown in Fig. 5.33. This device is part of a system designed to permit the accurate determination of the location of the endpoint of a robot, described by Leifer and

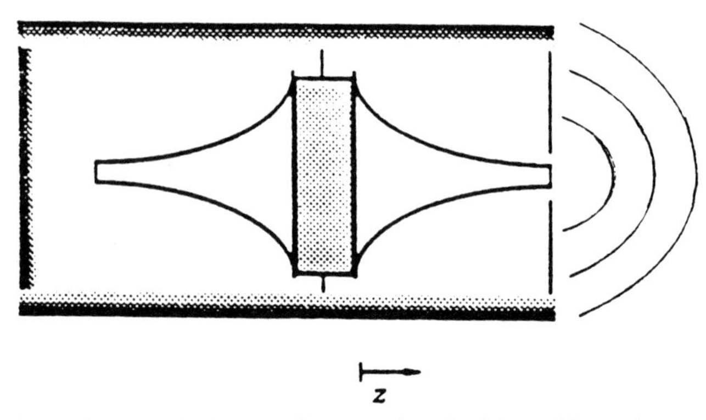

FIGURE 5.33. Robotic endpoint sound source described by Leifer and Busch-Vishniac [37].

Busch-Vishniac [37]. The system places the ultrasonic source shown in the figure at the robot endpoint. Microphones at fixed locations in the workspace detect the time of arrival of the pulse of sound emitted by the source. The differences in arrival times are used to determine the source location. The ultrasonic transducer uses a piezoelectric ceramic operated at its resonance frequency. The disk has two inverted solid exponential horns glued to its faces and tuned so that the entire system resonance has a high Q-factor. The purpose of the horn is to focus the sound to a point. This yields a high-frequency sound source which is nearly omnidirectional, rather than highly direction dependent. Only one of the horns is exposed to the environment. The other is present to reinforce the system resonance but is not permitted to radiate sound into the air.

5.5.3 Example of a Transducer Using Multiple Nonparallel Piezoelectric Elements

Many transducers are complicated combinations of multiple simple sensors and actuators in which the signals are combined in a manner which produces a desired effect. In this subsection, we provide an example of such a complicated transducer which is based on simple piezoelectric elements.

Figure 5.34, taken from Allocca and Stuart [38], is an example of a transducer which uses multiple piezoelectric elements which are not oriented parallel to one another. This device is a phonograph cartridge and uses two bender bimorph elements, one oriented at $+45^\circ$ and one at -45°. The stylus rides in the grooves of a phonograph and excites bending in the elements, each of which corresponds to a single channel of a stereo recording. This bending produces a set of output voltages which are an electrical representation of the phonograph signal. Such cartridges, referred to as ceramic cartridges, have been commercially available for many years (although phonographs are now nearly obsolete!). They are generally heavier than conventional, high-quality cartridges but they are very rugged and reliable.

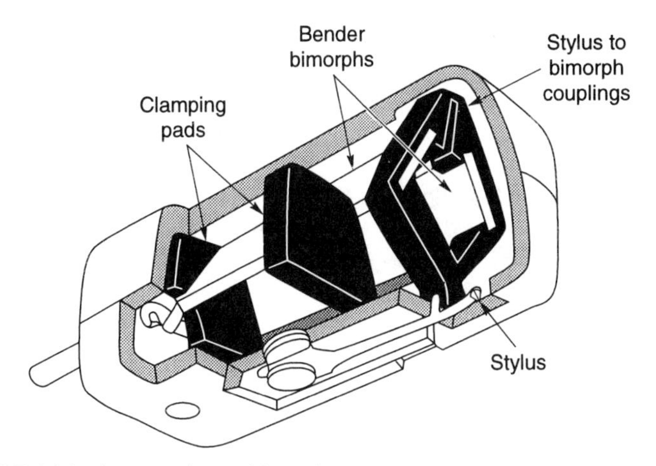

FIGURE 5.34. Phonograph cartridge using ceramic bimorph elements (from Allocca and Stuart [38]).

5.5.4 *Examples of Transducers which Are Unconventional*

The examples of piezoelectric transducers provided so far are for conventional sensors and actuators, primarily used in measurement or control systems. In addition, piezoelectric transducers may be found in a wide range of unconventional operations, such as cigarette lighters. This subsection provides a couple of examples of this sort of piezoelectric transducer.

Figure 5.35 is one example of an unusual transducer built from piezoelectric material. This device serves as a high-voltage source for gasoline engine ignition. It is built by Vernitron [35]. A cam driven by the engine actuates a lever which applies varying axial force to two PZT ceramic cylinders. Compression of the stack causes a voltage sufficiently high to generate a spark between two electrodes. A device of this type has been used successfully in small lawn-mowers, developing voltages up to 20,000 V. Further, a smaller version of this sort of device is found in many throw-away piezoelectric cigarette lighters and serves to generate the spark in response to human supplied power.

Figure 5.36 is another example of an unconventional piezoelectric transducer. It is a hydrophone, described in Bennett and Chambers [39], that not only detects the presence of an acoustic source and its amplitude, but can determine the distance to the source by dynamically varying the shape of the piezoelectric element. The piezoelectric material used is PVDF, which is readily available in sheets varying in thickness from about 4 μm to 1 mm. Thin PVDF is very flexible, which markedly distinguishes it from the conventional piezoelectric materials. In the transducer shown in Fig. 5.36 a vacuum is used to shape the PVDF membrane into a parabolic dish. This dish preferentially responds to sound incident from the focus. Thus, by dynamically varying the dish radius of curvature, and monitoring the received sound level, the transducer permits the source to be localized.

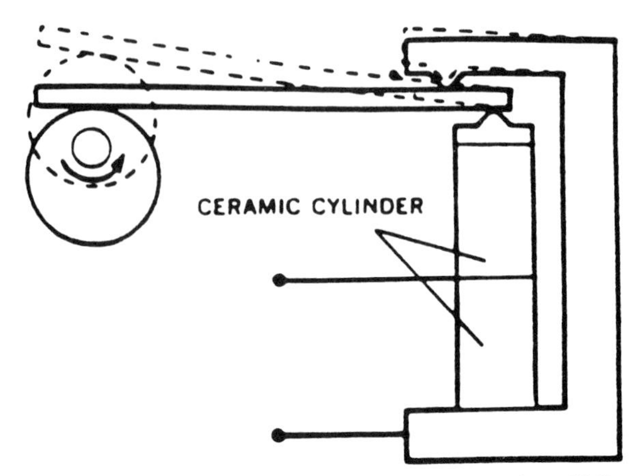

FIGURE 5.35. Piezoelectric ignition built by Vernitron [35].

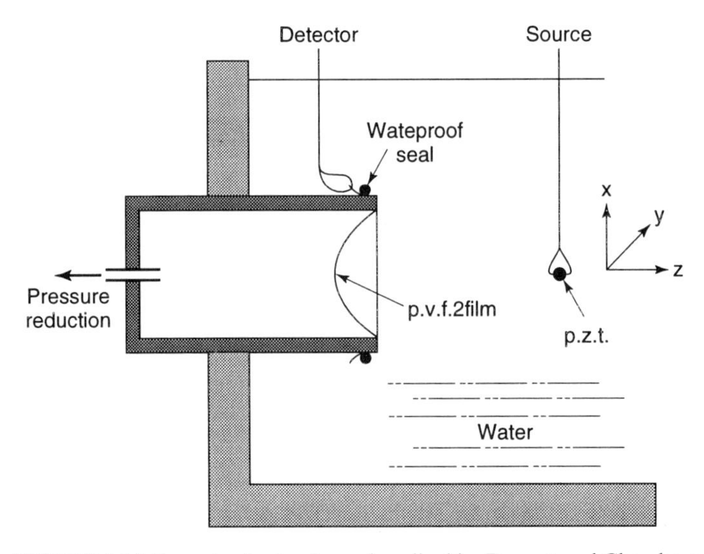

FIGURE 5.36. Scanning hydrophone described by Bennett and Chambers [39].

Figure 5.37 shows an inchworm motor® made from three piezoelectric tubes and manufactured by Burleigh Instruments Inc. [16]. In operation, a voltage is applied to the first element, which expands to clamp the internal rod. This transducer segment utilizes primarily the d_{31} piezoelectric coefficient of the material. Next, a voltage is applied to the second segment, which extends without touching the rod (using primarily the d_{ii} expansion mode). Next the third segment, which is identical to the first, is activated to clamp the rod. The first and second segments are then deactivated, so that the entire structure moves to the right. The process can be iteratively repeated to get steps as small as the nanometer level. These motors are commonly used in microscopy.

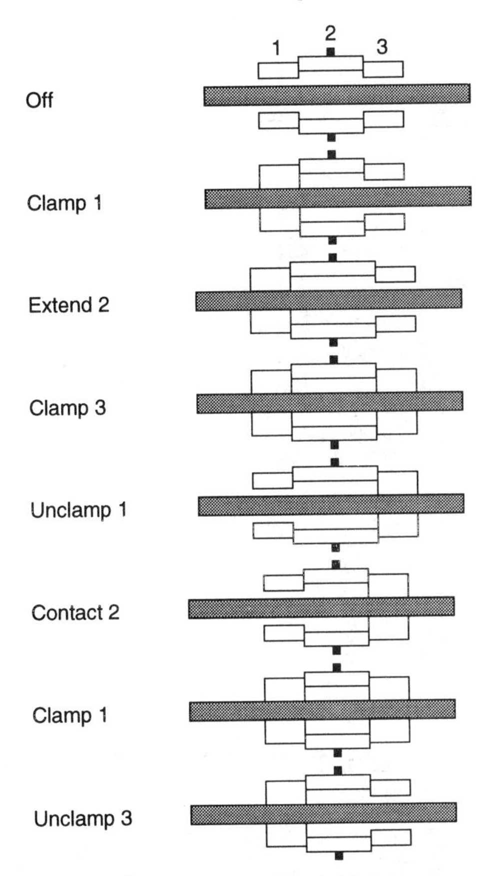

FIGURE 5.37. Inchworm motor of Burleigh Instruments, Inc. [16].

Figure 5.38 is an unusual transducer primarily because of the way it is operated. This transducer, described in Sinclair [40], detects rotational speed by allowing an eccentric cam to strike a piezoelectric element once per rotation. In response to the strike, the piezoelectric generates a voltage signal. The rotational speed is determined by simply counting the number of spikes in a given time period. Thus, this device essentially is operated in a digital mode, with a threshold detector determining whether a spike has occurred or not.

Another example of an unconventional piezoelectric transducer is unusual because of its size. Shown in Fig. 5.39 is one of the first microphones made using the techniques and scales of microelectronics. This microphone (later called "mike on a chip" by Motorola [41]) was built by Royer et al. [42]. It uses a 3–5 μm thick zinc oxide (ZnO) membrane as the piezoelectric material and sandwiches it between layers of silicon dioxide and thin aluminum electrodes. This structure sits on a 30 μm thick silicon diaphragm of 3 mm in diameter. The sensor was found to be fairly insensitive (50–250 μV/Pa) but to have a flat frequency response ($\pm$5 dB) from 10 Hz to 10,000 Hz.

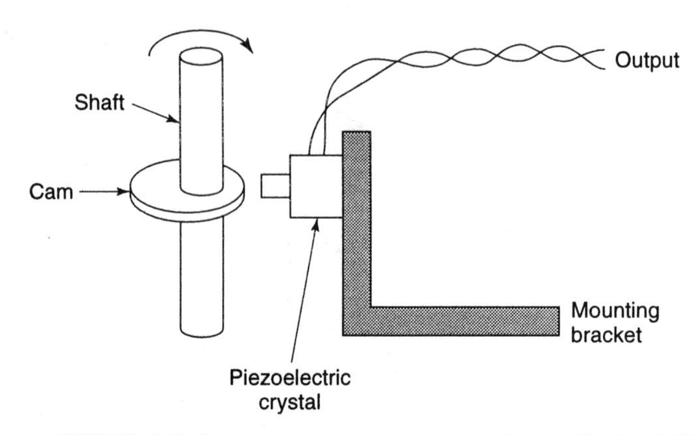

FIGURE 5.38. Rotational speed detector described by Sinclair [40].

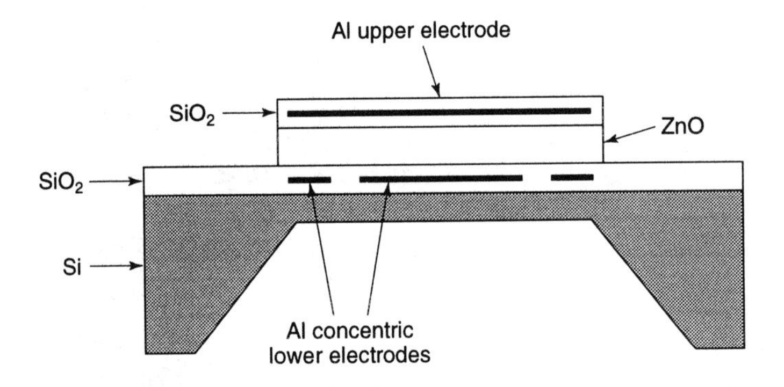

FIGURE 5.39. Piezoelectric microphone designed by Royer et al. [42].

Another microphone relying on piezoelectricity for transduction is shown in Fig. 5.40 and described in Reid et al. [43]. This microphone also uses zinc oxide as the piezoelectric material. The active membrane is segmented to permit temperature compensation and to allow for better performance control.

One final solid-state transducer incorporating piezoelectric transduction is shown in Fig. 5.41, taken from Shoji and Esashi [44]. This transducer is a three-way microvalve which is actuated using a stack of piezoelectric elements. Unlike the two solid-state microphones previously described, this transducer is not made monolithically. Instead it is assembled from two separate pieces.

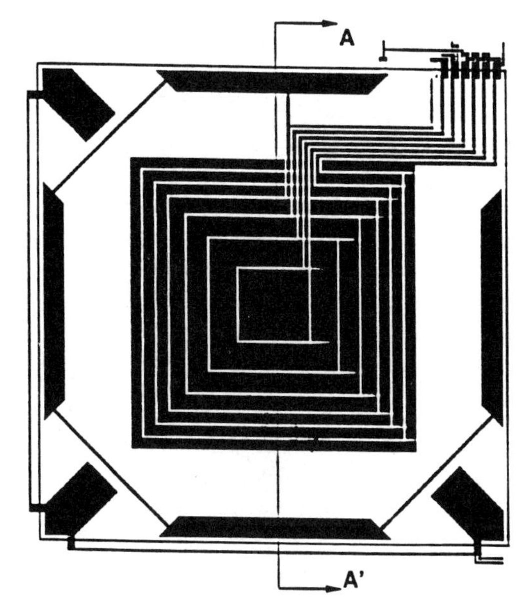

FIGURE 5.40. Piezoelectric microphone described by Reid et al. [43].

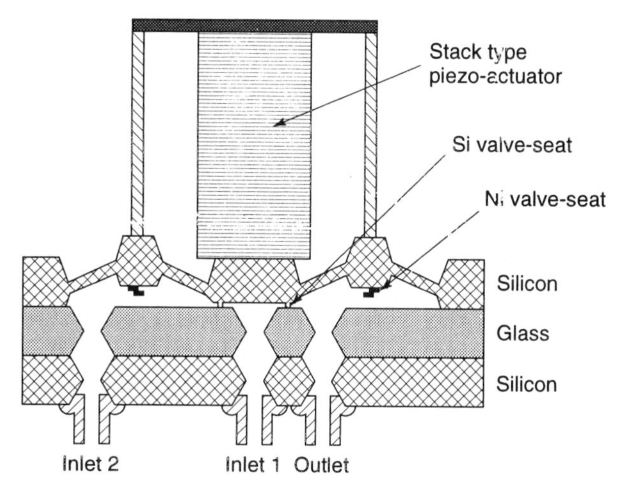

FIGURE 5.41. Piezoelectrically actuated microvalve described by Shoji and Esashi [44].

5.6 Pyroelectricity

Up to this point in the chapter we have assumed that the transducers under study operate at a fixed temperature. This has permitted us to ignore the coupling between thermal variations and the electric field. However, if we return to Eq. (22), we may now consider the thermoelectrical coupling described by the pyroelectric constants $p_m^{T,H}$. Pyroelectricity was first observed in tourmaline crystals in

TABLE 5.6. Pyroelectric coefficients of a variety of materials (from Marcus [45]).

Material	Pyro. Coef. (μC/m^{2}K)
PZT-5	300
BaTiO3	200
PVDF	−40
TGS	300
LiTaO3	190
PLZT	760

the eighteenth century. Later, as piezoelectric crystals were studied, they were found to be pyroelectric as well. Today, more than 1000 pyroelectric materials are known. Among the best (most pyroelectric activity) are triglycine sulfate (TGS) and lithium tantalate ($LiTaO_3$). Unfortunately (or fortunately, depending on your specific application) most piezoelectric materials also exhibit strong pyroelectric behavior. Table 5.6, with data taken from Marcus [45], shows the pyroelectric coefficients for a variety of materials.

Models of the fundamental pyroelectric behavior of materials have been developed (see, e.g., Marcus [45] and Fraden [46]). Generally, the view is that the materials consist of minute crystallites, each of which behaves as a dipole if the material is below the Curie temperature. The dipoles are randomly oriented or as randomly oriented as is possible within the context of retaining crystalline or semicrystalline structure. A change in temperature is thought to induce one of two sets of effects. Primary pyroelectricity is the result of a temperature change causing shortening or elongation of the dipoles, or a reduction in the randomness of their orientation. Secondary pyroelectricity is identified as thermoelectric coupling which results when a temperature change causes a dimensional change, and this dimensional change induces electrical activity via piezoelectricity.

While it is difficult to find materials which permit the separation of pyroelectric and piezoelectric behavior, let us assume now that such materials exist and focus only on pyroelectricity. Then (22) reduces to the following:

$$\begin{aligned} D_m &= \epsilon_{mk}^{T,H,\theta} E_m + p_m^{T,H}\, d\theta, \\ d\sigma &= p_m^{T,H} E_m + \frac{\rho c^{E,H,T}}{\theta} d\theta. \end{aligned} \tag{39}$$

In this set of equations, we have neglected the thermal expansion of the material, as it serves to make the following discussion more complicated. However, the thermal expansion could be included without the qualitative results changing. Equations (39) show that in pyroelectricity there is a thermoelectric coupling. More specifically, *changes* in temperature or entropy induce an electric field or vice versa.

Just as in the case of piezoelectricity, we may manipulate the equations above to get them in the form in which we use the energy variables (in this case displacement variables: charge and entropy) as independent variables, and the power conjugate variables (in this case effort variables: voltage and temperature) as dependent variables. We do this by solving the equations above for $d\theta$ and E_m in terms of D_m and $d\sigma$, and then rewriting the electric field parameters in terms of charge and voltage. The result is

$$e_m = \frac{\rho c l}{(\epsilon_{mk}\rho c - p_m^2\theta)A} q_m - \frac{\theta p_m l}{\epsilon_{mk}\rho c - p_m^2\theta} d\sigma,$$

$$d\theta = -\frac{\theta p_m}{(\epsilon_{mk}\rho c - p_m^2\theta)A} q_m + \frac{\epsilon_{mk}\theta}{\epsilon_{mk}\rho c - p_m^2\theta} d\sigma, \qquad (40)$$

where the superscripts on the constants have been eliminated for clarity. Note that these equations are not linear because the right-hand side retains a reliance on the temperature. It is customary to approximate θ by θ_0, the nominal operating temperature. Then (40) are linear coupled constitutive equations relating the voltage and temperature change to the charge and entropy change. As such, they define an essential 2-port capacitance element, quite analogous to that defined by piezoelectric relations.

More often than not, pyroelectricity is viewed as an undersired contaminant inherent to piezoelectric sensors and actuators. However, from (40), there are three relationships that could be exploited in the development of pyroelectric sensors and actuators: relating voltage to change in entropy, relating change in temperature to charge, and relating voltage to change in temperature. The first two of these three options pose the interesting problem of how to deal with a quantity which is virtually impossible to measure—namely, entropy. In order to avoid this problem, the most logical use of pyroelectricity in thermoelectric transducers, eliminates $d\sigma$ from the constitutive equations, and relates voltage and temperature.

Figure 5.42 shows an example application of pyroelectric transduction. In this device, reported in Bergman et al. [47], a lamp shines through an object slide and lens onto PVDF. The PVDF is oriented so that the heated region develops a negative charge, which makes toner concentrate in the unheated region. Thus an image of the slide is made.

Figure 5.43 shows another application of a pyroelectric sensor. In this device, made by Gallantree and Quilliam [48], a 50 element array of electrodes is patterned onto PVDF and used to measure the power profile of a laser. Assuming a uniform thermal absorption layer on the PVDF, the incident power is directly related to the absorbed heat and consequent temperature change.

5.7 Summary

Piezoelectric transducers are among the most common electromechanical transducers available. They take advantage of a material property which couples the mechanical and electrical fields. Thus, a mechanical strain is accompanied by the establishment of an electric field, and the application of an electric field produces a

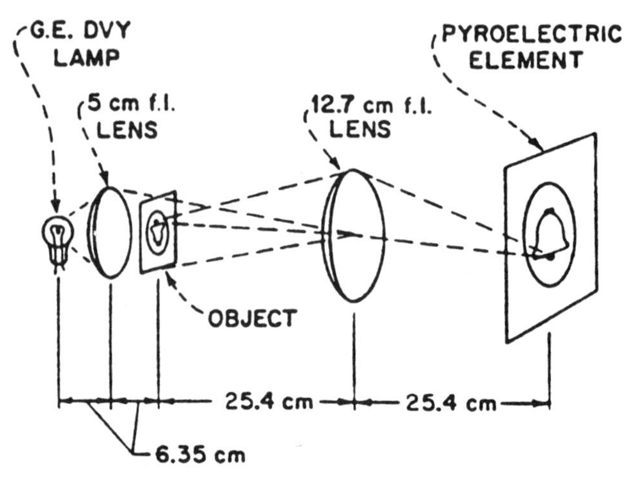

FIGURE 5.42. Prototype pyroelectric copier. Reprinted with permission from J. G. Bergman, G. R. Crane, A. A. Ballman, and H. M. O'Bryan, Jr., *Pyroelectric Copying Process*, Appl. Phys. Lett. **21**, 497 (1972). Copyright 1972 American Institute of Physics.

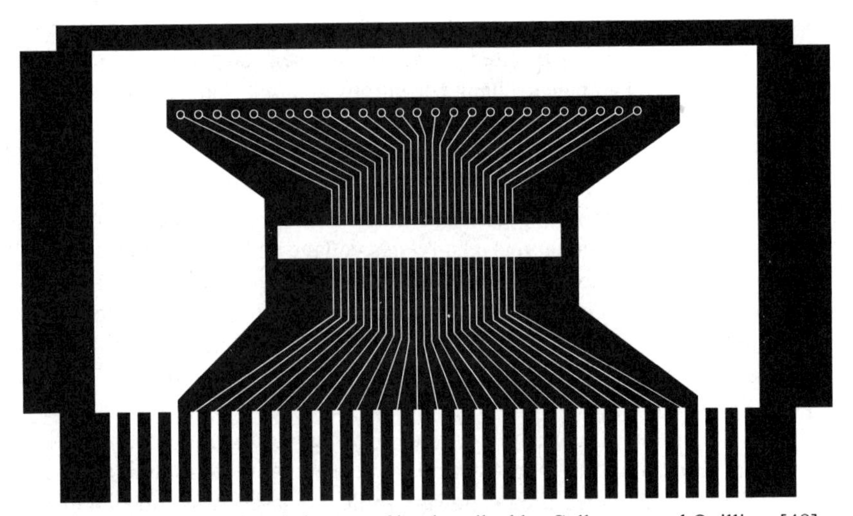

FIGURE 5.43. Pyroelectric laser profiler described by Gallantree and Quilliam [48].

strain of the material. Pyroelectricity is a coupling between thermal and electrical behavior and often accompanies piezoelectricity.

Unlike the transducers discussed previously, the electromechanical coupling in piezoelectric transducers is fundamentally linear. This means that there is no need to apply a mechanical or electrical bias to the material in order to obtain linear operation.

Piezoelectric transducers take many forms. They are commonly used as accelerometers, phonograph cartridges, ultrasonic cleaner vibrating elements, and spark ignitions. They are also used in invasive catheters and force heads.

In the next chapter, we discuss linear magnetic transduction principles. While there is a transduction mechanism directly analogous to piezoelectricity, namely piezomagnetism, there are few identified piezomagnetic materials. Thus, the next chapter will focus on magnetic transduction mechanisms which are linear due to geometrical considerations, rather than material properties.

5.8 References

1. P.-J. Curie and J. Curie, *Crystal Physics—Development by Pressure of Polar Electricity in Hemihedral Crystals with Inclined Faces* (in French), Acad. Sci. (Paris) C. R. Hebd. Seances **91**, 294 (1880). A translation appears in R. B. Lindsay (ed.), *Acoustics: Historical and Philosophical Development* (Dowden Hutchinson and Ross, Stroudsburg, PA, 1973).
2. P. Langevin, *Improvements Relating to the Emission and Reception of Submarine Waves*, British Patent No. 145,691, accepted July 28, 1921. Reprinted in I. D. Groves (ed.), *Acoustic Transducers* (Hutchinson Ross, Stroudburg, PA, 1981).
3. R. B. Gray, US Patent No. 2,486,560, issued Nov. 1, 1949.
4. B. Jaffe, US Patent No. 2,708,244, issued May 10, 1955.
5. H. Kawai, *The Piezoelectricity of Poly(vinylidene Fluoride)*, Jpn. J. Appl. Phys. **8**, 975 (1969).
6. R. E. Newnham, D. P. Skinner, and L. E. Cross, *Connectivity and Piezoelectric-Pyroelectric Composites*, Mat. Res. Bull. **13**, 525 (1978).
7. A. Halliyal, A. Safari, A. S. Bhalla, R. E. Newnham, and L. E. Cross, *Grain-Oriented Glass-Ceramics for Piezoelectric Devices*, J. Am. Cer. Soc. **67**, 331 (1984).
8. A. Safari, G. Sa-Gong, J. Giniewicz, and R. E. Newnham, *Composite Piezoelectric Sensors*, in C. A. Rosen, B. V. Hiremath, and R. E. Newnham, (eds.), *Piezoelectricity* (American Institute of Physics, New York, 1992).
9. Q. Y. Jiang and L. E. Cross, *Effects of Porosity on Electric Fatigue Behavior in PLZT and PZT Ferroelectric Ceramics*, J. Mat. Sci. **28**, 4536 (1993).
10. D. A. Berlincourt, D. R. Curran, and H. Jaffe, in W. P. Mason (ed.) *Physical Acoustics, Principles and Methods, Vol. I, Part A* (Academic Press, New York, 1964).
11. O. B. Wilson, *An Introduction to the Theory and Design of Sonar Transducers* (Deptartment of the Navy, Washington, DC, 1985).
12. M. Rossi, *Acoustics and Electroacoustics*, P. R. W. Roe (trans.), (Artech House, Norwood, MA, 1988).
13. W. A. Smith, *The Key Design Principles for Piezoelectric Ceramic/Polymer Composites,* in Proc. Conf. on Recent Advances in Adaptive and Sensory Materials and Their Applications, Blacksburg, VA, April 1992.
14. Materials Systems Inc., 521 Great Road, Littleton, MA 01460, USA.

15. C. Z. Rosen, B. V. Hiremath, and R. E. Newnham (eds.), *Piezoelectricity* (American Institute of Physics, New York, 1992).
16. Burleigh Instruments Inc., 9 Burleigh Park, Fisher, New York 14453-0755, USA.
17. G. Binnig and H. Rohrer, *The Scanning Tunneling Microscope*, Scientific American, **50** (August 1985).
18. W. G. Cady, *Piezoelectricy* (McGraw-Hill, 1946). Also available in Dover reprint (Dover, New York, 1964).
19. W. P. Mason, *An Electromechanical Representation of a Piezoelectric Crystal Used as a Transducer*, IRE Proc., **23**, 1252 (1935). Reprinted in I. D. Groves (ed.), *Acoustic Transducers* (Hutchinson Ross, Stroudburg, PA, 1981).
20. W. P. Mason, *Electromechanical Transducers and Wave Filters*, 2nd ed. (Van Nostrand, New York, 1948).
21. W. P. Mason, *Piezoelectric Crystals and Their Application to Ultrasonics* (Van Nostrand, New York, 1950).
22. I. J. Busch-Vishniac and H. M. Paynter, *Bond Graph Models of Acoustical Transducers*, J. Franklin Inst., **328**, 663 (1991).
23. W. Moon and I. J. Busch-Vishniac, *A Finite-Element Equivalent Bond Graph Modeling Approach with Application to the Piezoelectric Thickness Vibrator*, J. Acoust. Soc. Am., **93**, 3496 (1993).
24. W. Moon and I. J. Busch-Vishniac, *Modeling of Piezoelectric Ceramic Vibrators Including Thermal Effects.* Part I. *Thermodynamic Property Considerations*, J. Acoust. Soc. Am., **98**, 403 (1995).
25. W. Moon and I. J. Busch-Vishniac, *Modeling of Piezoelectric Ceramic Vibrators Including Thermal Effects.* Part II. *Derivation of Partial Differential Equations*, J. Acoust. Soc. Am., **98**, 413 (1995).
26. W. Moon and I. J. Busch-Vishniac, *Modeling of Piezoelectric Ceramic Vibrators Including Thermal Effects.* Part III. *Bond Graph Model for One-Dimensional Heat Conduction*, J. Acoust. Soc. Am., **101**, 1398 (1997).
27. W. Moon and I. J. Busch-Vishniac, *Modeling of Piezoelectric Ceramic Vibrators Including Thermal Effects.* Part IV. *Development and Experimental Evaluation of a Bond Graph Model of the Thickness Vibrator*, J. Acoust. Soc. Am., **101**, 1408 (1997).
28. Dytran Instruments Inc., 21592 Marilla Street, Chatsworth, CA 91311, USA.
29. Kistler Instrument Corporation, 75 John Glenn Drive, Amherst, NY 14228-2119, USA.
30. PCB Piezotronics, 3425 Walden Avenue, Depew, NY 14043-2495, USA.
31. R. J. Bobber, *Underwater Electroacoustic Measurements* (Peninsula, Los Altos, CA, 1988).
32. P. Dario, D. DeRossi, R. Bedini, R. Francesconi, and M. G. Trivella, *PVF_2 Catheter–Tip Transducers for Pressure, Sound and Flow Measurement*, in P. M. Galletti, D. E. DeRossi, and A. S. DeReggi (eds.),

Medical Applications of Piezoelectric Polymers (Gordon and Breach, NY, 1988).

33. K. Uchino *Electrostrictive Actuators: Materials and Applications*, in C. Z. Rosen, B. V. Hiremath, and R. Newnham (eds.), *Piezoelectricity* (AIP Press, New York, 1992).

34. M. R. Keeling, *Ink Jet Printing*, in C. Z. Rosen, B. V. Hiremath, and R. Newnham (eds.), *Piezoelectricity* (AIP Press, New York, 1992).

35. Vernitron, 1601 Precision Park Lane, San Diego, CA 92173, USA.

36. W. J. Toulis, *Flexural-Extensional Electromechanical Transducer*, US Patent 3,277,433, Oct. 4, 1966.

37. J. Leifer and I. J. Busch-Vishniac, *An Ultrasonic Source Incorporating a Solid Webster Horn for Three-Dimensional Position Monitoring in Robotics*, ASME Winter Annual Meeting, Dallas, Nov. 1990.

38. J. A. Allocca and A. Stuart, *Transducers: Theory and Applications* (Reston, VA, 1984).

39. S. D. Bennett and J. Chambers, *Novel Variable-Focus Ultrasonic Transducer*, Electron Lett., **13**, 110 (1977).

40. I. R. Sinclair, *Sensors and Transducers*, 2nd ed. (Newnes, Oxford, UK, 1992).

41. Motorola Inc., 1303 E. Algonquin Rd., Schaumburg, IL 60196, USA.

42. M. Royer, J. O. Holmen, M. A. Wurm, O. S. Aadland, and M. Glenn, *ZnO on Si Integrated Acoustic Sensor*, Sensors and Actuators, **4**, 357 (1983).

43. R. P. Reid, E. S. Kim, D. M. Hong, and R. S. Muller, *Piezoelectric Microphone with On-Chip CMOS Circuits*, J. MEMS, **2**, 111 (1993).

44. S. Shoji and M. Esashi, *Microflow Devices and Systems*, J. MEMS, **4**, 157 (1994).

45. M. A. Marcus, *Ferroelectric Polymers and Their Applications*, Fifth International Meeting on Ferroelectricity, State College, PA, Aug. 17–21, 1981.

46. J. Fraden, *AIP Handbook of Modern Sensors: Physics, Design and Applications* (AIP Press, New York, 1993).

47. J. G. Bergman, G. R. Crane, A. A. Ballman, and H. M. O'Bryan, Jr., *Pyroelectric Copying Process*, Appl. Phys. Lett. **21**, 497 (1972).

48. H. R. Gallantree and R. M. Quilliam, *Polarized Poly(vinylidene Fluoride)—Its Application to Pyroelectric and Piezoelectric Devices*, Marconi Review, **189** (fourth quarter, 1976).

6 Linear Inductive Transduction Mechanisms

Given that magnetism was discovered earlier than electricity, it should come as no surprise that the first magnetic (inductive) transducers predate the earliest electric (capacitive) transducers. Linear magnetomechanical coupling is among the most common transduction mechanisms, particularly for actuators, because magnetic devices have energy advantages on large scales.

The previous chapter of this book described transduction mechanisms in which there is a linear coupling between the mechanical (or thermal) and electrical fields in certain materials. In this chapter we describe linear transduction mechanisms which link magnetic fields and mechanical (or thermal) fields. While there is a mechanism which is directly analogous to piezoelectricity, and another directly analogous to pyroelectricity, there are additional linear inductive electromechanical coupling methods that do not have a strict capacitive analog. It is these additional coupling methods which are most frequently exploited in the design of inductive transducers. As will be shown, the most common of these mechanisms does not rely on changes in the energy stored in the magnetic field, but instead uses the interaction between charged particles to generate mechanical forces and torques. Electromechanical transducers which take advantage of this coupling include linear velocity transducers, loudspeakers, and fluid flow sensors.

In this chapter, we begin by presenting those linear inductive mechanisms which are directly analogous to the linear capacitive mechanisms described in Chapter 5. Then we discuss transduction based on interactions between charged particles and Hall effect transducers.

6.1 Piezomagnetism

The material property which links magnetic fields and mechanical fields and is directly analogous to piezoelectricity is referred to as piezomagnetism. In fact, (5.21) and (5.22) define the piezomagnetic constants and include piezomagnetic effects as well as piezoelectric effects. In transducers which seek to take advantage primarily of piezomagnetic coupling, we would assume that the piezoelectric and thermal effects are small. The resulting equations linking the mechanical and magnetic domains are

$$
\begin{aligned}
S_i &= s_{ij}^H T_j + d_{mi} H_m, \\
B_m &= d_{mi} T_i + \mu_{mk}^T H_k,
\end{aligned}
\tag{1}
$$

where the d_{ij} are now piezomagnetic constants measured at constant electric field and temperature.

Equations (1) are written with the stress field and the magnetic field as the independent variables, and the strain and magnetic flux density as dependent variables. Clearly, given that the equations are linear, one could easily rewrite them with independent variables of strain and magnetic field, strain and magnetic flux density, or stress and magnetic flux density. In each case the result would be a pair of linear equations describing the magnetomechanical coupling. Of the four possible representations of piezomagnetism, the set of equations in which strain and magnetic flux density are the independent variables is the most useful. This is because it permits us to write constitutive equations of the standard form, with power conjugate variables expressed as functions of energy variables. In this form we write

$$
\begin{aligned}
\mathbf{T} &= [c^B]\mathbf{S} - [h]^T\mathbf{B} \\
\mathbf{H} &= -[h]\mathbf{S} + [\gamma^S]\mathbf{B},
\end{aligned}
\tag{2}
$$

where $[c^B]$ is the stiffness matrix determined with fixed magnetic flux density, $[h]$ is a matrix of piezomagnetic constants, and $[\gamma^S]$ is the matrix whose inverse is the permeability matrix. The components of $[h]$ are related to the piezomagnetic d constants by $d_{mi} = \mu_{nm}^S s_{ji}^H h_{nj}$.

From Eqs. (2) we produce the standard constitutive equations by using the following relations:

$$
\begin{aligned}
\mathbf{x} &= [l]\mathbf{S}, \\
\mathbf{F} &= [A]\mathbf{T}, \\
\phi &= [A^*]\mathbf{B}, \\
\mathbf{M} &= [l^*]\mathbf{H}.
\end{aligned}
\tag{3}
$$

Here $[l]$ is the 6×6 diagonal matrix whose components are the equilibrium lengths that scale the strain, $[A]$ is the 6×6 diagonal matrix whose components are the areas that scale the stress, $[A^*]$ is the 3×3 diagonal matrix whose elements describe the cross-sectional area of the magnetic flux paths, and $[l^*]$ is the 3×3 diagonal matrix whose elements are the magnetic path lengths. The result is as follows:

$$\mathbf{F} = [A][c^B][l]^{-1}\mathbf{x} - [A][h]^T[A^*]^{-1}\phi,$$
$$\mathbf{M} = -[l^*][h][l]^{-1}\mathbf{x} + [l^*][\gamma^S][A^*]^{-1}\phi. \tag{4}$$

Equations (4) relate mechanical effort variables and magnetic effort variables, each to mechanical displacement and magnetic displacement variables. Thus this set of equations defines a 9-port capacitive element in which six of the ports are mechanical, and three are magnetic. In general, we are concerned only with the magnetic field in one of the principal directions. Also, the piezomagnetic matrix tends to be sparse. Thus, it is possible to simplify the model to a multiport capacitance with far fewer than nine ports (and a minimum of two ports).

From (4) it is clear that there are three distinct ways to exploit piezomagnetism in electromechanical transduction: by relating force and magnetic flux, by relating magnetomotance (or current) and displacement, and by relating magnetomotance and force. In each case, since the constitutive equations are fundamentally linear, the transduction relations would be linear.

Up to this point, the discussion of piezomagnetic behavior has parallelled piezoelectricity precisely. However, there is an important distinction between piezoelectricity and piezomagnetism, namely, that there are naturally occurring piezoelectric materials (e.g. quartz), whereas there are no known naturally occurring piezomagnetic materials. What this means is that piezomagnetism is simply used as a means of modeling what is really magnetostrictive behavior. The advantage offered by the piezomagnetic model is that it is linear rather than quadratic, and that it permits a quick comparison of so-called *piezomagnetic* materials with piezoelectric materials. Using the piezomagnetic model presented here, one can define the piezomagnetic d constants for various materials. The results, which are dependent on the compressive mechanical bias, are compiled from [1]–[3] and presented in Table 6.1 with nearly maximal values shown for d in each case. Note that these results can be compared directly to the d values cited for piezoelectric materials in Chapter 5. On this basis, nickel and alfenol are roughly equivalent to quartz, Terfenol is roughly equivalent to PVDF, and Metglas 2605SC to PZT.

There is a great deal of work currently being done by the magnetic materials research community on new magnetostrictive materials. In particular, great emphasis is being placed on thin-film magnetostrictive materials that might be deposited

Material	d_{33} (nm/A)	k
Nickel	3	0.15–0.31
Alfenol	7	0.25–0.32
Terfenol-D	15	0.7–0.75
Metglas 2605CO	70–88	0.7
Metglas 2605SC	300–400	0.6–0.9

TABLE 6.1. Piezomagnetic properties of some materials [1]–[3]. Alfenol is 13% aluminum and 87% iron by weight. Metglas is made by Allied Corporation.

onto silicon. Using the piezomagnetic model of magnetostrictive materials which are compressively prestressed and operated in a linear regime, it is reasonable to anticipate significant material breakthroughs in the near future.

6.2 Pyromagnetism

The material property that links magnetic and thermal variables and is directly analogous to pyroelectricity is called pyromagnetism. From Eqs. (5.21) and (5.22), we describe pyromagnetic interactions as

$$\begin{aligned} B_m &= \mu^\theta_{mk} H_k + i_m \, d\theta, \\ d\sigma &= i_m H_m + \frac{\rho c^H}{\theta} d\theta, \end{aligned} \tag{5}$$

where we have ignored thermoelectrical, magnetoelectrical, and thermomechanical interactions. This set of equations is precisely analogous to that given for pyroelectricity in (5.39).

With the analogy between pyroelectricity and pyromagnetism, it is possible to immediately use (5.40) to produce constitutive equations in the standard form for piezomagnetism. This requires the following transformations $e_m \to M_m$, $\epsilon_{mk} \to \mu_{mk}$, $p_m \to i_m$, and $q_m \to \phi_m$, and produces

$$\begin{aligned} M_m &= \frac{\rho c l}{(\mu_{mk} \rho c - i_m^2 \theta) A} \phi_m - \frac{\theta i_m l}{\mu_{mk} \rho c - i_m^2 \theta} d\sigma, \\ d\theta &= -\frac{\theta i_m}{(\mu_{mk} \rho c - i_m^2 \theta) A} \phi_m + \frac{\mu_{mk} \theta}{\mu_{mk} \rho c - i_m^2 \theta} d\sigma. \end{aligned} \tag{6}$$

These constitutive equations are nonlinear, because of the appearance of θ on the right-hand side. One may linearize them by approximating θ by θ_0, the nominal operating temperature.

From (6) it is clear that there are three ways that we might exploit pyromagnetic coupling: by relating magnetomotance (or current) and entropy change by relating temperature change and magnetic flux; or by relating magnetomotance and temperature change. The first two of these present the nearly insurmountable problem of coping with a need to know (or restrict) the entropy change. Hence, the only logical way to use pyromagnetism is to relate magnetomotance and temperature change by eliminating entropy change from the constitutive relations.

Pyromagnetic transducers are very rare, as materials with high pyromagnetic coefficients are difficult to find. One example of a transducer which takes advantage of pyromagnetism is described by Seki et al. [4]. In this photomagnetic device, a radiant absorber is deposited over a pyromagnetic semiconductor made from Mn–Cu ferrite powder, RuO, and binders. Light striking the transducer is converted to heat by the radiant absorber and then to an electrical signal through the pyromagnetic semiconductor.

6.3 Charged Particle Interactions

The most common linear magnetic transducers are based on the interaction between charged particles. This mechanism has no capacitive analog. We begin the description of these transducers by consideration of the motion of a single charged particle in a magnetic field. If the velocity of the charge is described by **V** and the magnetic flux density field imposed in the environment is **B**, then the force on the particle is given by

$$\mathbf{F} = q\mathbf{V} \times \mathbf{B}, \tag{7}$$

where q is the particle charge. This force is referred to as the Lorentz force and is named after the scientist who first described it. Note that (7) shows that the force is orthogonal to the velocity. Hence the mechanical power, which is given by $\mathbf{F} \cdot \mathbf{V}$ is zero. In other words, the charged particle is subjected to a force, but no power is converted from the magnetic domain to the mechanical. It is this property which results in the transduction being linear as opposed to quadratic, as in the transduction based on changes in the magnetic field energy. Of course, in order to use the generated force to produce a mechanical motion of a structure (rather than motion of a charged particle), power is required on the mechanical side.

Figure 6.1 shows an example of a charged particle moving in a magnetic flux density field. In this figure the particle has velocity components only in the directions normal to the flux density field. Hence we have

$$\begin{aligned} F_1 &= qBV_2, \\ -F_2 &= qBV_1. \end{aligned} \tag{8}$$

Note that these equations can be viewed as the defining relations for a gyrator, with the gyrator modulus defined as the product qB. If the magnetic flux density varies in time, then the relations define a modulated gyrator. A modulated gyrator is shown in bond graph notation using a dashed line to the gyrator modulus. The dashed line indicates signal flow, and its origin is the junction or bond which describes the modulating agent. This approach to modeling will be discussed further as we proceed through this chapter.

The Biot–Savart law provides the means for generalization of the interaction of a charged particle in a magnetic field to a current-carrying wire in a magnetic field. We imagine that there is a wire in which charge is moving with velocity **V**. Then the force caused by each infinitesimal of charge, dq, is

$$\mathbf{dF} = dq\mathbf{V} \times \mathbf{B}. \tag{9}$$

Since the charge is confined to the wire, we may rewrite **V** as $\mathbf{dl}/dt$, where $\mathbf{dl}$ is an infinitesimal segment of the wire. Then the contribution from a segment is given by

$$\begin{aligned} \mathbf{dF} &= dq\frac{\mathbf{dl}}{dt} \times \mathbf{B} \\ &= i\mathbf{dl} \times \mathbf{B}. \end{aligned} \tag{10}$$

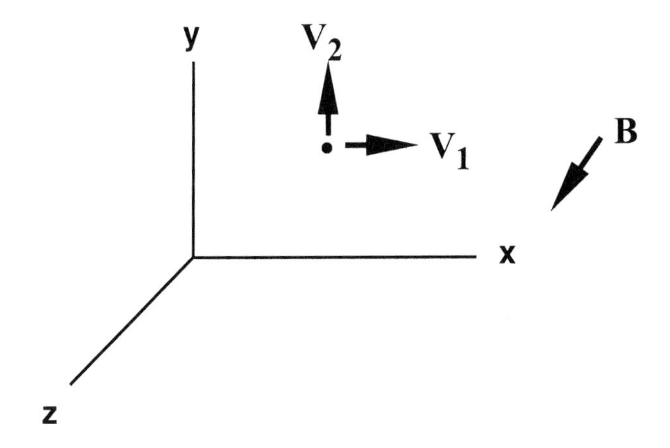

FIGURE 6.1. Lorentz force due to motion of a charged particle in a magnetic field.

The total force is obtained by integrating over the length of the wire. If, as is often the case, the wire is maintained orthogonal to the magnetic field direction, then the result is simply

$$F = Bli. \tag{11}$$

Again, this can be viewed as a defining gyrator relation between the mechanical and electrical domains. Since power is neither stored nor dissipated, we must have the complementary relation

$$e = -BlV, \tag{12}$$

where the minus sign exists by convention and clearly depends on the definition of potential ground.

Equations (11) and (12) are constitutive equations for the energy transformation part of an electromechanical transducer based on the motion of charged particles in a magnetic field. They suggest two types of transduction relations: relating force and current, and relating voltage and velocity. Note that the constitutive equations developed up to this point have had *three* means by which they could be exploited in transduction. This transduction mechanism is an exception because energy is conserved going from the electrical to mechanical port or vice versa. This means that there is no way to use the stored energy to establish a relation between the power conjugate variables. Instead, that relation is established by the rest of the system of the transducer.

In keeping with the preceeding chapters of this book, we will model transducers which exploit the interaction between moving charges using a bond graph approach. However, we could also develop circuit models of these transducers quite easily. In general, for a transducer, in which the coupling is based on the motion of charged particles in a magnetic field, we may write

$$F = Bli + Z_m^i V, \tag{13}$$

and

$$e = Z_e^V i - BlV. \tag{14}$$

Here Z_m^i is the mechanical impedance, measured with no current in the transducer and Z_e^V is the electrical impedance measured with no velocity of the current-carrying wires. Together they represent all of the system besides the transduction portion (i.e., that which would be modeled in the rest of the bond graph or circuit model). Z_m^i is often referred to as the open circuit mechanical impedance and $Z_m^i V$ is a measure of the force required to move the wire in the absence of a current. Z_e^V is referred to as the clamped electrical impedance and $Z_e^V i$ is a measure of the voltage drop induced by a current in the absence of motion. At a minimum it accounts for the fact that materials are imperfect conductors.

Note that in (13) and (14), we have represented what might be a quite complicated system using very simple linear equations. These equations should be thought of as nothing more than a shorthand notation. The actual physical system might include elements with fundamentally nonlinear behavior, rendering an impedance representation inappropriate.

6.3.1 Examples of Transducers Based on the Motion of Charged Particles in a Magnetic Field

Because of the simplicity and linearity of the transduction relations ((11) and (12)), associated with the motion of charged particles in a magnetic field, there are large numbers of transducer types which exploit the relations.

An example of a common transducer which uses the motion of charged particles in a magnetic field for transduction is a loudspeaker, such as shown in Fig. 6.2. The magnetic flux density is established by a permanent magnet. A solenoid, usually referred to as a voice coil, is positioned so that the current flow is perpendicular to the magnetic flux direction. Current running through the coil produces a force on it in the direction orthogonal to both the current flow and magnetic flux. This produces a motion of the speaker diaphragm attached to the voice coil and the motion causes a sound wave to propagate.

Figure 6.3 is an elementary bond graph model of the loudspeaker shown in Fig. 6.2. Typically a voltage signal is supplied to the speaker. The voice coil is modeled as a lumped electrical resistance and inductance. Taken together these items are the components that would be included in Z_e^V. The gyrator represents the electromechanical coupling which results in the speaker diaphragm motion. The diaphragm and coil mass and the suspension stiffness are the primary elements included in the mechanical impedance, Z_m^i.

The strict reciprocal of the loudspeaker is the moving coil, or dynamic, microphone. This device is quite similar to the loudspeaker, but instead of a voltage drive producing a velocity, a velocity is used to generate a voltage. Additionally, there are geometric changes to accommodate the desire for high sensitivity as a microphone.

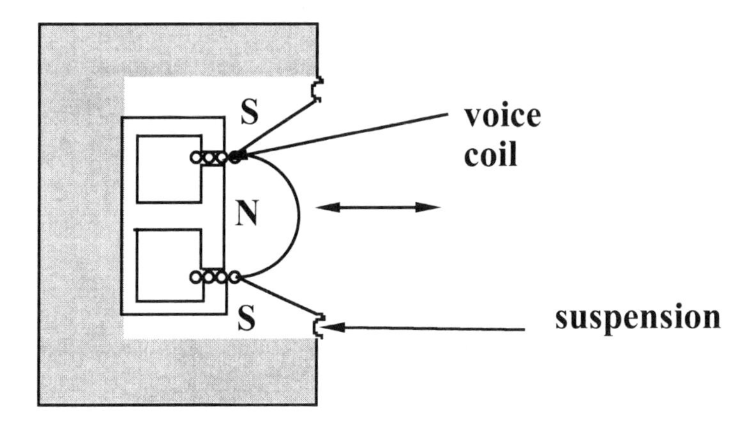

FIGURE 6.2. Simplified view of a loudspeaker.

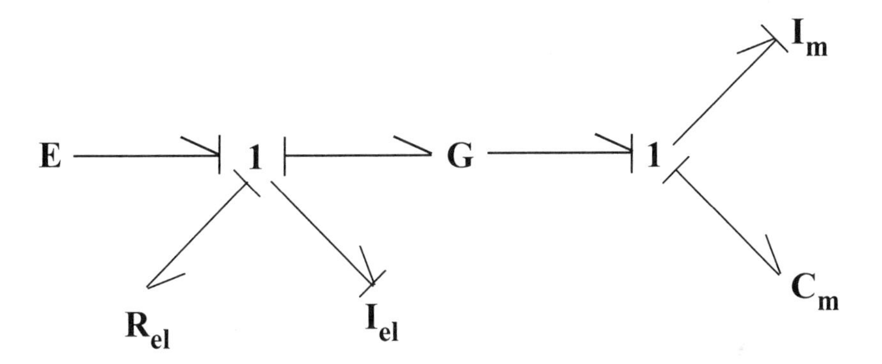

FIGURE 6.3. Bond graph model of a loudspeaker.

There are a number of actuators which exploit the same transduction mechanism and geometry as that of the loudspeaker. Because of this, these transducers typically are referred to as *voice coil actuators*. For example, Fig. 6.4 shows a voice coil actuator which is sold by BEI Sensors and Systems Company [5]. This transducer uses a curved coil to generate a radial force proportional to current. Similar actuators with other geometries are available for the generation of linear force or of torque.

An example of an application of a voice coil actuator is a magnetic disk drive such as shown in Fig. 6.5, taken from Mee and Daniel [6]. In standard magnetic and magneto-optical disk drives, the disk heads are moved by using a system nearly identical to that of a loudspeaker. This produces repeatable, linear motion with high resolution and rapid positioning speed.

In the voice coil actuators described so far, the transduction mechanism is used to produce a force rather than a velocity. An example of a transducer in which the motion of charged particles in a magnetic field is used to produce a velocity is shown in Fig. 6.6. This linear liquid velocity transducer was first proposed in

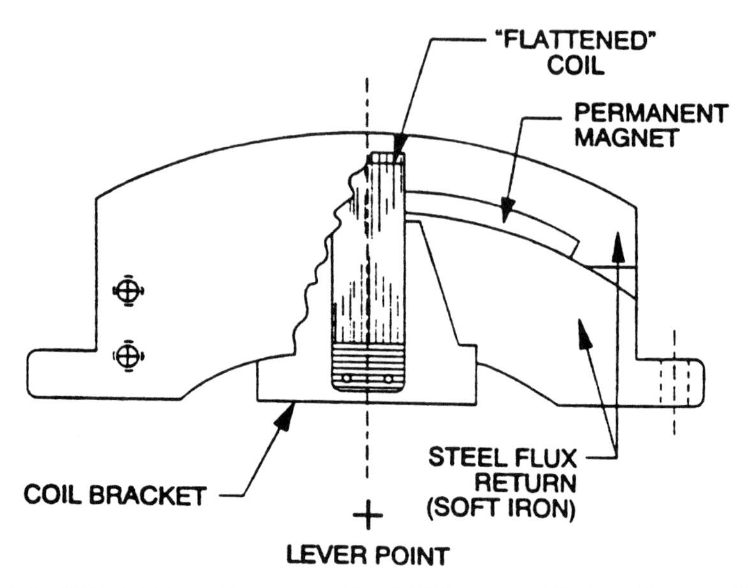

FIGURE 6.4. Voice coil actuator sold by BEI Sensors and Systems Company [5].

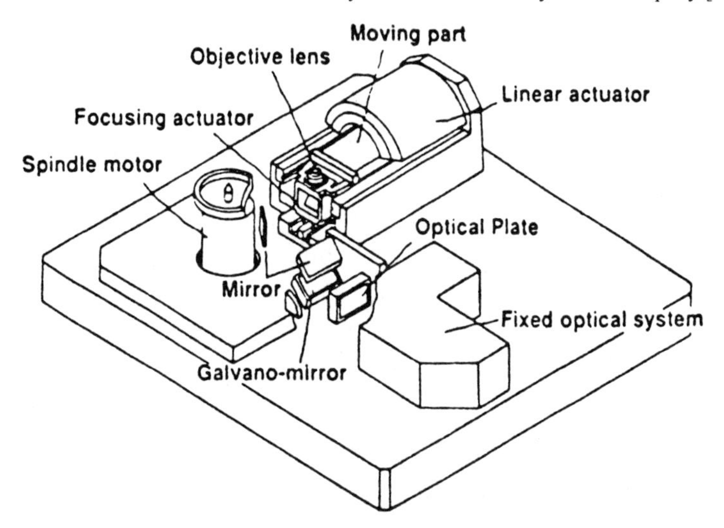

FIGURE 6.5. Magneto-optical disk drive, taken from C. D. Mee and E. D. Daniel (eds), **Magnetic Recording Handbook, Technology and Applications**, (McGraw-Hill, NY, 1990). Reproduced with permission by The McGraw-Hill Companies.

1887. In the device, a magnetic flux density is established by a permanent magnet. A duct containing conducting fluid is located between the north and south poles of the magnet, with its axis aligned orthogonal to the magnetic field. Application of a voltage between the top and bottom plates of the duct induces fluid motion through the duct. Alternatively, motion of the fluid can be used to induce a potential which is sensed as the transduction signal.

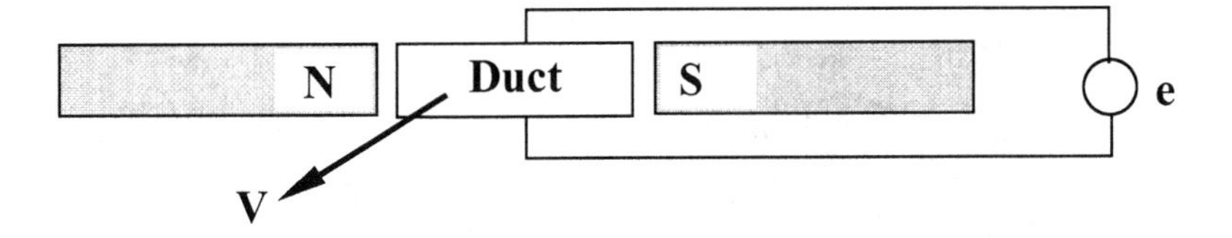

FIGURE 6.6. Liquid level velocity transducer.

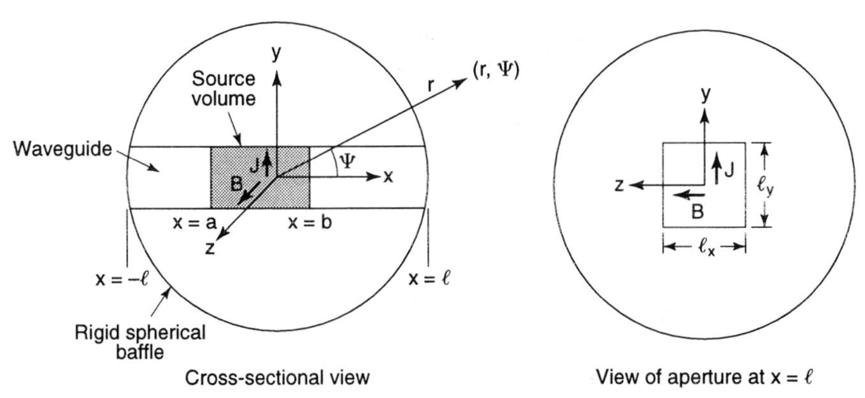

FIGURE 6.7. Underwater sound source from Schreppler and Busch-Vishniac [8]. Reprinted with permission from *A Magnetohydrodynamic Underwater Acoustic Transducer*, J. Acoust. Soc. Am. **89**, 927 (1991). Copyright 1991 American Institute of Physics.

Another realization of a velocity transducer, based on the motion of charged particles in a magnetic field that has been investigated, is an underwater sound source [7] and [8]. Figure 6.7, from Schreppler and Busch-Vishniac [8], shows such a device. Here a duct filled with salt water is sealed with rubber whose acoustic impedance closely matches that of water. The duct is surrounded on the top and bottom with permanent magnets. An oscillating current is applied to the duct sides, producing an oscillating velocity that generates sound in the water.

Another realization of this type of transducer is a magnetic flowmeter, which is a sensor designed to measure the flow of conducting liquid. In conducting fluids we can measure the fluid velocity by imposing a constant magnetic field and monitoring the voltage induced in the orthogonal direction as a result of the motion of the charged particles. This device is the reciprocal of the sound source described previously, as the voltage drive generating a velocity is replaced by a velocity drive generating a voltage. Figure 6.8 shows such a flowmeter marketed by Omega Engineering, Inc. [11]. In this device, the magnetic field is vertical. Horizontal flow of a conducting fluid results in a voltage which is measured across electrodes orthogonal to both the magnetic field and the flow. Figure 6.9 shows a similar flowmeter but with a slightly different geometry. In this flowmeter by Schlumberger Industries, Inc.[9] the flow moves a rotor which drives fluid in a side branch. This causes a magnet-containing float to raise and lower and induces a voltage in a fixed solenoid.

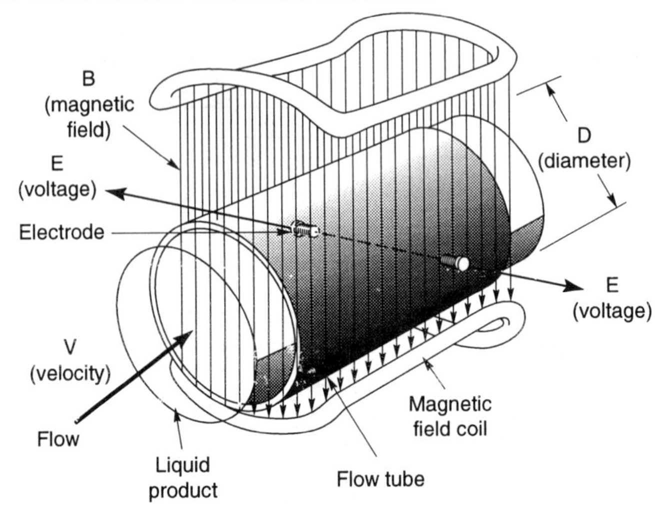

FIGURE 6.8. Magnetic flowmeter sold by Omega Engineering, Inc. [9]. Copy right 1995 Omega Engineering, Inc. All rights reserved. Reproduced with the permission of Omega Engineering, Inc., Stamford, CT 06907.

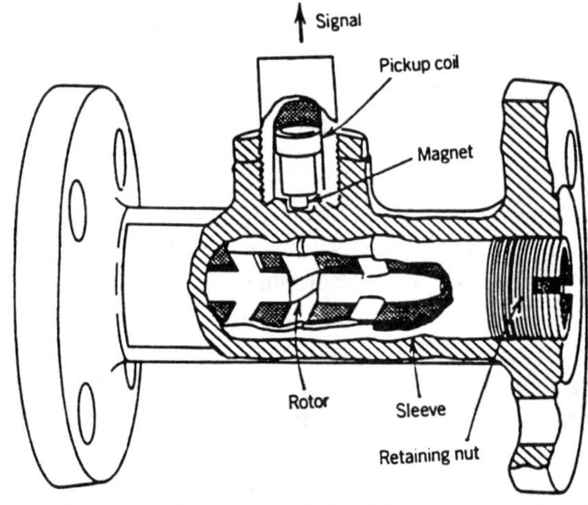

FIGURE 6.9. Turbine flowmeter sold by Schlumberger Industries, Inc. [9].

One final example of a device which exploits transduction based on the motion of charges in a magnetic field is a linear velocity transducer (often written LVT). In LVTs, the motion of a permanent magnet is used to generate a voltage according to (12). An example of the structure is shown in Fig. 6.10. Here, a pair of fixed coils responds to the motion of a permanent magnet. The generated voltages are added out of phase so as to add constructively. Such a design is typical of the LVTs sold by Trans-Tek, Inc., e.g., [10]. Note that this design differs from the loudspeaker in that the solenoid is fixed while the magnet moves. What matters from the transduction viewpoint is the relative motion between these two physical components. The advantages of LVTs are that they do not require excitation, they

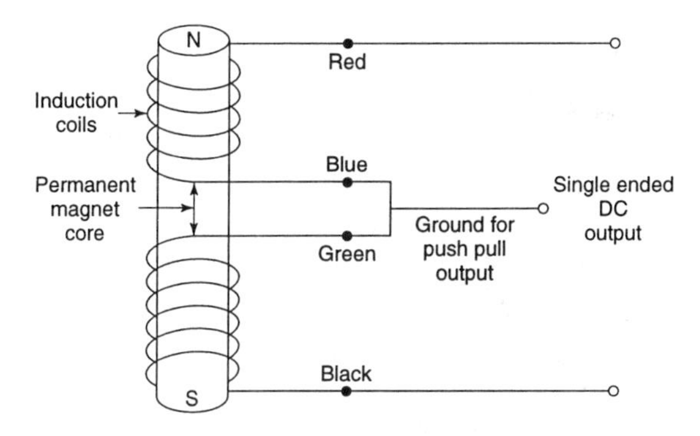

FIGURE 6.10. Linear velocity transducer configuration for Trans-Tek LVTs [10].

generate a dc voltage linearly proportional to velocity, and they are very simple. They are very similar to LVDTs, but they use motion of a magnet rather than motion of a ferrous material. It is striking that this seemingly simple change could have such a large impact on device operation.

Figure 6.11 shows a complete LVT made by Bell and Howell [12]. In this device, the permanent magnet is supported between two springs. The magnet is surrounded by a chrome-plated stainless steel sleeve. Threaded retainers seal the mechanical assembly. In operation, the entire transducer is attached to the vibrating object and the output is proportional to the velocity of the case relative to the magnet.

A simple bond graph of this LVT is shown in Fig. 6.12. Here I_c represents the mass of the case, I_m the magnet mass, C_{springs} the compliance of the springs, R the damping due to friction, and G the transduction from mechanical to electrical energy. Note that the case mass is a dependent energy storage element because its velocity is necessarily that of the body to which it is attached. Thus the model is a second-order system. We can use this model to examine qualitatively the high- and low-frequency limits of behavior. For instance, at low frequencies, the magnet mass admittance (since it is on a 0-junction) goes as the inverse of frequency. Similarly the spring impedance (since it is on a 1-junction) also goes as inverse frequency. Thus we would expect the low frequency behavior to be dominated by the mass and springs. At high frequencies, the magnet mass admittance is low so the spring velocity is close to that of the case, and the output voltage is almost proportional to the case velocity. It is in this regime that we would like to operate the sensor, and this mitigates the choice of magnet mass and spring stiffness.

6.4 Hall Effect Transducers

Another linear inductive transduction mechanism commonly used in electromechanical sensors is known as the Hall effect. Hall effect transducers actually combine two transduction mechanisms: electromechanical coupling based on the

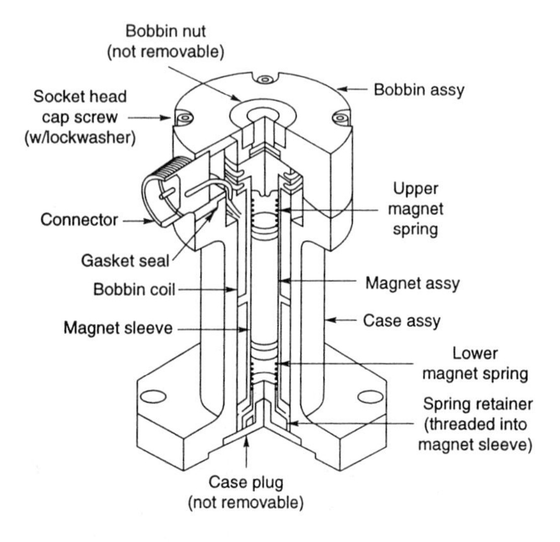

FIGURE 6.11. Bell and Howell Co. LVT [12].

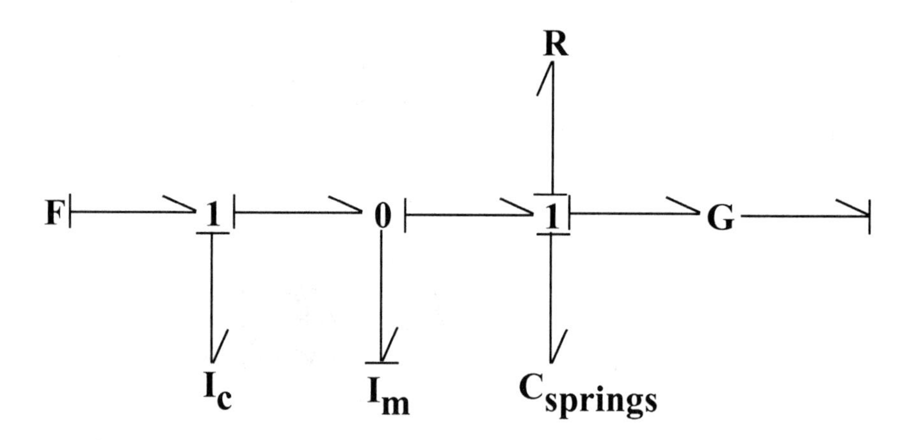

FIGURE 6.12. Bond graph model of a Bell and Howell Co. LVT.

motion of charges in a magnetic field, and magnetoelectric coupling via the Hall effect. The result is used to couple voltage to the position of a permanent magnet.

The Hall effect is named for its discoverer, Dr. Edwin Hall, who first noted the effect in 1879 while working on his Ph.D. dissertation. Its first practical application came in the 1950s, when it was exploited as a microwave power sensor. In 1968 the first solid-state Hall effect sensors were produced. These devices, which were used in keyboards, were highly competitive with alternatives because they were mass produced at low cost and small size. Today, Hall effect devices may be found in computers, sewing machines, cars, aircraft, machine tools, and medical equipment.

When a current-carrying conductor has a spatially uniform current density, there is no voltage generated normal to the current flow direction. When a magnetic flux density is added orthogonal to the current, then we know from Section 6.3 that a force is developed on the moving charges, with $F = Bli$. As shown in Fig. 6.13, taken from Honeywell's Micro Switch Division product literature [13], this tends to bend the current, inducing a nonuniform component to it. What Edwin Hall observed is that such a spatially nonuniform current induces a voltage in the direction orthogonal to both the magnetic flux and the current. This voltage is given by

$$e = \frac{R_H B i}{t} \tag{15}$$

$$= \frac{R_{\mathrm{H}}}{lt} F_{\mathrm{Lorentz}}, \tag{16}$$

where R_H is the Hall coefficient of the material, t is the thickness of the conductor, and l is the conductor length.

From (16) we see that there is a linear relation between voltage and force. This would be represented in a model using a transformer. Such a representation suggests that there is another expression which linearly relates the velocity and orthogonal current:

$$v = \frac{R_{\mathrm{H}}}{lt} i. \tag{17}$$

Hall effect devices can thus be used as velocity sensors, although they are rarely employed in this manner.

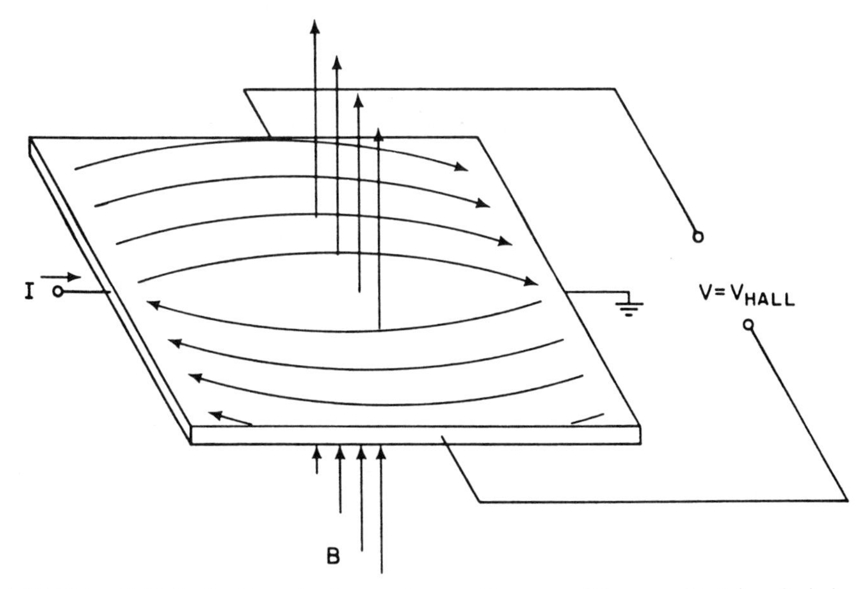

FIGURE 6.13. Hall effect transduction mechanism. Courtesy of Honeywell's Micro Switch Division [13].

Another interpretation of the Hall effect is possible, and explains Hall's results. Equation (15) may be thought of as a resistance expression since it relates a voltage to a current. However, the resistance so defined is not the conventional resistance because, in this case, the current flows in a direction orthogonal to the direction in which the potential arises. With this caveat, the effective Hall resistance is given by $R_{\mathrm{H}}B/t$. As we might expect, the resistance increases with decreasing thickness of the material. It also increases with increasing Hall coefficient and increasing magnetic field strength.

It is the Lorentz force which accelerates the charge carriers. This produces a current density described by

$$j = \frac{nq_e c}{B}E, \tag{18}$$

where n is the carrier density (a material property), q_e is the charge of an electron, and c is the speed of light. Here E and j are magnitudes of the electric field and current density, respectively. They are oriented in orthogonal directions. This permits us to define the material resistivity as follows:

$$\rho = \frac{B}{nq_e c}. \tag{19}$$

The resistance of a material is generally taken as $\rho l/A$, where A is the area normal to the current flow and l is the length of the material over which voltage is developed. In this case, A is defined normally (as the product tw, where w is the material width). However, the length is not the length of the material in the direction of current flow, but instead is the width of the material since the voltage is determined in the width direction. Thus, the resistance is given by

$$R = \frac{B}{nq_e ct} = \frac{R_{\mathrm{H}}B}{t}, \tag{20}$$

from (15). This permits us to solve for the Hall coefficient:

$$R_{\mathrm{H}} = \frac{1}{nq_e c}, \tag{21}$$

which shows that the coefficient is indeed a constant for a given material.

There is a very interesting aspect to (15) and (16) which accounts for the commerical success of Hall effect sensors—namely, that the output voltage generated increases with the inverse of the conductor thickness. This suggests that the thinner the conductor, everything else remaining fixed, the greater the sensitivity of the sensor. Microelectronics fabrication techniques make it possible to produce very thin conducting sheets, and it is this which permits Hall effect sensors to be mass produced at relatively low cost. Even with microelectronics techniques, the sensitivity of Hall effect sensors is quite low, with roughly 10 μV produced for a 1 G magnetic field [13].

In operation, it is not terribly useful to exploit (16) to relate the Lorentz force and voltage. Instead, it is typical to use a permanent magnet to produce the magnetic flux density, and to vary the magnet position so as to vary B. Because the magnetic flux density is a nonlinear function of the distance from a permanent magnet, this introduces a nonlinearity into the transduction relation. In other words, (15) becomes

$$e = \frac{R_{\mathrm{H}} B i}{t} = \frac{R_{\mathrm{H}} i}{t} B(x), \tag{22}$$

where $B(x)$ is the magnetic flux density as a function of position x. This expression defines a modulated gyrator. In other words, it linearly relates a voltage to a current, but the modulus $(R_H t/B)$ is a function of another variable, x. Hence the gyrator can be used to sense the variable that is causing the modulus to vary, i.e., the modulating agent.

There are a number of geometries for the permanent magnet and conducting sheet which can be exploited. Some of these are taken from Honeywell's Micro Switch Division publication [13] and shown in Figs. 6.14 to 6.17 below, along with their associated variation of the magnetic flux density as a function of position. Yet another approach uses a fixed magnet and fixed conductor sheet, with a moving vane made of a ferrous material located between them. This is shown in Fig. 6.18, also taken from Reference [13].

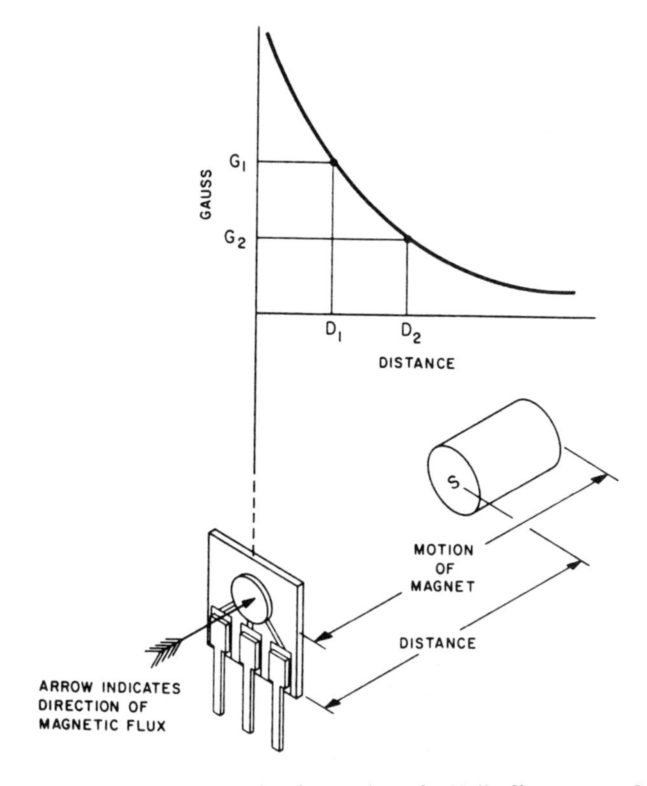

FIGURE 6.14. Unipolar head-on mode of operation of a Hall effect sensor. Courtesy of Honeywell's Micro Switch Division [13].

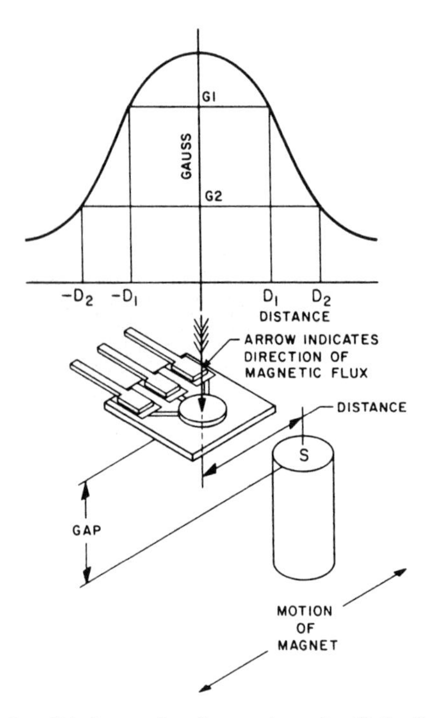

FIGURE 6.15. Unipolar slide-by mode of operation of a Hall effect sensor. Courtesy of Honeywell's Micro Switch Division [13].

6.4.1 *Example Hall Effect Sensors*

Because they are reasonably inexpensive and small, Hall effect sensors have been used for a wide variety of applications including position measurement, joy sticks, flowmeters, relays, proximity detectors, and pressure sensors. A few of the transduction applications are shown here.

Figure 6.19, used courtesy of Honeywell's Micro Switch Division, shows a Hall effect transducer operating as a level/tilt sensor. In this application, the sensor is fixed while the permanent magnet is attached to a part which swings freely (like a pendulum). The part is level when the sensor output is at a maximum.

Figure 6.20 shows a Hall effect sensor used in a beverage gun (taken from [13]). In this device, the sensor is operated in the unipolar head-on mode. A spring (not shown in the figure) restores the button to its undisturbed position.

Figure 6.21 shows a geartooth sensor manufactured by American Electronic Components, Inc. [14]. In this sensor, the steel gear teeth serve as the vane which is located between the sensor and a fixed permanent magnet (not seen in the figure).

Figure 6.22 shows a flowmeter based on the Hall effect. In this device, sold by Gems Sensors [15], fluid flow rotates a rotor which contains magnets in its vanes. Each time a magnet passes the fixed Hall effect sensor, a voltage pulse is sent. Then the number of such pulses per second is easily converted to a fluid flow rate.

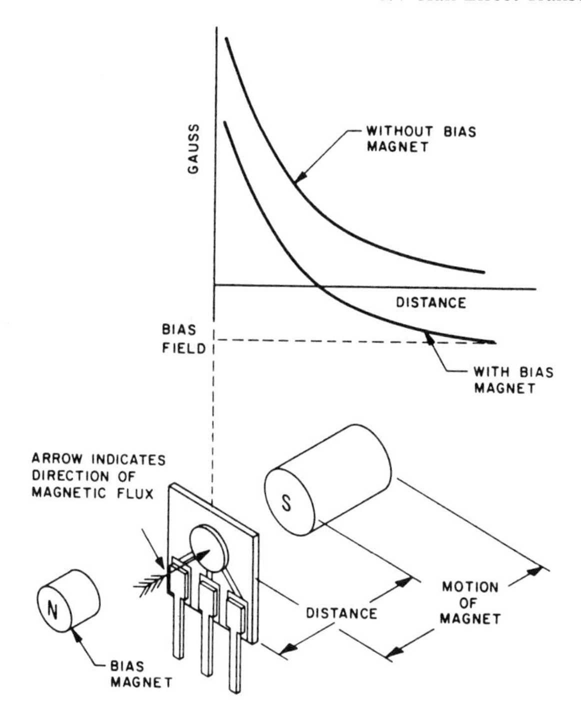

FIGURE 6.16. Bipolar head-on mode of operation of a Hall effect sensor. Courtesy of Honeywell's Micro Switch Division [13].

Figure 6.23 shows a bond graph model of this device. In this model, losses at the input and output are included in R_{in} and R_{out}, respectively. A transformer converts pressure to torque, and the rotor inertia and bearing resistance are modeled by I and R. The Hall effect device is modeled as a gyrator, in which a current is supplied (flow source). The modulus of the gyrator is affected by the displacement of the rotor. Hence there is a signal line (dashed arrow) from the 1-junction, which defines the rotor velocity, to the gyrator.

A last set of interesting Hall effect devices are pressure sensors. For instance, Fig. 6.24, taken from Webster [16], shows a pressure sensor designed for biomedical application. In this device, a magnet is embedded into a Teflon diaphragm which deforms. The Hall effect sensor responds to the changing distance from the magnet. Figure 6.25 shows a more conventional pressure sensor. In this device, described in Popovic [17], pressure causes deformation of a diaphragm which has a magnet suspended below it. The magnet is in a bipolar slide-by configuration to which the Hall effect sensor responds.

In the last two decades there has been a great deal of research on the Hall effect in two-dimensional electron gases at low temperature and under a high magnetic field. Under these conditions, the Hall effect has been shown to be quantized. This research has led to very accurate and useful laboratory measuring instruments.

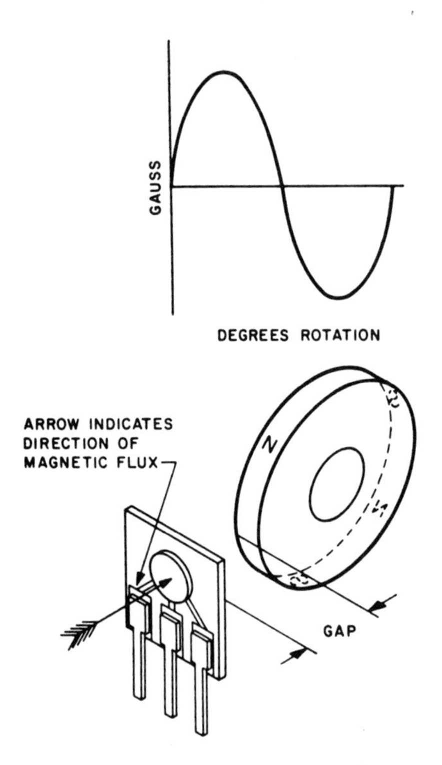

FIGURE 6.17. Bipolar slide-by ring magnet operation of a Hall effect sensor. Courtesy of Honeywell's Micro Switch Division [13].

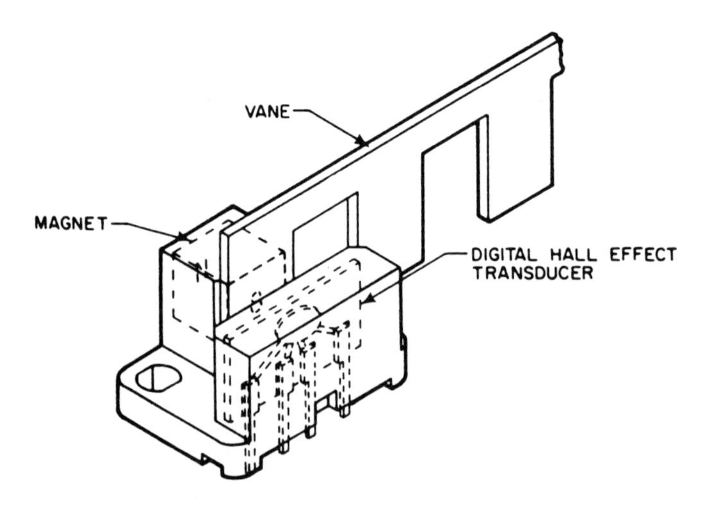

FIGURE 6.18. Hall effect sensor using a moving vane located between a fixed permanent magnet and a conductor. Courtesy of Honeywell's Micro Switch Division [13].

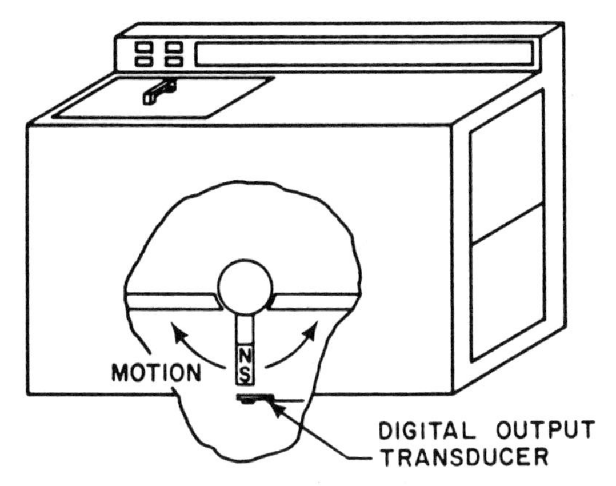

FIGURE 6.19. Hall effect level/tilt sensor. Courtesy of Honeywell's Micro Switch Division [13].

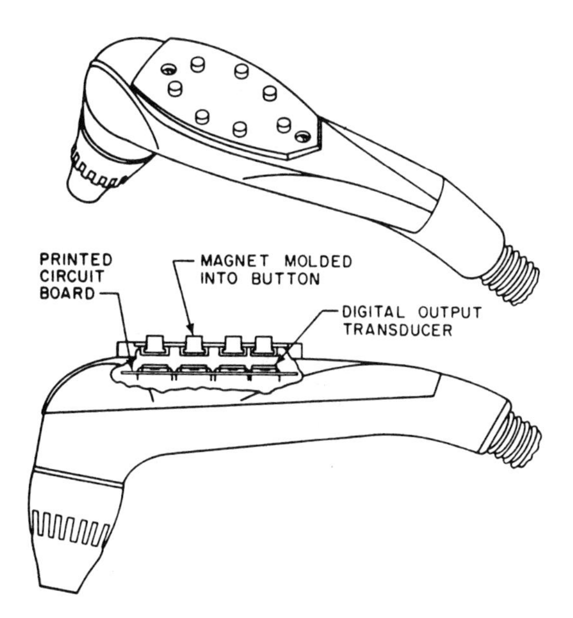

FIGURE 6.20. A Hall effect beverage gun sensor. Courtesy of Honeywell's Micro Switch Division [13].

6.5 Summary

This chapter has discussed transduction mechanisms which are inherently inductive and linear. While there are transduction mechanisms directly analogous to piezoelectricity and pyroelectricity, two other transduction mechanisms are far more common. Transduction which relies on the motion of charged particles in the presence of a magnetic field has been shown not to affect the system energy.

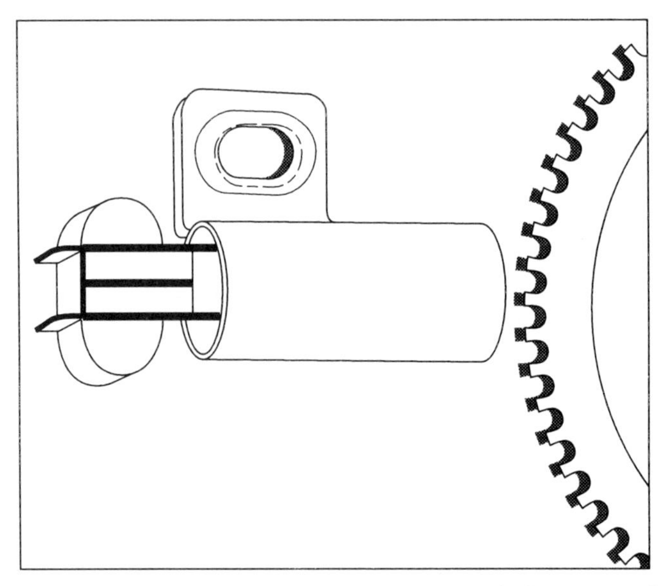

FIGURE 6.21. Geartooth sensor sold by American Electronic Components, Inc. [14].

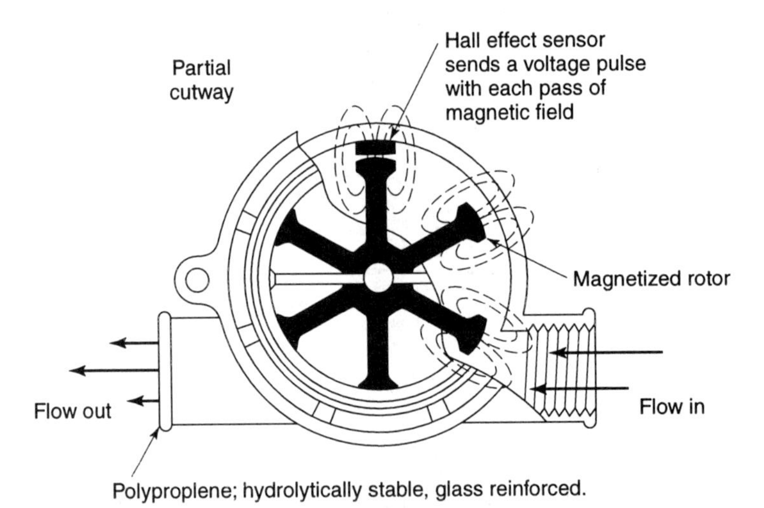

FIGURE 6.22. Flowmeter sold by Gems Sensor [15].

This transduction is found in common transducers such as loudspeakers, disk drives, and rotameters. Hall effect sensors are also common, linear inductive devices, although their typical use incorporates a nonlinearity related to variation of the magnetic flux density field as a function of distance from a permanent magnet. Hall effect sensors are commonly found in automobiles and a wide variety of sensors.

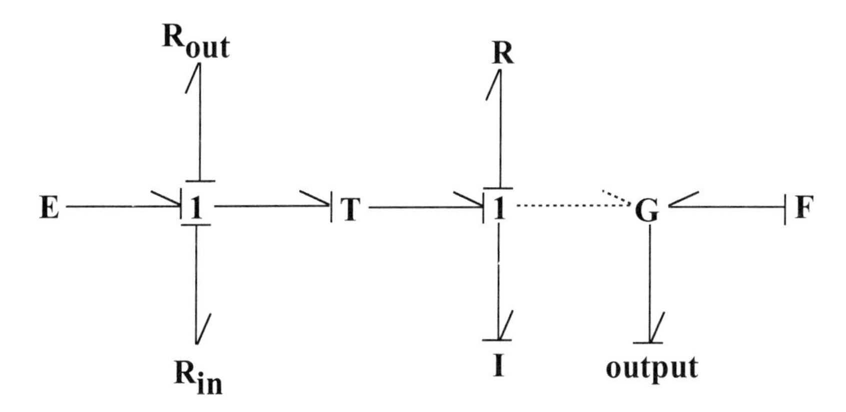

FIGURE 6.23. Bond graph model of the flowmeter sold by Gems Sensor [15].

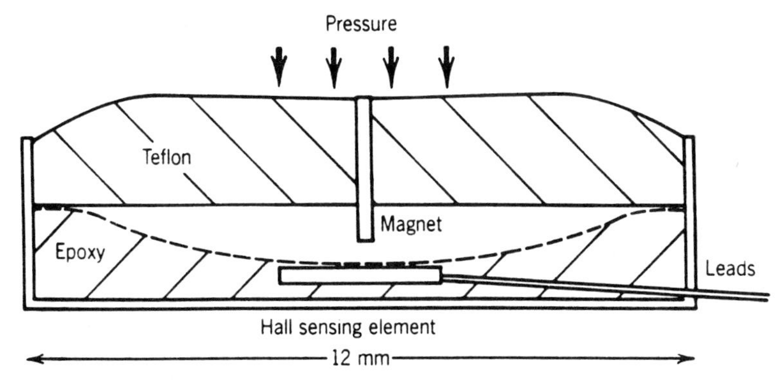

FIGURE 6.24. Biomedical pressure sensor. This figure reprinted by permission of John Wiley and Sons, Inc. from J. G. Webster (ed), **Tactile Sensors for Robotics and Medicine**, (Wiley, NY, 1988).

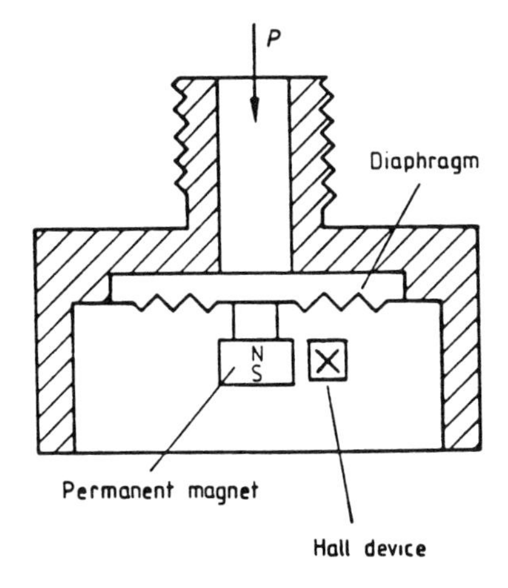

FIGURE 6.25. Conventional pressure sensor (taken from Popovic [17]).

6.6 References

1. D. A. Berlincourt, D. R. Curran, and H. Jaffe, *Piezoelectric and Piezomagnetic Materials and Their Function in Transducers*, in W. P. Mason (ed.), *Physical Acoustics: Principles and Methods*, Vol. 1, Part A (Academic Press, New York, 1964).
2. S. W. Meeks and J. C. Hill, *Piezomagnetic and Elastic Properties of Metallic Glass Alloys* $Fe_{67}CO_{18}B_{14}Si_1$ *and* $Fe_{81}B_{13.5}Si_{3.5}C_2$, J. Appl. Phys. **54**, 6584 (1983).
3. J. L. Butler, *Application Manual for the Design of Etrema Terfenol-*D_{tm} *Magnetostrictive Transducers* (Etrema, Ames, IA, 1988).
4. K. Seki, T. Takahashi, J. Shida, H. Matsuki, and K. Murakami, *Photo-Magnetic Semiconductor Using Pyromagnetic Effect*, IEEE Trans. Magnetics, **27**, 5256 (1991).
5. BEI Sensors and Systems Company, Kimco Magnetics Division, 804-A Rancheros Dr., San Marcos, CA 92069, USA.
6. C. D. Mee and E. D. Daniel (eds.), *Magnetic Recording Handbook, Technology and Applications* (McGraw-Hill, New York, 1990).
7. H. Adjisaka and E. L. Hixson, *Transduction Principles of a Liquid State Electromagnetic Transducer*, J. Acoust. Soc. Am., **77**, 1933 (1985).
8. S. C. Schreppler and I. J. Busch-Vishniac, *A Magnetohydrodynamic Underwater Acoustic Transducer*, J. Acoust. Soc. Am., **89**, 927 (1991).
9. ________, *Hall Effect Transducers* (Honeywell's Micro Switch Division, Dept. 722, 11 W. Spring St., Freeport, IL. 61032, 1982).
10. Schlumberger Industries, Measurements Division, 1310 Emerald Rd., Greenwood, SC 29646, USA.
11. Trans-Tek Inc., PO Box 338, Route 83, Ellington, CT 06029, USA.
12. Omega Engineering Inc., PO Box 4047, Stamford, CT, 06907, USA.
13. Bell and Howell Co., 5215 Old Orchard Rd., Skokie, IL 60077.
14. American Electronic Components, Inc., 1010 North Main St., PO Box 280, Elkhart, IN 46515, USA.
15. Gems Sensors, 1 Cowles Rd., Plainville, CT 06062-1198, USA.
16. J. G. Webster (ed.), *Tactile Sensors for Robotics and Medicine*, (Wiley, New York, 1988).
17. R. S. Popovic, *Hall Effect Devices* (Adam Hilger, Bristol, UK, 1991).

7

Transduction Based on Changes in the Energy Dissipated

In the preceding chapters of this book we have discussed transduction mechanisms associated with energy storage, capacitive or inductive, and energy transformation (linear magnetic transduction). In all of these cases the transduction mechanism is fundamentally reciprocal in that it can be used either to produce an electrical signal in response to a mechanical input, or to produce a mechanical signal in response to an electrical input. In this chapter we discuss the broad class of transduction which is based on changes in the resistance of the device. Given that resistance causes the irreversible transformation of energy from a form that could be used to perform work into that which cannot perform work, these transducers are not reciprocal. In other words, the process has only one way it can work: typically to produce an electrical signal in response to a mechanical input.

Resistive transducers come in a very wide variety of types, but fundamentally all of them take advantage of the resistance between electrodes changing in response to some input variable. In this chapter, we divide these resistive transducers into five basic types: those which take advantage of the coupling between temperature and resistance (thermoresistive); those which exploit the coupling between strain and resistance (piezoresistive); those which directly couple thermal and electrical behavior (thermoelectric); those which move a slide in a potentiometer, and all others. Of these classes, the first three take advantage of material properties present, to varying extents, in common materials. The potentiometric class of transducers uses a direct geometrical coupling between the mechanical and electrical domains. The remaining resistive transducers tend to work either as electrical switches or to look at resistance changes due to the medium between electrodes changing in some fundamental way. In this chapter we describe the fundamental operation of all of these sensors, and present commercially available examples. We begin with the simplest resistive sensors—conductive switches.

7.1 Conductive Switches

In conductive switches a change in the mechanical state of a system causes a switch to change either from open to closed, or closed to open. Thus, the change in resistance is a dramatic shift between two states: one with nearly infinite resistance and one with nearly zero resistance. The change in resistance is easily detectable by observation of a change in the output voltage typically from zero to some preset level. As such, these sensors behave like digital devices. Their output signal is one of two levels indicating that the sensed quantity is either above or below the threshold.

In general, we may model a conductive switch as a nonlinear resistance. As the resistance depends on some mechanical variable, there must be at least one mechanical port. Further, since the resistance is fundamentally an electrical resistance, there must be an electrical port. Hence, a reasonable model of a conductive switch is a multiport R element.

There are two general classes of conductive switches. In the first class, motion of some component in the sensor either pushes the button of a switch or forces electrodes into (or out of) contact. In the second class, a set of electrodes is fixed, and a conductive material (usually liquid) moves in response to changes in the mechanical variable we desire to sense. When the conducting material connects the electrodes, the switch is closed. When the conducting material is not between the electrodes, which are then separated by air or vacuum, then the switch is open.

Figure 7.1 shows a pressure switch which is an example of the first class of conductive switches. In this device, sold by Dwyer Instruments, Inc. [1], a large diaphragm is exposed to high pressure on one side, and low pressure on the other. Motion of the diaphragm in response to the pressure differential can cause a switch button to be extended. A spring in the sensor makes it possible to adjust the threshold pressure which will trigger the switch.

Figure 7.2 is a bond graph model of the pressure switch shown in Fig. 7.1. In this second-order model, I is the mass of the diaphragm, C represents the spring, and the 2-port R is the switch itself. For static pressure applications, the inertance can be ignored, and the system reduces to first order.

A similar pressure switch sold by Micro Pneumatic Logic, Inc. [2] is shown in Fig. 7.3. Used as a vacuum switch, the adjustment screw is set so that the electrodes are touching. Then when the vacuum pressure is below a certain value, the contacts will move apart, breaking the electrical contact between the output terminals. As a pressure sensor, the adjustment screw is set so that the electrodes are initially not in contact. When the pressure at the pressure port exceeds some value, the contacts will be pushed together, and the output terminals shortcircuited.

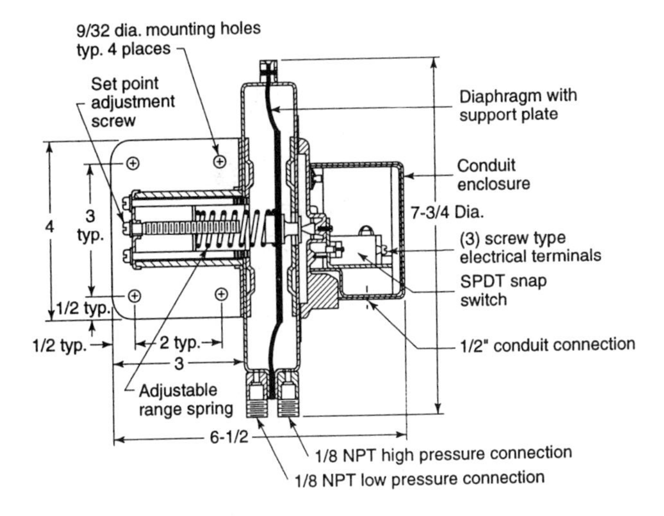

FIGURE 7.1. Pressure switch sold by Dwyer Instruments, Inc. [1].

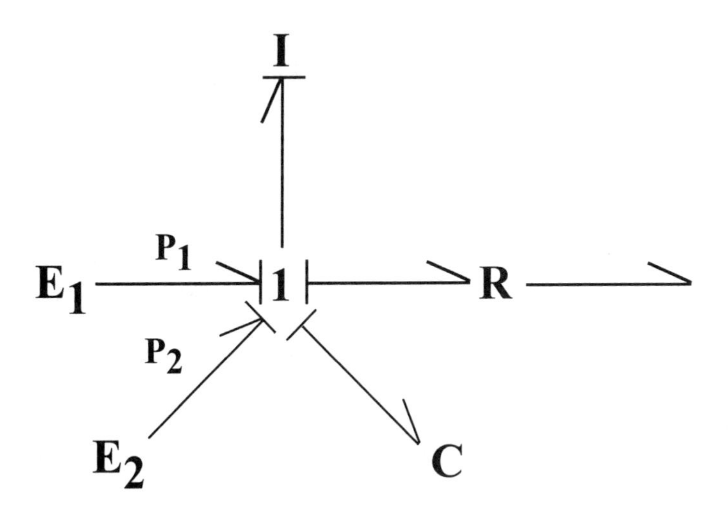

FIGURE 7.2. Bond graph model of the Dwyer pressure switch.

Figure 7.4 shows an example of a sensor which relies on the motion of a conducting fluid to close the circuit between fixed electrodes. This level sensor is manufactured by Gems Sensors [3] and is useful for regulating the height of a conducting fluid in a controlled process. In this application, the fluid level is permitted to rise until it touches both the high and low electrodes, shortcircuiting the output terminals. Then the pump is turned on and the water level lowered. A precipitation detector by Qualimetrics, Inc. [4] works on an identical principle.

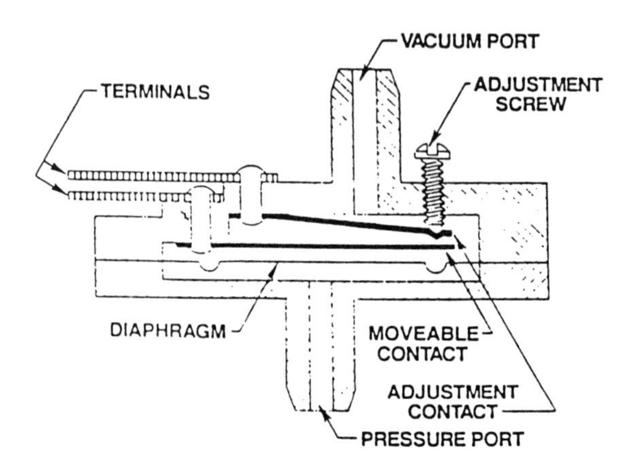

FIGURE 7.3. Pressure or vacuum switch sold by Micro Pneumatic Logic, Inc. [2].

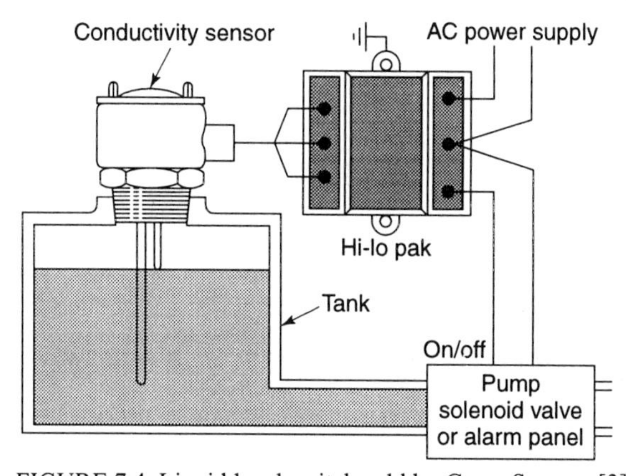

FIGURE 7.4. Liquid level switch sold by Gems Sensors [3].

Figure 7.5 is another example of the class of conductive switches in which a liquid is used to complete the circuit between two electrodes. This switch, sold by American Electronic Components, Inc. [5] uses mercury as the conducting fluid, and can be configured either to respond to translational accelerations (Fig. 7.5(a)) or to tilting (Fig. 7.5(b)).

7.2 Continuously Variable Conductivity Transducers

The conductive switches described previously are digital devices which essentially monitor whether a circuit is open or closed. There are many devices which resemble these switches but which operate so as to provide resistance information on a continuum scale. Typically these devices produce a change in resistance due to a change in the medium between electrodes, or due to a change in the distance

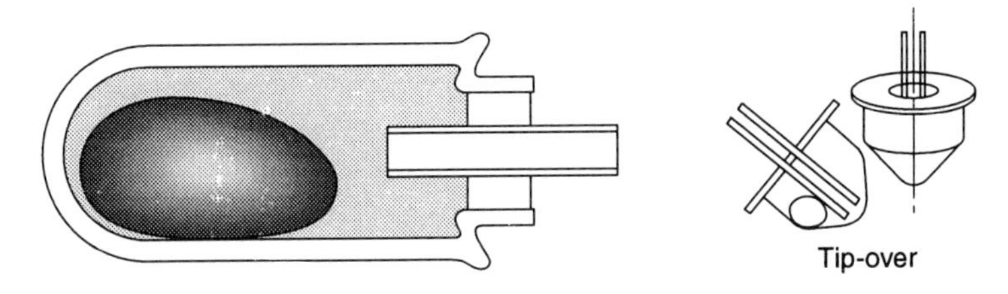

FIGURE 7.5. Mercury switch sold by American Electronic Components, Inc. [5].

between the electrodes. Examples of each of these transducers are provided below. The models appropriate for them are discussed in the context of the examples.

Figure 7.6, taken from Rossi [6], shows the carbon button microphone that was the workhorse of all telephones in the United States for many years. This microphone has a metal diaphragm which serves as a moving electrode and responds to sound pressure. Carbon between the diaphragm and the fixed electrode provides electrical resistance. As the diaphragm moves, the distance between the electrodes varies, thus changing the resistance. This change is monitored using circuitry to produce a voltage proportional to the change in resistance. Figure 7.7 shows a bond graph model of the microphone excluding the circuitry. The sound pressure is an effort source delivered to the diaphragm which has both inertia, I, and compliance, C. The carbon has compliance, C_c, and is modeled as a 2-port resistance, with one port mechanical and one electrical.

A very different sort of variable conductance transducer is the tunneling displacement transducer used in tunneling microscopes. Figure 7.8, taken from Coombs [7], shows the tunneling transducer. This device monitors a tunneling current from the tip to the surface under study. The closer the surface atoms, the stronger the current (i.e., the lower the resistance). Based on the current, then, a map of the surface profile on very fine scales can be generated.

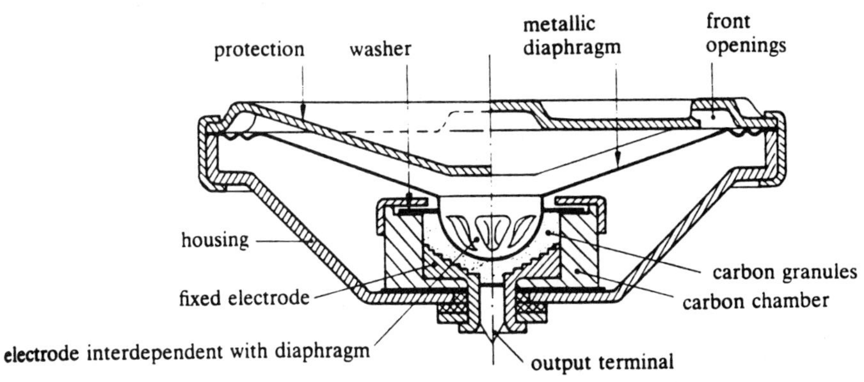

FIGURE 7.6. Carbon button microphone. Reprinted with permission from Acoustics and Electroacoustics, by M. Rossi, Artech House, Inc., Norwood, MA, USA. http://www.artech-house.com [6].

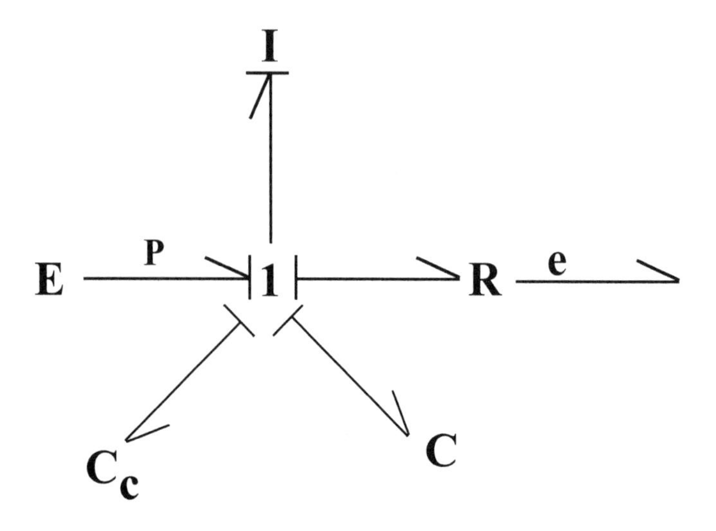

FIGURE 7.7. Bond graph model of the carbon button microphone.

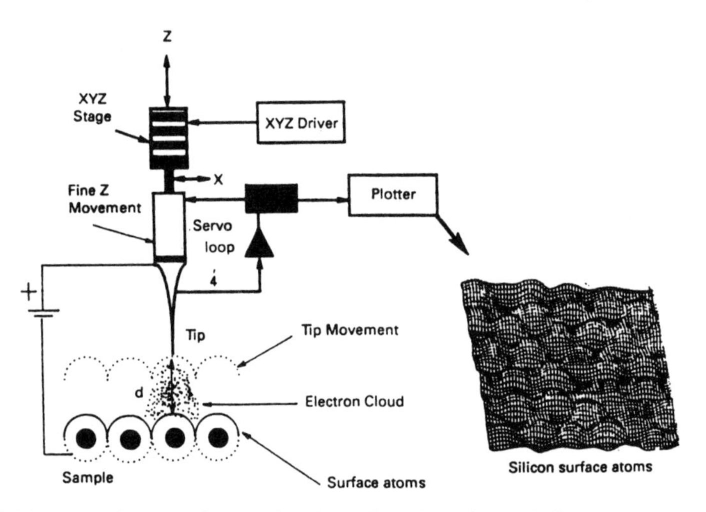

FIGURE 7.8. Tunneling transducer, taken from C. F. Coombs, Jr. (ed), **Electronic Instrument Handbook**, 2nd ed., (McGraw-Hill, NY, 1995). Reproduced with permission by The McGraw-Hill Companies.

Figure 7.9, taken from Fraden [8], is a humidity sensor which operates by exploiting a material whose electrical properties change as its water content varies. In particular, although the distance between the electrodes is fixed, the material resistivity varies as a function of the water absorbed by the hygroscopic layer. Thus, by monitoring this resistance, we can determine the relative humidity of the surrounding environment. The important part of this transducer is merely an electrical resistance. Clearly there is no direct link in the mechanical and electrical

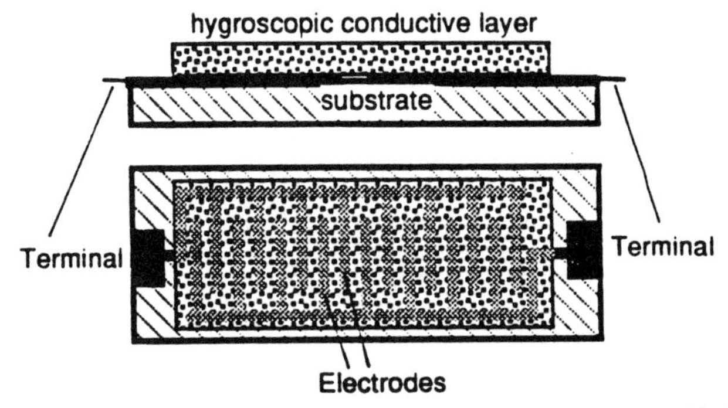

FIGURE 7.9. Relative humidity sensor. This figure originally appeared in Fraden [8].

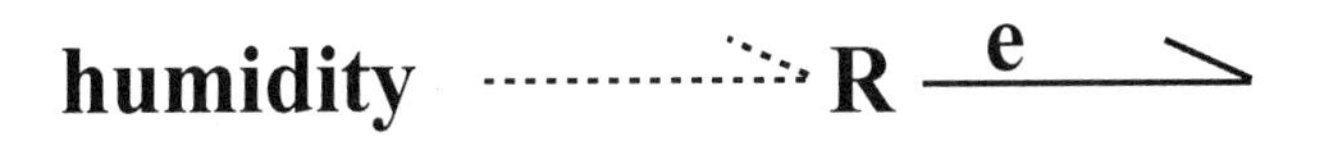

FIGURE 7.10. Modulated resistance model of the humidity sensor.

domain in this resistance. Instead, the linear electrical resistance is *modulated* by the amount of water present. We model this using a modulated resistance, as shown in Fig. 7.10. The element is a 1-port resistance, but its resistance value depends on the humidity signal, shown as a dashed line to the R.

The idea of a resistance path of fixed length but variable medium is exploited in a number of sensors. For instance, the liquid level switch shown in Fig. 7.4 does not have to operate as a binary switch. It could easily be used to correlate the liquid level with the resistance. Also, we could detect the presense of various liquids or gases as long as their resistivities are substantially different from the nominal medium they displace.

7.3 Potentiometric Devices

Potentiometric devices are very simple resistive transducers in which the potentiometer slide is moved in response to some mechanical signal. This produces a change in the voltage across the potentiometer which is linearly related to the slide displacement. Hence, a linear conversion from a mechanical signal to an electrical signal is accomplished.

Figure 7.11 is a simple view of the potentiometric transducer. In this device a fixed voltage input, E_{in} is supplied and the output voltage, e_{out} is monitored as the voltage drop across the measurement resistor R_m. The electrical resistance is divided into two portions, R_1 and R_2, the sum of which is fixed. As shown in Fig. 7.11 the voltage drop across R_1 is the monitored signal.

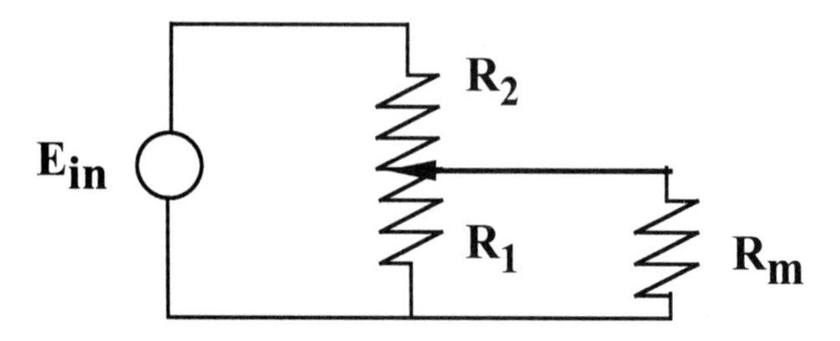

FIGURE 7.11. Simple model of a potentiometer.

The model of the slide potentiometer can be analyzed quite simply. The current through the resistor is given by

$$E_{\text{in}} = R_2 i + R_1 i', \tag{1}$$

where $i - i'$ is the current through the measuring arm. The output voltage is given by

$$e_{\text{out}} = R_m(i - i') = R_1 i'. \tag{2}$$

We can use (2) to determine the ratio of i' to i.

$$\frac{i'}{i} = \frac{R_m}{R_1 + R_m}. \tag{3}$$

Combining Eqs. (1) – (3) we have

$$\frac{e_{\text{out}}}{E_{\text{in}}} = \frac{R_1}{R_1 + R_2\left(\frac{R_1+R_m}{R_m}\right)}. \tag{4}$$

We can simplify this equation if it is safe to assume that the measurement resistance is very large compared to R_1 (i.e. normal operating conditions). Then we have

$$\frac{e_{\text{out}}}{E_{\text{in}}} = \frac{R_1}{R_1 + R_2} = \frac{x}{L}. \tag{5}$$

Potentiometric transducers can be used to measure either a translational position or an angular position. Figure 7.12 shows an example of a potentiometric transducer designed for measurement of the angular position. This device functions in a manner precisely analogous to that shown in Fig. 7.11. The defining relationship, assuming a uniform resistance, is

$$\frac{e_{\text{out}}}{E_{\text{in}}} = \frac{\theta}{\theta_{\text{tot}}}, \tag{6}$$

where θ_{tot} is the total angular extent of the resistance potentiometer.

Figure 7.13 shows an example of a potentiometric transducer described in Doebelin [9]. This transducer is used to indicate the attitude of an aircraft and uses a rotational potentiometer to distinguish the angular displacement. While potentiometric devices generally are noisy and not very accurate, they tend to be quite rugged, a distinct advantage for an aircraft application.

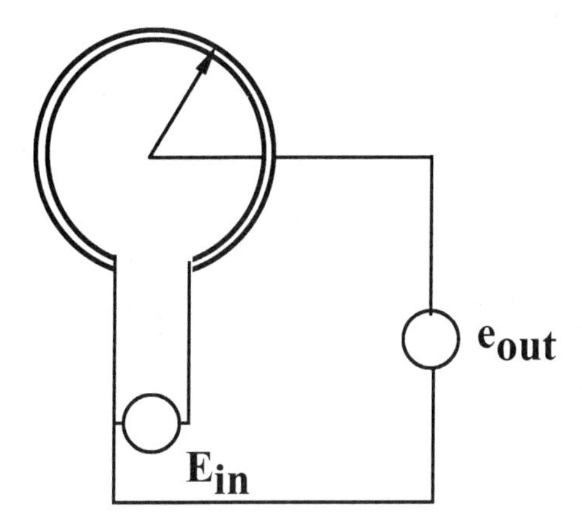

FIGURE 7.12. An example of a potentiometric angular displacement transducer.

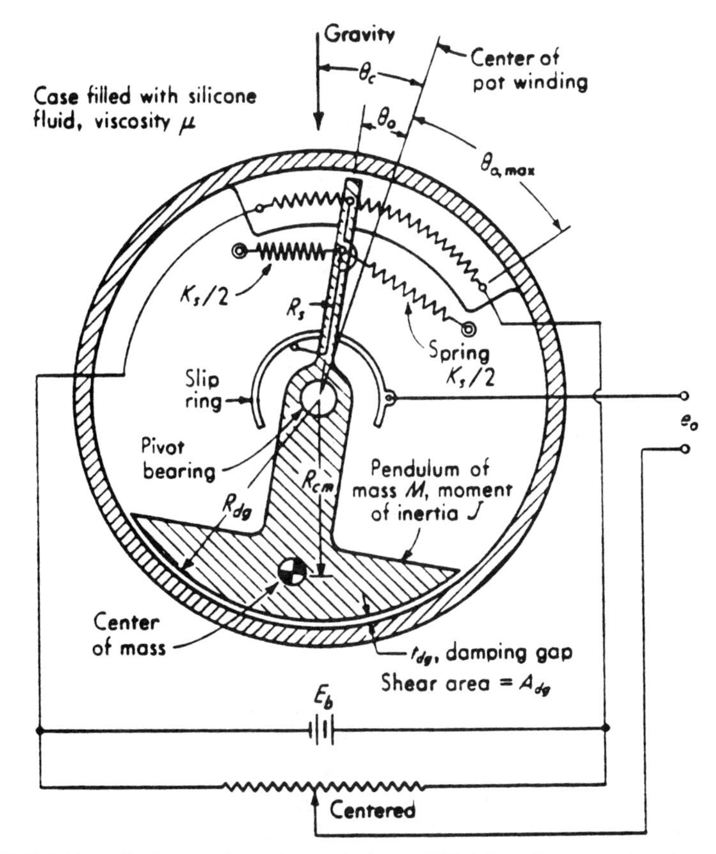

FIGURE 7.13. Attitude transducer for an airplane which is based on a rotational potentiometer. Figure from E. O. Doebelin, **Measurement Systems: Application and Design**, 3rd ed., (McGraw-Hill, NY, 1983). Reproduced with permission by The McGraw-Hill Companies.

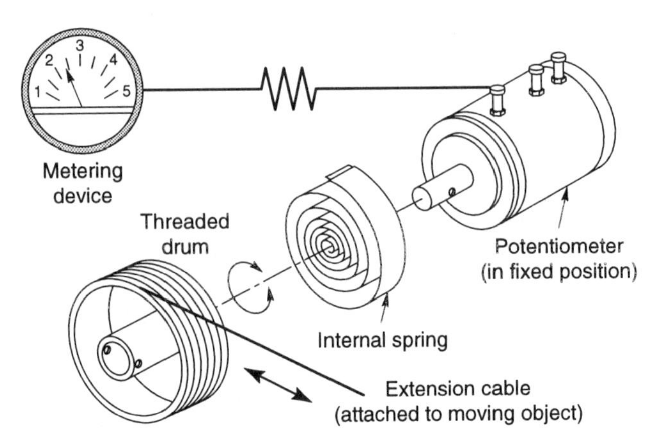

FIGURE 7.14. Position sensing potentiometric transducer sold by SpaceAge Control, Inc. [10].

Figure 7.14 shows another potentiometric device available commercially. This position sensor, sold by SpaceAge Control, Inc. [10], uses a cable attached to a moving object to convert translational motion into rotational motion of a slide on a plastic or metal/plastic hybrid potentiometer. Note that an internal spring tends to resist motion of the drum so that the cable can rewind if the translational motion reverses.

The major advantage of potentiometric sensors is their relatively low cost. However, they tend to be fairly low in sensitivity, and because the slide moves repeatedly over the resistive surface, wear is an important issue. To some extent, these negatives can be overcome through innovative use of materials. For example, Novotechnik U.S., Inc. [11] now sells a line of potentiometric sensors made from conductive plastic. In materials such as this, the resistivity continuously varies as a function of distance from the electrode, rather than varying in small steps set by the resistor structure. Further, the polymer surface suffers from friction less than a typical metal surface.

7.4 Piezoresistivity

Piezoresistivity is a material property in which the electrical resistance of a material is a function of its mechanical properties. In particular, crystalline materials when strained often suffer changes in their crystal lattice which may be observed as a change in the electrical resistivity. By exploiting the change in electrical resistivity, it is possible to build sensors which directly measure the strain in a structure. We call such sensors strain gauges.

The origin of the first strain gauge is not well documented. However, the phenomenon of strain-induced changes in electrical resistance was certainly known as early as the 1850s, when Lord Kelvin reported on the effect of strain on the resistance of wires. In 1908, St. Lindeck investigated this phenomenon further using thin-walled brass tubes with fine manganin wire wound around them. He is thought to be the first to suggest exploiting the resistance variation with strain in a pressure sensor [12]. The heyday of strain gauge sensors began in earnest with the creation of small metal wire strain gauges by E. E. Simmons of California Institute of Technology and A. C. Ruge of MIT in the 1940s [13]. In their strain gauges, a fine wire was glued to an elastic element which could then be attached to the structure whose strain was to be monitored. The change in resistance was monitored as a change in the voltage drop across the resistor. The first industrial strain gauge-based load and torque sensors were marketed in the early 1940s.

In the 1950s, wire strain gauges gave way to metallic foil strain gauges. These sensors offered the advantages of lower mass, so they were far less likely to load the structure whose strain they were to monitor. Further, the metallic foil strain gauges could be produced easily and at low cost. They could be backed with a variety of materials, including adhesives making it possible to simply peel the strain gauge off their package and attach it to a structure. Metal foil strain gauges are still a major segment of the strain gauge market today.

In the 1960s the first strain gauges using materials other than metal resistors emerged. In particular, the discovery of a strong piezoresistive effect in semiconductor materials made them very attractive for strain gauge sensors. These strain gauges can be made extremely small and tend to have high sensitivity compared to metal strain gauges. The first semiconductor strain gauges were made of single crystals. However, work dating from the early 1970s has concentrated on semiconductor strain gauges of polycrystalline thin films. The first strain gauges of this type were reported by Grigorovici et al. [14] and Johanessen [15].

7.4.1 *Material Description*

If a material is subjected to a stress, it experiences a strain. This has two effects. First, there is a change in the length and cross-sectional area of the material that causes a change in its electrical resistance. This is a purely geometric effect. Second, if the material is significantly piezoresistive, straining the crystal lattice produces a change in the resistivity of the material. The change in resistivity may be expressed as

$$\begin{aligned}\frac{\Delta\rho_{\alpha\beta}}{\rho} &= \sum_{k,l} \gamma_{\alpha\beta kl}\epsilon_{kl} \\ &= \sum_{k,l} \pi_{\alpha\beta kl}\tau_{kl}. \end{aligned} \tag{7}$$

Here, ρ is resistivity, ϵ is strain, τ is stress, the γ coefficients are referred to as elastoresistance coefficients, and the π coefficients are the piezoresistance coefficients.

Given that symmetry arguments result in the restriction that $\epsilon_{\alpha\beta} = \epsilon_{\beta\alpha}$ and $\tau_{\alpha\beta} = \tau_{\beta\alpha}$, we may simplify the piezoresistive expressions using precisely the

TABLE 7.1. Piezoresistive coefficients for silicon and germanium, as first published by Smith [16].

Material	ρ (Ω cm)	$\pi_{11} \times 10^{-7}$ cm^2/N	$\pi_{12} \times 10^{-7}$ cm^2/N	$\pi_{44} \times 10^{-7}$ cm^2/N	γ_{11}	γ_{12}	γ_{44}
n-Ge	1.5	−2.3	−3.3	−69.0	−9.5	−10.3	−93.0
n-Ge	5.7	−2.7	−3.9	−68.4	−15.1	−16.1	−92.0
n-Ge	9.9	−4.7	−5.0	−69.0	−16.6	−16.8	−93.0
n-Ge	16.6	−5.2	−5.5	−69.4	−18.3	−18.5	−93.0
p-Ge	1.1	−3.7	3.2	48.4	0.2	−5.8	65.0
p-Ge	15.1	−10.6	5.0	49.0	−7.5	6.0	66.0
n-Si	11.7	−102.2	53.4	−6.8	−72.6	86.4	−10.8
p-Si	7.8	6.6	−1.1	−69.0	10.5	2.7	109.0

same notation as employed in piezoelectricity, i.e., with the following index transformations: $11 \to 1$, $22 \to 2$, $33 \to 3$, $23 \to 4$, $13 \to 5$, and $12 \to 6$. Then we have

$$\frac{\Delta\rho_i}{\rho} = \sum_{j=1}^{6} \gamma_{ij}\epsilon_j = \sum_{j=1}^{6} \pi_{ij}\tau_j. \tag{8}$$

As an example, Table 7.1 shows the piezoresistive coefficients for silicon and germanium as first published by Smith in 1954 [16]. For these semiconductor materials, only three coefficients need be given because the ii components are the same for $i = 1, 2, 3$, the i, j components are the same for $i, j = 1, 2, 3$ and $i \neq j$, and the ii components are the same for $i = 4, 5, 6$.

Considering the piezoresistive effect only (and ignoring the dimensional change for now), one typically considers two types of conditions: hydrostatic pressure and uniaxial strain. In hydrostatic pressure, an isotropic material or cubic crystal can be assumed to strain so that $\epsilon_1 = \epsilon_2 = \epsilon_3 = \epsilon$. Then, the relative change in resistivity is the same in each principle direction:

$$\frac{\Delta\rho_i}{\rho} = \gamma_{11}\epsilon + \gamma_{12}\epsilon + \gamma_{13}\epsilon. \tag{9}$$

Recalling that $\gamma_{12} = \gamma_{13}$ for metals and cubic crystals, we can simplify this equation to

$$\frac{\Delta\rho_i}{\rho} = (\gamma_{11} + 2\gamma_{12})\epsilon. \tag{10}$$

The volumetric change of the material under hydrostatic pressure conditions is given by

$$\frac{\Delta U}{U} = (1 + \epsilon)^3 - 1. \tag{11}$$

Given that the strain is typically very small (no larger than 0.001), we may approximate (11) by

$$\frac{\Delta U}{U} = 3\epsilon = \chi \Delta P, \tag{12}$$

where χ is the compressibility of the material and ΔP is the gauge pressure. Finally, combining (10) and (12) we can determine the piezoresistive sensitivity of the material:

$$\psi = \frac{\Delta \rho_i}{\rho \Delta P} = \frac{\chi(\gamma_{11} + 2\gamma_{12})}{3}. \tag{13}$$

A similar consideration of uniaxial strain can be carried out. For metals and cubic crystals undergoing uniaxial strain we have

$$\begin{aligned} \epsilon_1 &= \epsilon, \\ \epsilon_2 &= -\nu\epsilon, \\ \epsilon_3 &= -\nu\epsilon, \end{aligned} \tag{14}$$

where ν is Poisson's ratio. Now there are two resistivity changes to consider: that in the direction of strain (longitudinal), and that in the directions normal to the strain (transverse). These are given by the following:

$$\begin{aligned} \frac{\Delta \rho_1}{\rho} &= \gamma_{11}\epsilon - \nu\gamma_{12}\epsilon - \nu\gamma_{13}\epsilon \\ &= (\gamma_{11} - 2\nu\gamma_{12})\epsilon, \end{aligned} \tag{15}$$

from which we define the longitudinal coefficient to be $\gamma_l = \gamma_{11} - 2\nu\gamma_{12}$, and

$$\begin{aligned} \frac{\Delta \rho_2}{\rho} &= -\nu\gamma_{22}\epsilon + \gamma_{12}\epsilon - \nu\gamma_{23}\epsilon \\ &= \left[\gamma_{12} - \nu(\gamma_{12} + \gamma_{11})\right]\epsilon \quad , \end{aligned} \tag{16}$$

from which we define the transverse elastoresistance coefficient to be $\gamma_t = [\gamma_{12} - \nu(\gamma_{11} + \gamma_{12})]$.

Table 7.2, with data taken from Belu-Marian et al. [12], shows the longitudinal and transverse coefficients for p and n type silicon and germanium with various crystal orientations. For metals, these same coefficients tend to vary from 0.1 to 3.8, a decrease in magnitude (and thus in piezoresistive sensitivity) by at least one order of magnitude. However, silicon and germanium exhibit a much greater sensitivity to temperature changes than do the metal strain gauges (by two to four orders of magnitude), so the needs of a specific application tend to dictate which material choice is best.

7.4.2 *Strain Gauge Structures*

The previous analysis has focused on the change in electrical resistivity in response to strain. In strain gauges, we must consider both of the mechanisms which effect

TABLE 7.2. Longitudinal and transverse elastoresistance constants for n and p germanium and silicon (from Belu-Marian et al. [12]).

Material	γ_l 110	γ_t 110	γ_l 100	γ_t 100	γ_l 111	γ_t 111	γ_l Random	γ_t Random
n-Ge	−96	40	−60	39	111	42	−87	40
p-Ge	63	−28	37	−27	74	50	58	−28
n-Si	−81	47	−108	40	−72	50	−87	46
p-Si	120	−54	77	−51	135	−54	111	−53

resistance, namely, the dimensional change and piezoresistivity. In metals, the dimensional change typically dominates, while in semiconductors, the piezoresistive effect is dominant. In what follows we consider a general material undergoing a strain and examine both effects. We use the gauge factor, k_g, as the figure of merit for the transduction process. The gauge factor is defined as

$$k_g = \frac{\text{fractional increase in resistance}}{\text{fractional increase in length}}. \tag{17}$$

Consider a material of length l and cross-sectional area A, as shown in Fig. 7.15. We stretch the material so that the strain along its axis, ϵ_1, is positive. Then due to Hooke's law interactions, the strains in the orthogonal directions will be negative. Equation (14) provides the strain relations. The electrical resistance of the material, R, is given by

$$R = \frac{\rho l}{A}. \tag{18}$$

From this we can determine dR/dl

$$\frac{dR}{dl} = \frac{\rho}{A} - \frac{\rho l}{A^2}\frac{\partial A}{\partial l} + \frac{l}{A}\frac{\partial \rho}{\partial l}. \tag{19}$$

We multiply this equation by the quantity dl/R to get

$$\frac{dR}{R} = \frac{dl}{l} - \frac{dA}{A} + \frac{d\rho}{\rho}. \tag{20}$$

The first term on the right-hand side of (20) is the strain in the axial direction, and the third is the piezoresistivity contribution which may be found directly from (15). It remains then to manipulate the middle term on the right-hand side of the equation.

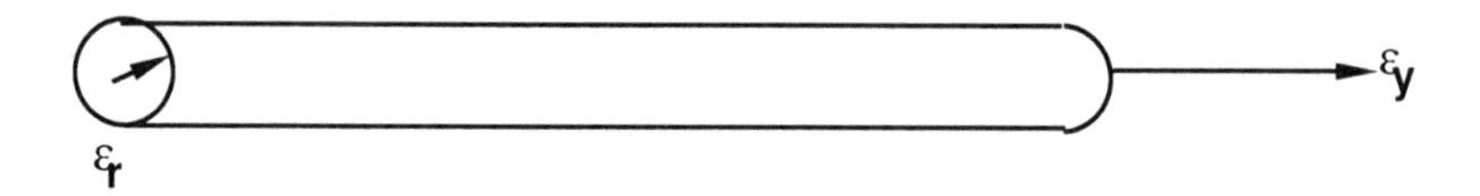

FIGURE 7.15. Cylindrical material under uniaxial strain.

The fractional change in cross-sectional area can be written as

$$\frac{dA}{A} = (1 - \epsilon\nu)^2 - 1. \tag{21}$$

We use a linear approximation to this equation, valid because the strain is very small. This yields

$$\frac{dA}{A} = -2\epsilon\nu. \tag{22}$$

Substituting this result into Eq (20) we have

$$\frac{dR}{R} = (1 + 2\nu)\frac{dl}{l} + \frac{d\rho}{\rho}. \tag{23}$$

Finally, we use the definition of the gauge factor, and (15), to produce

$$k_g = (1 + 2\nu) + \gamma_l. \tag{24}$$

The first term in (24) is that associated with the dimensional change of the conductor. The second term on the right-hand side is that associated with piezoresistivity. For metals, where the second term varies from 0.1 to 3.8, the gauge factor ranges from 1.7 to 5.4. For silicon and germanium, which are highly piezoresistive, k_g varies between 70 and 135. For cermets, which are ceramic–metal mixtures that can be fashioned into thin films, the gauge factors vary between 5 and 50.

Piezoresistive strain gauges made of silicon have a number of advantages when compared to metal strain gauges relying solely on changes in electrical resistivity due to dimensional variations. The semiconductor strain gauges are higher sensitivity, as the gauge factors indicate. Further, they tend to be smaller with typical dimensions of 0.5 mm length and 0.25 mm width. They have a very high fatigue life, with greater than 10^7 cycles producing no damage in the gauge. Semiconductor strain gauges also exhibit low hysteresis.

One of the main problems in using changes in the electrical resistance of a material as a transduction mechanism, is that such transduction tends to be sensitive to temperature changes. This is particularly true for single crystals of semiconductor materials, and was a key motivator for the development of polycrystalline semiconductor strain gauges. We typically compensate for these temperature effects by using a reference gauge which is exposed to the same temperature but not the applied mechanical strain.

Figures 7.16 to 7.18 show three examples of gauges to measure longitudinal strain. The first is a wire-type strain gauge using a metal wire bonded to a polymer or glass. The substrate is attached to a structure for measurement of the strain. Such strain gauges are relatively large, relatively stiff, and have high transverse sensitivity. Figure 7.17 shows an almost identical strain gauge but made of a metal foil and sold by BLH Electronics, Inc. [17]. This strain gauge, compared to the wire type, is more flexible, has better heat dissipation, and has lower transverse sensitivity. Figure 7.18 shows a single crystal semiconductor strain gauge on a silicon substrate. This gauge is more sensitive than the metal foil and wire types,

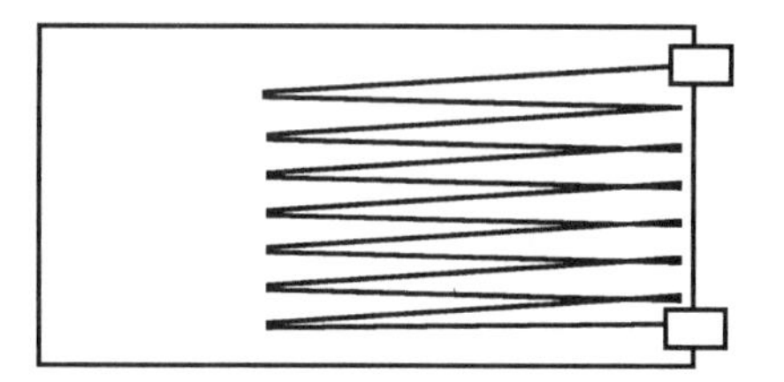

FIGURE 7.16. Wire strain gauge.

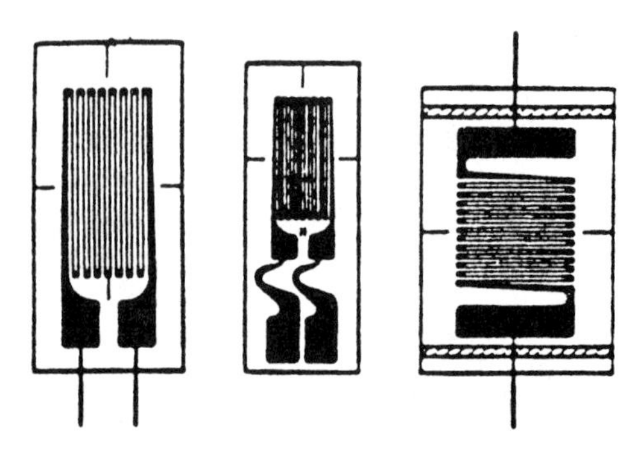

FIGURE 7.17. Metal foil strain gauge sold by BLH Electronics, Inc. [17].

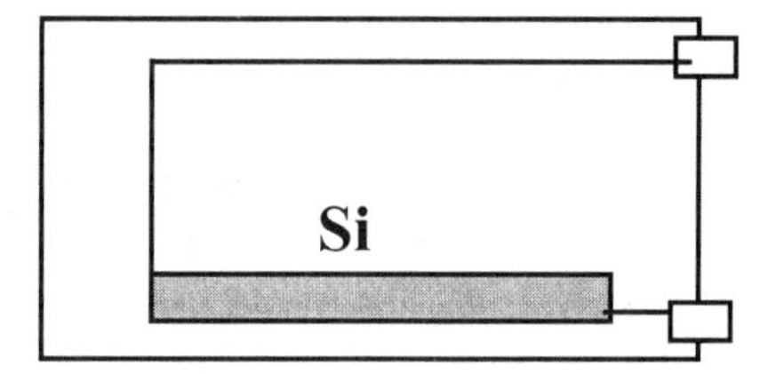

FIGURE 7.18. Semiconductor strain gauge on a silicon substrate.

and has better fatigue life, but has a greater sensitivity to temperature changes, is stiffer, and generally is more expensive.

Strain gauges are also available for the measurement of rotational strain or the simultaneous measurement of strains in multiple directions. Figures 7.19 and 7.20 are examples of this technology which are sold by BLH Electronics, Inc. [17].

Strain gauges are often used as the transduction element in tranducers not designed for the direct measurement of strain. For example, Fig. 7.21 shows a resistive load cell pictured in Doebelin [9]. In this device there are four strain gauges used. Gauges 1 and 2 provide output but are temperature dependent. Gauges 3 and 4 are placed skewed to the other two gauges at the angle at which there is no strain $\tan(\theta) = 1/\nu$. These strain gauges are used to compensate for thermal effects.

Figure 7.22 shows another load cell arrangement which uses strain gauges. In this sensor, sold by Entran Sensors and Electronics [18], a load button sits on a diaphragm which has four bonded strain gauges. A load applied to the button displaces the diaphragm, causing the strain gauges to strain.

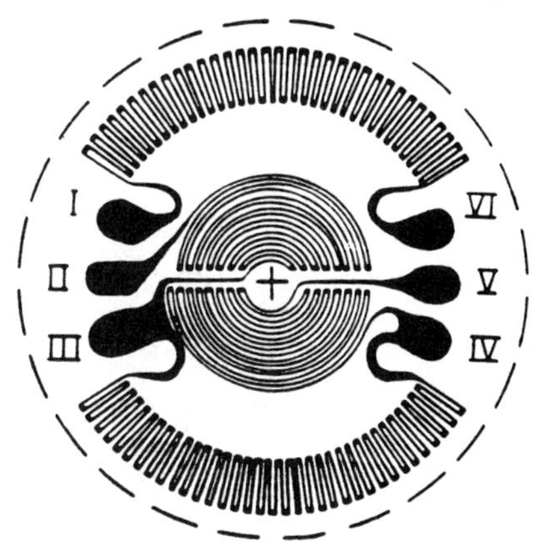

FIGURE 7.19. Rotational strain gauge sold by BLH Electronics, Inc. [17].

FIGURE 7.20. Three-directional strain measurement with a single gauge sold by BLH Electronics, Inc. [17].

Figure 7.23 shows an example of a strain gauge incorporation into a bolt torque transducer. In this sensor, sold by Advanced Mechanical Technology, Inc. [19], tightening the nut results in strain of an S-structure onto which strain gauges have been mounted. An additional S-structure is symmetrically placed in the system in order to maintain balanced forces.

A similar approach to the problem which eliminates the S-structure is a load bolt. Figure 7.24 shows a load bolt in which a strain gauge is embedded in the interior of a bolt. This device is sold by Omega Engineering, Inc. [20] and is a particularly good solution to the problem of sensing joint forces.

Figure 7.25 shows a silicon micromachined piezoresistive pressure sensor sold by Silicon Microstructures Inc. [21]. This device is 2mm × 2mm × 0.9 mm, operates with 10 V maximum excitation, and is available in versions with full scale ranges from 5 psi to 300 psi. It incorporates silicon resistors diffused into a substrate, much as shown schematically in Fig. 7.26 from Angell et al. [22]. Another silicon-based pressure sensor which uses strain gauges is sold by Omega Engineering [20]. In this device, the normal metal diaphragm to which strain gauges

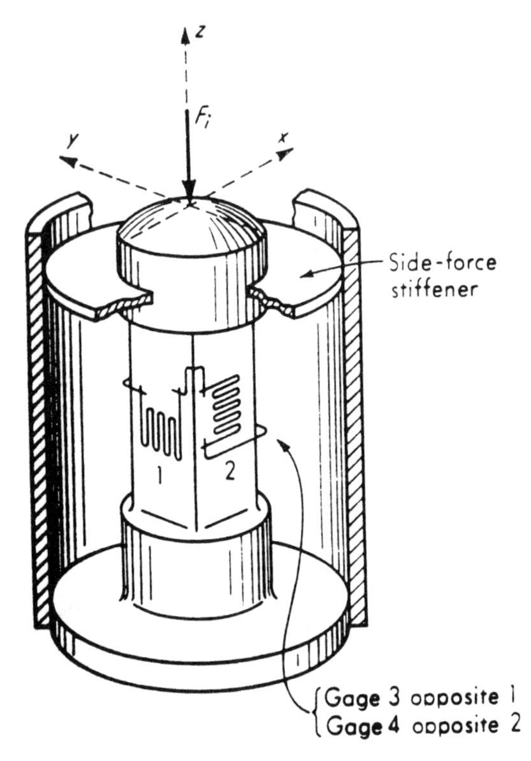

FIGURE 7.21. Resistive load cell (from Doebelin [9]) employing four strain gauges.

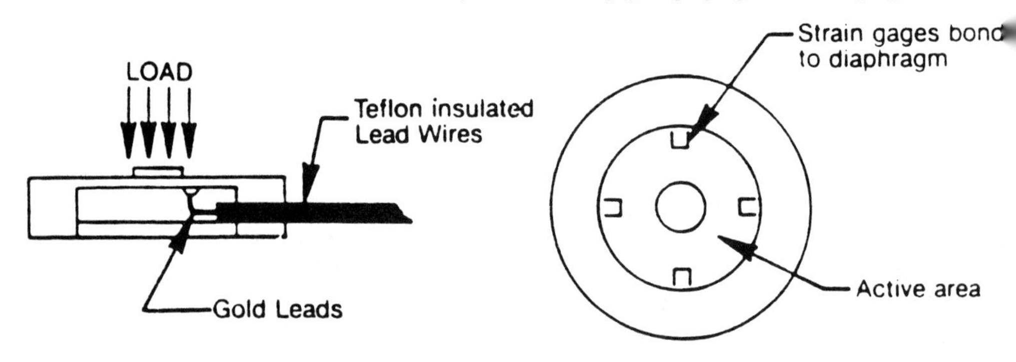

FIGURE 7.22. Load cell sold by Entran Sensors and Electronics [18].

are bonded is replaced by a circular silicon diaphragm into which piezoresistive stain gauges have been diffused. Yet another pressure sensor which exploits strain gauges, and is of more conventional size and design is shown in Fig. 7.27, taken from Fraden [8]. Here a metal corrugated diaphragm with attached plunger converts pressure into the linear deflection of a cantilever beam. The beam has strain gauges on it, so deflection produces an electrical output.

One final strain gauge example, is an accelerometer sold by Entran Sensors and Electronics [18] and pictured in Fig. 7.28. In this sensor, semiconductor strain

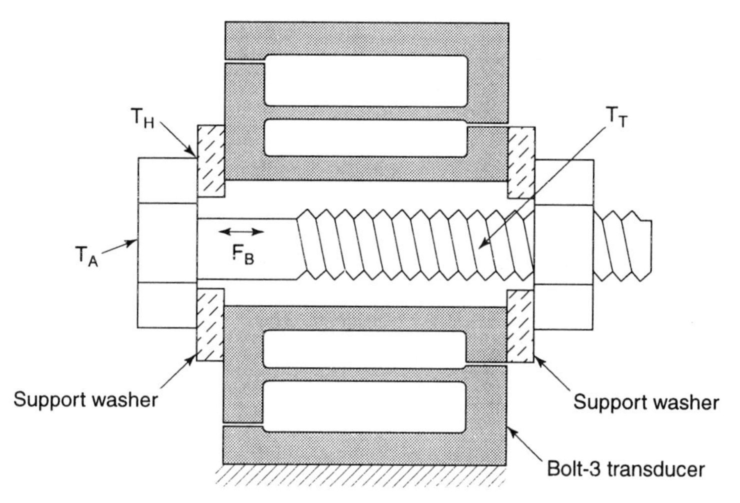

FIGURE 7.23. Bolt torque transducer made by Advanced Mechanical Technology, Inc. [19].

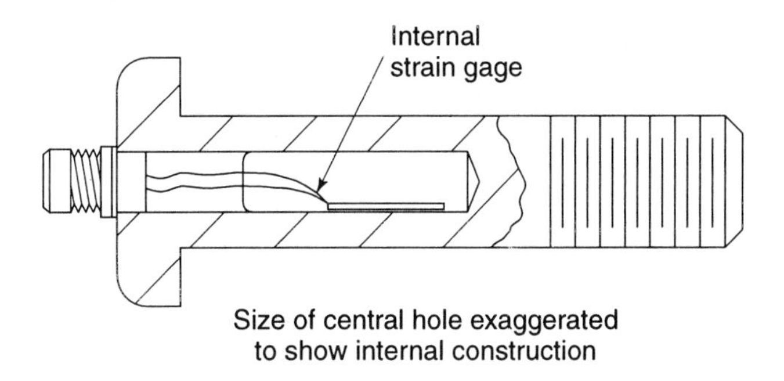

FIGURE 7.24. Load bolt sensor made by Omega Engineering, Inc. [20]. Copyright 1995 Omega Engineering, Inc. All rights reserved. Reproduced with the permission of Omega Engineering, Inc., Stmaford, CT 06907.

gauges are bonded to a cantilever beam which has a seismic mass at its free end. Acceleration of the case causes relative motion between the clamped end of the beam and the end with the mass, thus flexing the beam and straining the gauges.

7.4.3 Electrical Operation

In normal operation of a piezoresistive sensor, the sensor gauge element is one of the four resistances used to form a Wheatstone bridge. A Wheatstone bridge is shown in Fig. 7.29. For this sensor, an excitation voltage is supplied and a voltage difference across the two arms of the bridge measured. Assuming that three of the four resistors are fixed, any change in voltage difference can be linearly related to the change in the resistance of the gauge.

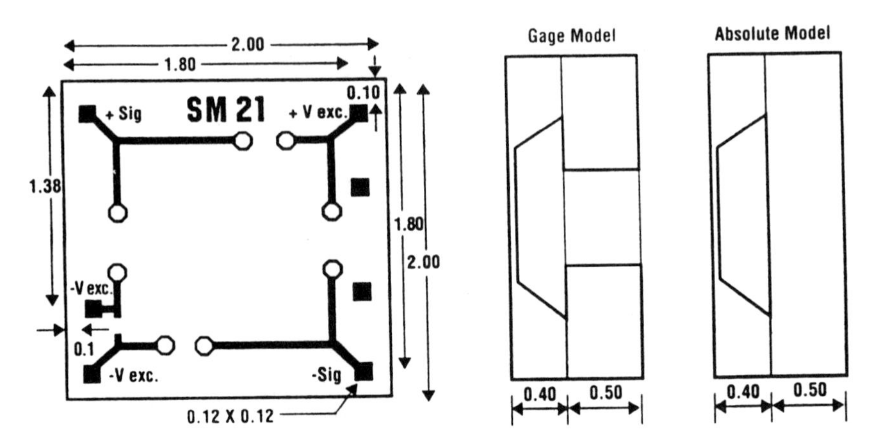

FIGURE 7.25. Piezoresistive pressure sensor made using silicon micromachining and sold by Silicon Microstructures Inc. [21].

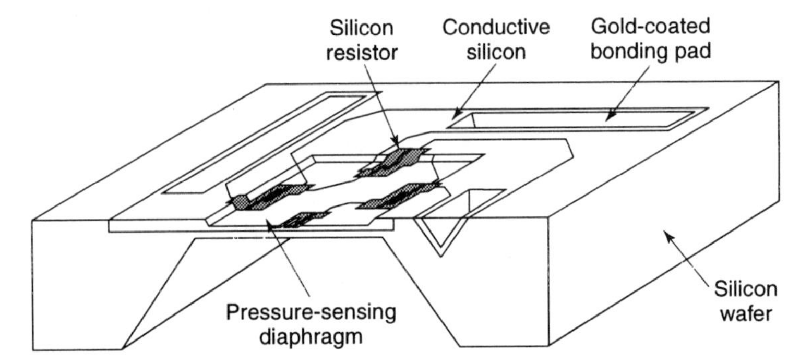

FIGURE 7.26. Piezoresistive pressure sensor schematic showing diffused silicon resistors. Figure taken from Angell et al. [22] .

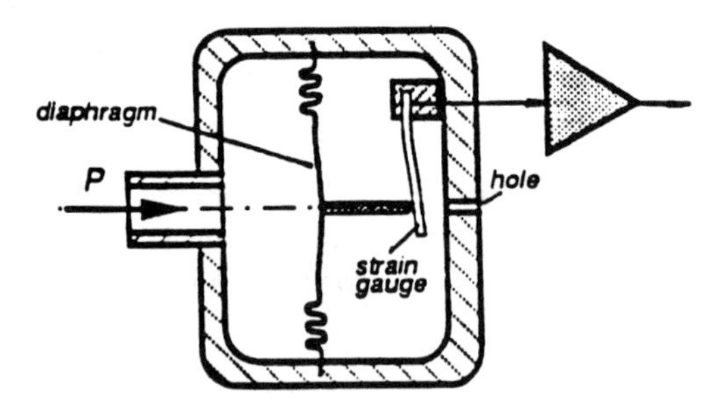

FIGURE 7.27. Conventional piezoresistive pressure sensor. Figure taken from Fraden [8].

The basic bridge configuration is useful for two reasons: it produces a voltage difference linearly proportional to the resistance change of the gauge, and it can self-compensate for temperature changes. This latter advantage assumes that the temperature coefficient of resistance is the same for all of the members of the bridge,

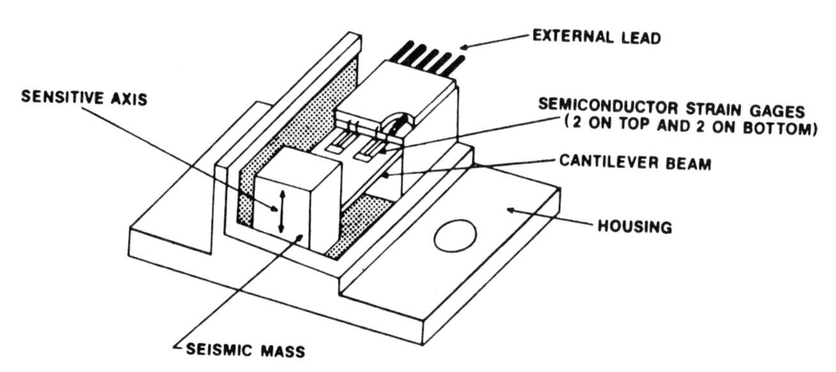

FIGURE 7.28. Piezoresistive accelerometer sold by Entran Sensors and Electronics [18].

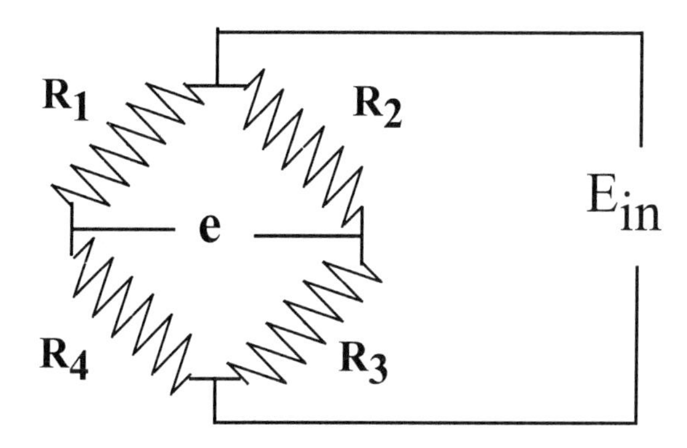

FIGURE 7.29. Wheatstone bridge. The smallest resistance in the bridge usually is the gauge.

so that their relative resistances are independent of temperature. In practice, it is not easy to achieve this ideal situation.

There are a number of variations of the basic Wheatstone bridge using a single strain gauge. Each offers advantages when operated in a particular mode (e.g., uniaxial tension). Many of these alternative electrical configurations are described in Belu-Marian et al. [12].

An alternative to a Wheatstone bridge is a simple voltage divider, shown in Fig. 7.30. In this voltage divider a voltage is supplied across a fixed resistance and a variable resistance which are in series. The transducer, which could be any resistance which varies with the desired measurand, is the variable resistance, R_{var}. The output voltage is

$$e_{\text{out}} = e_{\text{source}} \left(\frac{R}{R + R_{\text{var}}} \right). \tag{25}$$

Note that the voltage divider approach does not produce an output signal which varies linearly with the varying resistance. However, it is a perfectly reasonable approach to resistance detection when accuracy is less of an issue than cost.

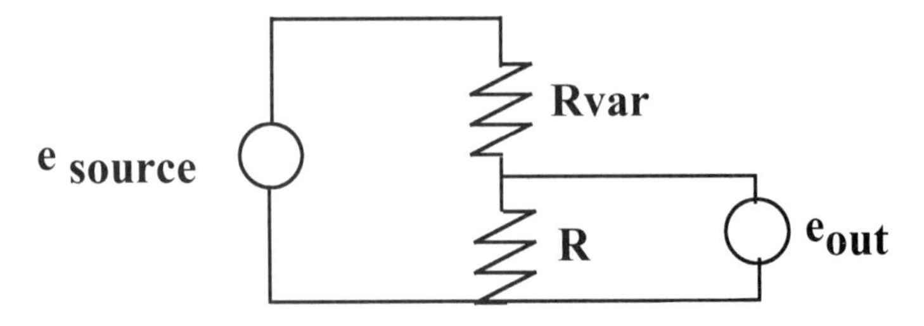

FIGURE 7.30. Voltage divider for strain gauge measurement.

7.5 Thermoresistivity

Thermoresistivity refers to the variation of electrical resistivity as a function of temperature. Historically, thermoresistivity was first noted in 1821 by Sir Humphry Davy who found that the resistance of metals depends on temperature. In 1871, Sir William Siemens described a platinum thermometer exploiting this temperature variation of resistivity, but it was not until 1887 that Hugh Callendar presented means of practically using platinum resistance thermometry. Since then, the field of thermoresistivity has progressed rapidly, as alternative materials to the very expensive platinum were found. These alternatives have included silicon and metal oxides.

To understand thermoresistive sensors, we must first consider that the electrical resistivity of virtually all materials is a function of temperature. Debye theory (see Popescu and Popescu [23]) models the resistivity as a function of T^n, with $n \approx 5$ for temperatures below the Debye temperature, and as a function of T^1 above the Debye temperature. The Debye temperature is given as $h\nu_{\max}/k$, where h is Planck's constant, $\nu_{\max}$ is the maximum vibrational frequency of the material, and k is Boltzmann's constant. For most metals, the Debye temperature is about 300 K, so the resistivity can be modeled as a linear function of temperature above room temperature. In this regime, it is useful to define a temperature coefficient of resistivity (TCR), α, which is approximately constant and given by

$$\alpha = \frac{1}{\rho}\frac{d\rho}{dT}. \tag{26}$$

Table 7.3, with data taken from Fraden [8], shows α for a variety of metals. Note that although most metals have a positive TCR, it is possible for α to be negative (e.g., carbon), indicating that the resistivity *decreases* with increasing temperature. Note also that Table 7.3 shows that α varies over at least two orders of magnitude for the metals shown. Clearly, it is those materials with higher values of α which are best suited for exploitation in temperature sensors.

Of the metals, platinum has been the most popular choice for resistive temperature sensing. This choice is prompted by its wide operating temperature (roughly $-220°$C to $1050°$C), its resistance to oxidation and corrosion, and its stability.

Silicon and other semiconductors have also been exploited in thermoresistive temperature sensing. Generally bulk silicon is used, with the resistivity and α

TABLE 7.3. Resistivity and temperature coefficient of resistivity (α) for various metals. Data taken from Fraden [8].

Material	ρ 10^{-8} Ω-m	α 10^{-3}/K
Aluminum	2.65	3.9
Beryllium	4.0	0.025
Brass	7.2	2.0
Carbon	3500	−0.5
Constantan	52.5	0.01
Copper	1.678	3.9
Gold	2.24	3.4
Iron	9.71	6.5
Lead	20.6	3.36
Manganin	44	0.01
Mercury	96	0.89
Nichrome	100	0.4
Nickel	6.8	6.9
Palladium	10.54	3.7
Platinum	10.42	3.7
Silver	1.6	6.1
Tantalum	12.45	3.8
Tin	11.0	4.7
Tungsten	5.6	4.5
Zinc	5.9	4.2

controlled, within limits, through the judicious use of dopants. Qualitatively, there is little difference between the use of silicon and the metals described in Table 7.3.

The materials which have shown the largest resistance change with temperature are oxides of nickel, manganese, cobalt, iron, and titanium. Typically these materials exhibit a nonlinear relationship between temperature and resistance which is modeled as follows:

$$R = R_0 e^{\beta(1/T - 1/T_0)}, \tag{27}$$

where β is some characteristic temperature of the material (usually in the range of 3000 to 5000 K) and T_0 is the temperature at which the resistance is found to be R_0. Sensors made from these materials are referred to as thermistors. We distinguish them from other thermoresistive materials by referring to transducers made using metals or semiconductors as resistive thermal detectors, RTDs. Note that (27) indicates that for thermistor materials, the resistance decreases as the temperature increases. Thus, for thermistors, the equivalent α is larger than that for RTDs and negative (although some positive temperature coefficient thermistors are now available). This is demonstrated in Fig. 7.31, taken from Thermometrics [24], which shows the resistance change as a function of temperature for a variety of thermistors and a platinum RTD. The advantage of thermistors relative to RTDs is

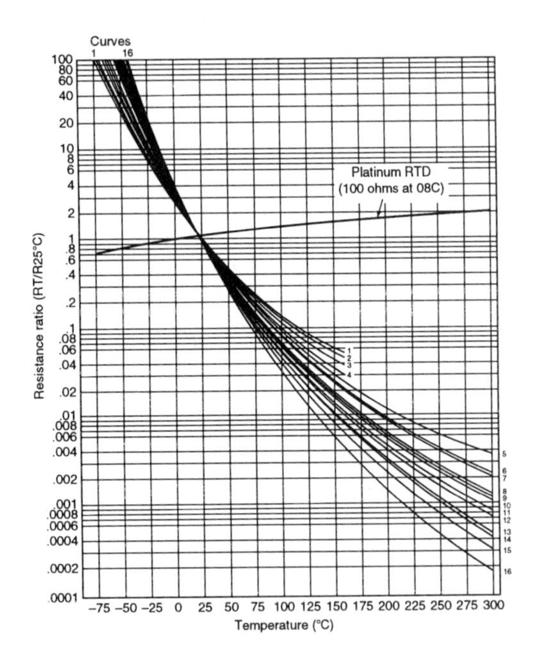

FIGURE 7.31. Resistance-temperature characteristics of thermistors compared to a platinum RTD (from Thermometrics [24]).

that they are generally faster, smaller, less expensive, and more sensitive. However, they are far less linear and have a narrower operating range of temperatures.

Before proceeding to discuss examples of the uses of thermoresistive sensors, let us consider how one would model thermoresistivity using bond graph techniques. Clearly, thermoresistive materials are those in which the electrical resistance is a strong function of temperature. Hence, it must be modeled with a multiport resistance with one electrical port and at least one thermal port. Indeed, (27) and (26) may be thought of as different forms of constitutive equations for the multiport resistance which describes thermoresistivity.

Thermistors and RTDs are widely available for the measurement of temperature, and gas or liquid flow rates. Figure 7.32, taken from Figliola and Beasley [25] shows the construction of a conventional platinum RTD. Note that a mica cross form supplies structural support for a platinum coil. The form and coil are embedded in a pyrex tube.

Another example of a thermoresistive sensor is shown in Fig. 7.33. This device is a thin-film thermoresistive temperature sensor. It uses a meandering electrode of nickel and measures thermal changes as changes in resistance. It is described in more detail in Popescu and Popescu [26]. The advantages of thin-film thermoresistive sensors over other types of thermoresistive sensors are twofold: the

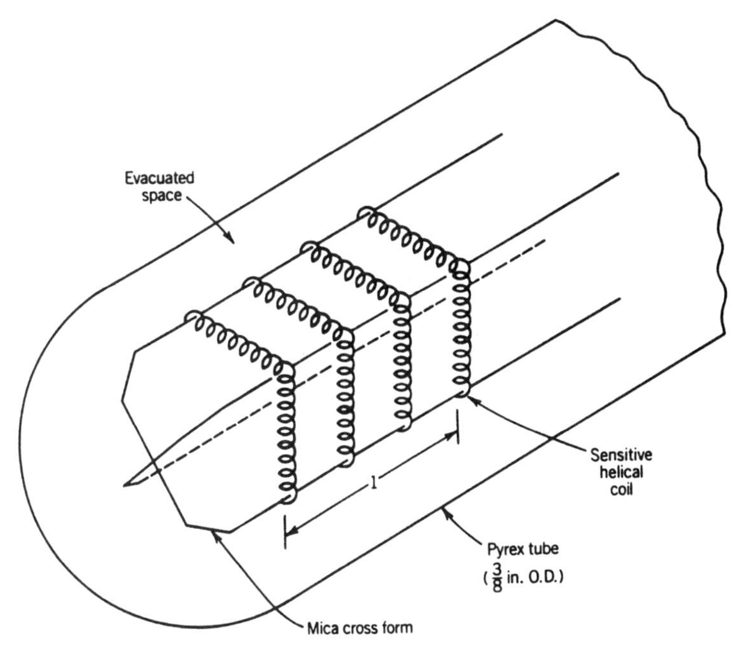

FIGURE 7.32. Schematic of a conventional platinum RTD described in Theory and Design for Mechanical Measurements, 2nd ed., R. S. Figliola and D. E. Beasley, 1995 [25]. Figure is reprinted by permission of John Wiley and Sons, Inc.

FIGURE 7.33. Thin-film thermoresistive sensor described in Popescu and Popescu [26].

electronics can be integrated with the sensor on a single chip, and the sensor can be made extremely small so its response time is relatively low.

Figure 7.34 shows a schematic of a thin platinum RTD manufactured on a chip and sold by Hy Cal Sensing Products [27]. Note the extremely small size of this device, which helps to give it a relatively small (fast) time constant of about one second on metal surfaces.

Figure 7.35 (from Dantec Measurement Technology [28]) shows as example of a thermistor used to determine flow rate. In such a device, usually called a hot-wire or hot-film anemometer (depending on the structure), the temperature changes are the result of heat lost by convection to the flowing fluid. These temperature changes

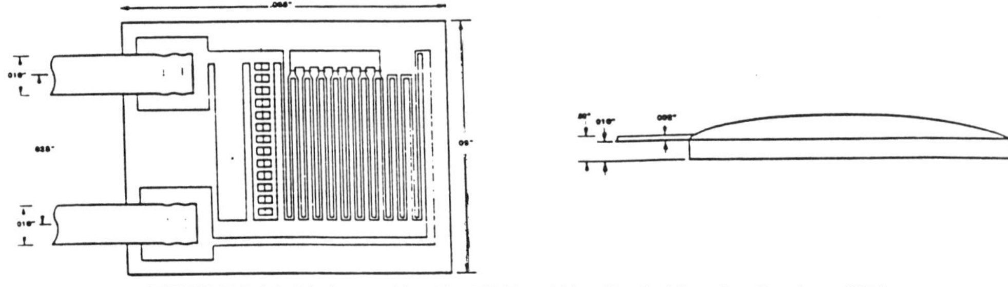

FIGURE 7.34. Platinum thin-film RTD sold by Hy Cal Sensing Products [27].

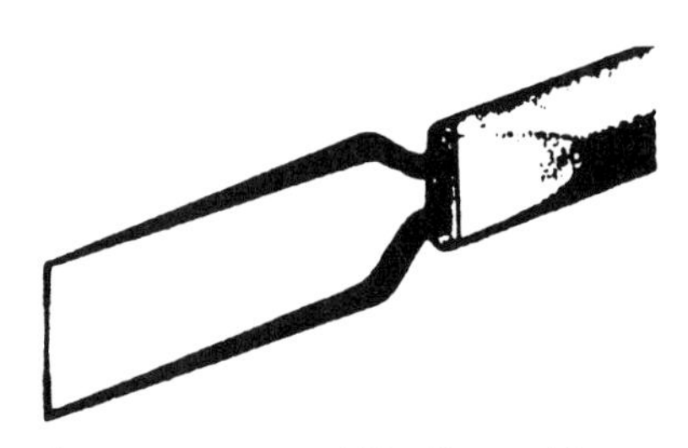

FIGURE 7.35. Hot wire anemometer sold by Dantec Measurement Technology [28].

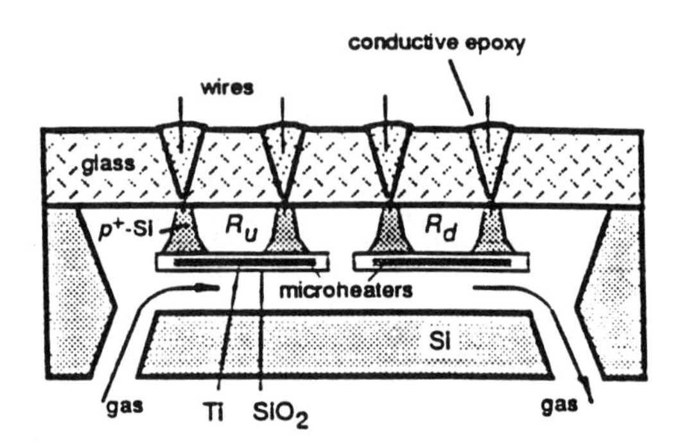

FIGURE 7.36. Microflow gas sensor using titanium RTDs. Figure is taken from Fraden [8].

can be sensed as resistance changes. (They can also be sensed by monitoring the power needed to maintain a given temperature.)

A number of solid state microflow sensors operate using thermoresistivity in a manner very similar to hot-wire anemometers. For instance, Fig. 7.36, taken from Fraden [8], shows a sensor with two self-heating titanium resistors. The difference in the resistances is related to the flow of gas through the channel. Quite similar approaches but different structural geometries have been described by Hocker et al. [29] (Fig. 7.37) and Betzner et al. [30] (Fig. 7.38).

As a final example, Fig. 7.39 shows an RTD system used to monitor liquid level. In this transducer, sold by Fluid Components International [31], RTDs are placed on two rods, one of which is close to a third rod with an embedded resistance heater. In the absense of a liquid, there is a relatively large temperature differential between the two RTDs. This translates to a significant resistance difference (and

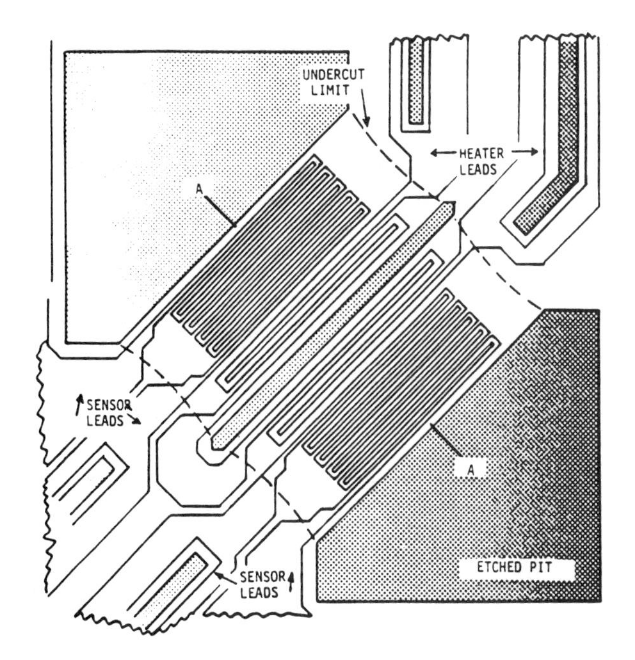

FIGURE 7.37. Microflow gas sensor described by Hocker et al. [29].

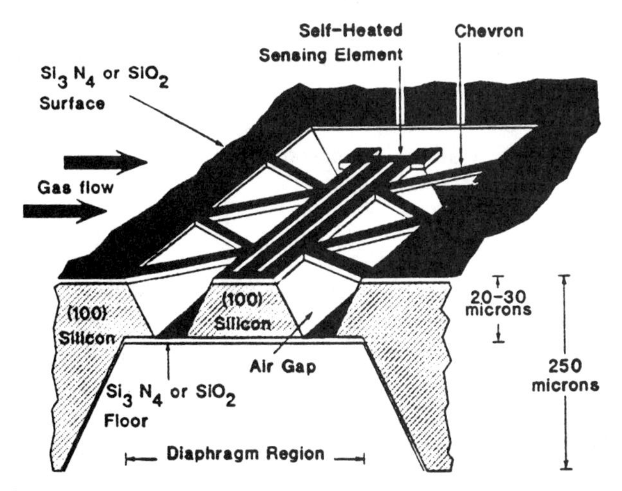

FIGURE 7.38. Microflow gas sensor described by Betzner et al. [30].

voltage difference) in the two RTD transducers. In the presence of liquid, this resistance difference drops dramatically, as conduction through the liquid tends to make the temperatures more uniform.

Thermistors and other thermoresistive sensors are operated in a manner quite similar to piezoresistive sensors. Usually the sensor is one arm of a Wheatstone

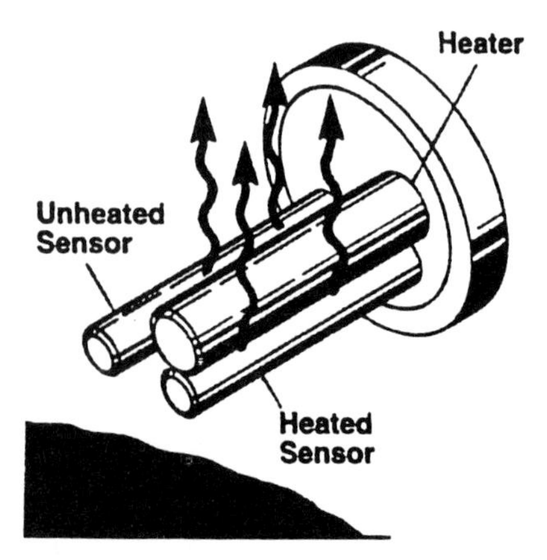

FIGURE 7.39. Liquid level sensor sold by Fluid Components International [31].

bridge network. Changes in the sensor cause a change in the voltage difference between the two sides of the bridge.

7.6 Thermoelectricity

Thermoelectricity is any phenomenon which involves the direct interchange of thermal and electrical energy. The most common thermoelectric effect is Joule heating in resistors. Three specific thermoelectric effects are reversible thermodynamically, and two of these are interesting from a transduction perspective. These thermoelectric effects are known by the names of their discoverers: the Seebeck effect, the Peltier effect, and the Thomson effect.

7.6.1 Seebeck Effect

The Seebeck effect is a thermoelectric coupling mechanism that is named for Thomas Johann Seebeck. In 1821 Seebeck observed a magnetic disturbance when he joined semicircular pieces of bismuth and copper. This magnetic disturbance was the result of the establishment of an electric field in response to the thermal gradient between the two junctions of the conductors. While A. C. Becquerel suggested using Seebeck's discovery in the development of a temperature sensor in 1826, it was not until almost 60 years later that Henry Le Chatelier produced the first thermoelectric sensor exploiting the effect [32]. These devices, which normally are referred to as thermocouples, are very common thermal sensors.

If a conductor has one hot end and one cold end, then there is heat flow from the hot end toward the cold. We have previously modeled this heat transfer using a 2-port thermal resistance. In many materials, the thermal gradient also establishes an electric field. The constitutive relation defining this generally is expressed as follows:

$$e = S_A dT, \tag{28}$$

where S_A is the absolute Seebeck coefficient of the material. We may view this relation as a constitutive relation between a temperature differential and a voltage. It is modeled using a 3-port resistance element. Two of the ports are thermal ports whose temperatures define the temperature difference. The third port is the electrical port.

Equation (28) suggests that we could exploit the absolute Seebeck effect in a pure material in order to determine a temperature difference. In fact, this is not true. The problem is that we would need to measure the potential difference between the two ends, and the act of making the measurement would either create a thermocouple, by adding a second material between the hot and cold regions, or would use the same conductor, in which case we can show that the potential difference would go to zero. Indeed, consider the materials shown in Fig. 7.40. Materials 1 and 2 are different. Junction A is at low temperature, T_A and junction B at high temperature, T_B. A *perfect* measuring device, with infinite impedance, is located in the loop with its contacts at ambient temperature, T_a. If we measure the potential starting with the ground shown, then traversing to junction A we have a voltage given by (28) with $S_A = S_1$ and dT given by the difference between T_a and T_A. Let us assume that S_1 is negative, so that the temperature drop causes a rise in the potential. This is shown in Fig. 7.41 and labeled as region 'i'.

Now imagine that we continue to traverse the circuit by moving along material 2 to junction B. From Eq. (28) we have a potential difference for this arm equal to $S_2(T_B - T_A)$. Let us suppose that S_2 is positive so that the potential rises as we traverse this arm. This is shown in Fig. 7.41 and labeled as region "ii". Finally,

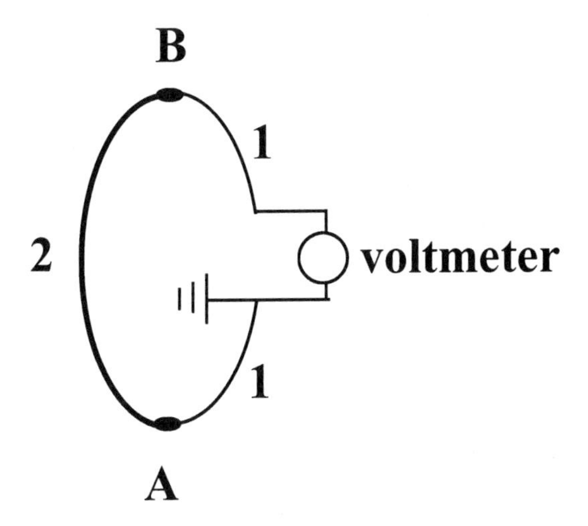

FIGURE 7.40. Structure of a thermocouple.

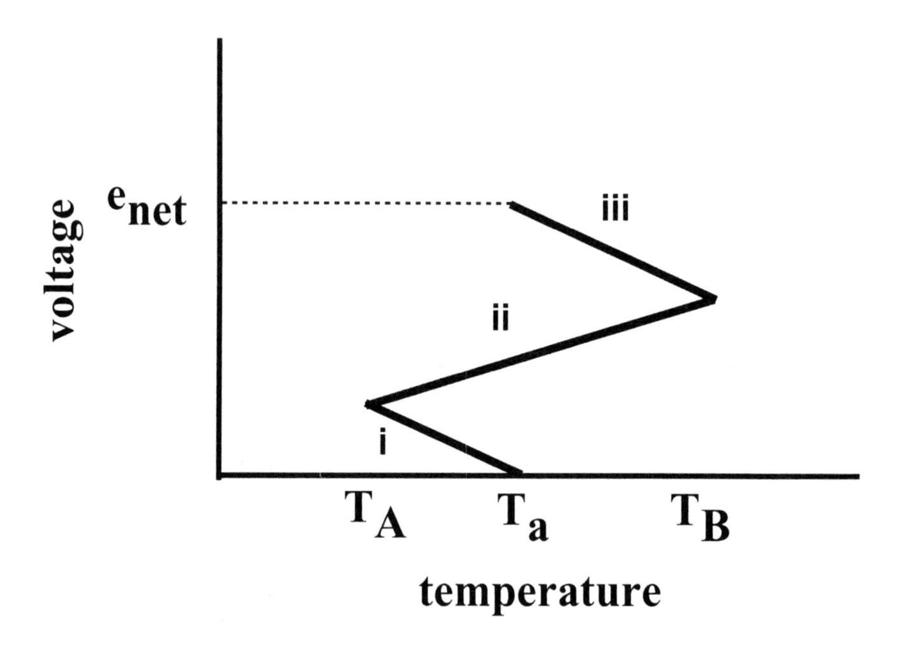

FIGURE 7.41. Voltage versus circuit position in a thermocouple.

we traverse back to the measuring device along material 1, producing another rise in potential, as shown in region "iii" in Fig. 7.41. Hence we measure a net rise of e_{net}, which is clearly related to the temperature difference at the junctions, but independent of the ambient temperature of the voltage measuring device.

From Fig. 7.41 it is clear that if the circuit were completed with a single material, the voltage would rise around half of the circuit and fall the same amount around the other half. The net result would be that the voltage monitor would see no voltage. Thus, while (28) is interesting, it is not useful directly in transduction. If, on the other hand, the materials are chosen as suggested above, with opposite signs for the absolute Seebeck coefficients, then the potential difference around the loop can be measurable. It is this need to couple two different materials that leads to the term *thermocouple*.

For two materials joined as shown in Fig. 7.40 the voltage drop between the junctions is given by

$$e = S_{12}\,dT = (S_1 - S_2)\,dT. \tag{29}$$

This expression shows that there is an advantage to using two materials with very different absolute Seebeck coefficients. The differential Seebeck coefficient, S_{12}, can be regarded as the sensitivity of the sensor. Figure 7.42, taken from Moffat [33], shows the Seebeck effect for various materials. The slopes of the lines define the absolute Seebeck coefficients. From this, it is clear that we might choose to

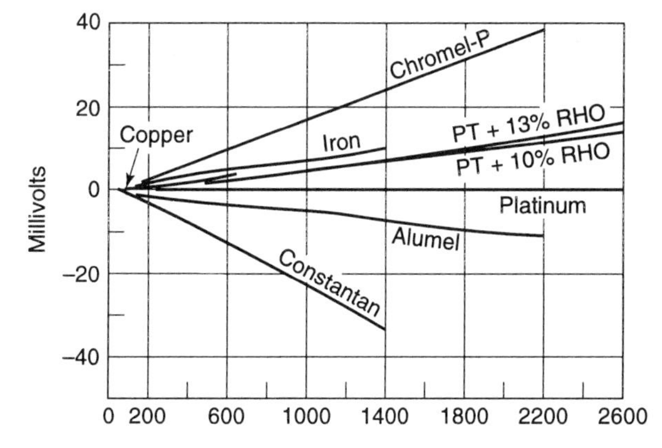

FIGURE 7.42. Seebeck effect in some materials. Figure taken from Moffat [33].

couple a material such as chromel with one such as constantan in order to achieve high sensitivity.

What is not revealed in (29) is that the absolute Seebeck coefficient, and thus the differential Seebeck coefficient, is typically a function of temperature. In commercial thermocouples the materials are chosen to produce high sensitivity and a fairly linear response as a function of temperature. Thus the S_{12} for a given pair of materials is typically modeled as

$$S_{12} = a_0 + a_1 T. \tag{30}$$

Table 7.4, with information taken from Fraden [8], shows the characteristics of common thermocouple materials. Figure 7.43 shows how the differential Seebeck coefficients vary with temperature for these common thermocouple types.

Thermocouples come in a large variety of shapes and sizes. For example, Fig. 7.44 shows a bare wire thermocouple element available from Nanmac Corp. [34]. Figure 7.45 shows a thermocouple using ribbons instead of wires, and Fig. 7.46 shows a trident style thermocouple using ribbons. Both of these sensors are available from Nanmac Corp. [34]. Still another type of thermocouple is shown in Fig. 7.47. This thermocouple uses thin foil attached to removable carrier material. It is sold by RdF Corp. [35]. The advantages of thermocouples compared

TABLE 7.4. Characteristics of various thermocouple types, from Fraden [8].

Materials	Sensitivity $\mu V/^{\circ}C$	Temperature range $^{\circ}C$	Designation
Copper/Constantan	40.9	−270 to 600	T
Iron/Constantan	51.7	−270 to 1000	J
Chromel/Alumel	40.6	−270 to 1300	K
Chromel/Constantan	60.9	−200 to 1000	E
Pt(10%)/Rh-Pt	6.0	0 to 1550	S
Pt(13%)/Rh-Pt	6.0	1 to 1600	R

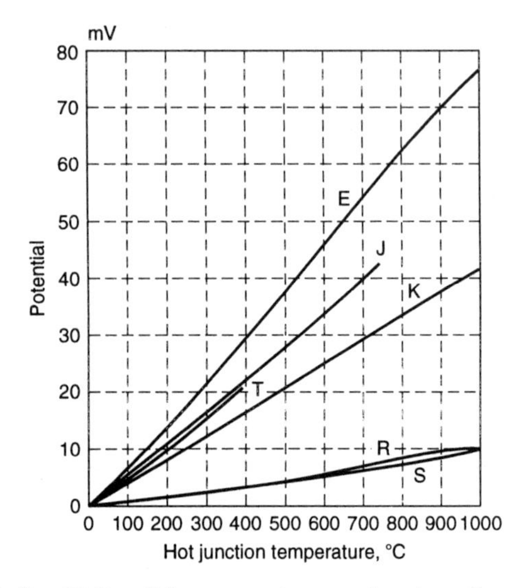

FIGURE 7.43. Sensitivity of thermocouples as a function of temperature. This figure is taken from Doebelin [9].

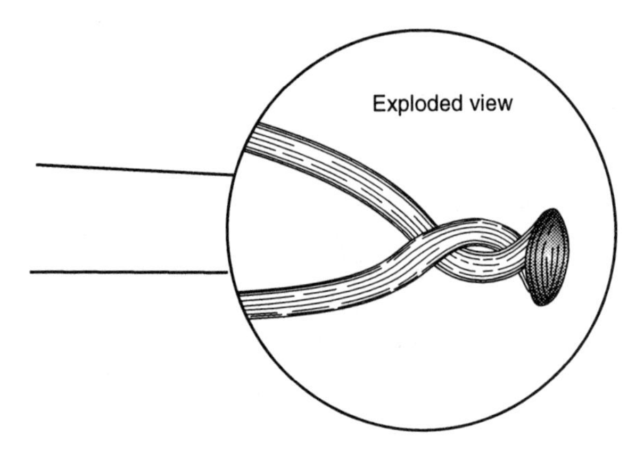

FIGURE 7.44. Thermocouple using twisted wires, sold by Nanmac Corp. [34].

to other temperature measuring sensors are their relatively high sensitivity and their potentially fast response times. As in most sensors, the system response time is highly dependent on the sensor size and material. The thermocouples shown here have response times in the 1–100 ms range.

While thermocouples used as temperature sensors are by far the most common devices exploiting the Seebeck effect, they are not the only transducers based on the Seebeck effect. One example of another transducer is the thermopile. The thermopile was invented by James Joule in an attempt to increase the output voltage of the thermocouple. It is essentially an array of thermocouples that are connected in series. An example of a thermopile, implemented on a solid-state sensor, is shown in Fig. 7.48, taken from van Herwaarden and Meijer [36].

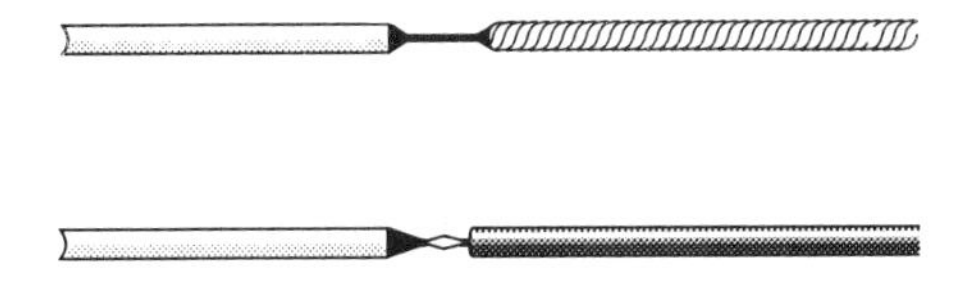

FIGURE 7.45. Thermocouple using ribbons sold by Nanmac Corp. [34].

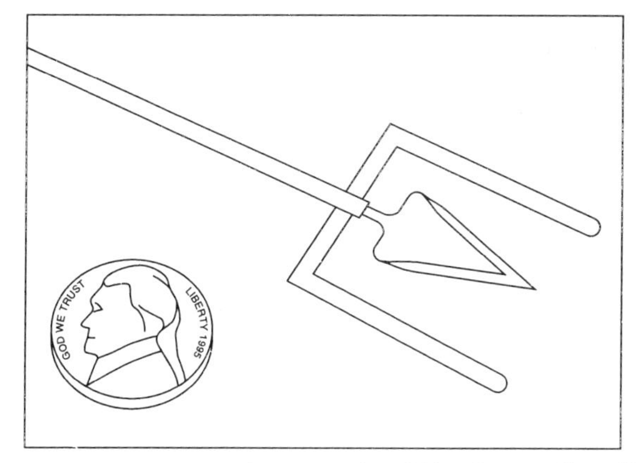

FIGURE 7.46. Trident thermocouple sold by Nanmac Corp. [34].

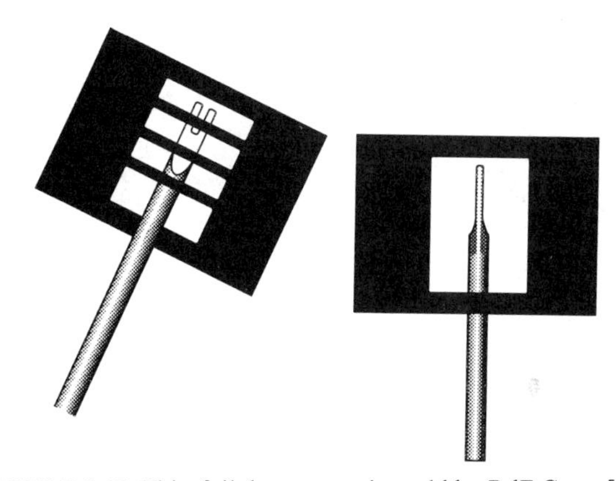

FIGURE 7.47. Thin-foil thermocouples sold by RdF Corp. [35].

Another use of thermocouples is obtained by combining thermocouples at different locations of a structure to monitor heat flux. Figure 7.49 shows a heat flux sensor created by RdF Corp. [35] in this manner.

Yet another application of thermocouples is in solid-state sensors of gas flow. For example, Fig. 7.50 shows a gas flow sensor incorporating thermopiles described by Robadey et al. [36]. This device is able to measure both flow velocity and direction, by electronically comparing the output signals of the four thermocouples.

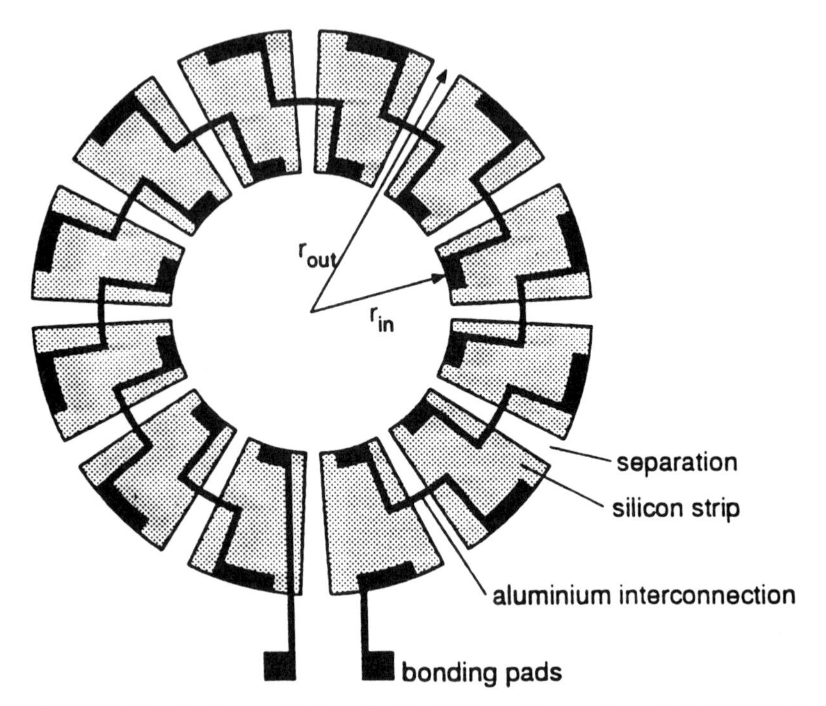

FIGURE 7.48. Circular thermopile described in A. W. van Herwaarden and G. C. M. Meijer, *Thermal sensors*, in S. M. Sze (ed), **Semiconductor Sensors**, (Wiley, NY, 1994). Reprinted by permission of John Wiley and Sons, Inc.

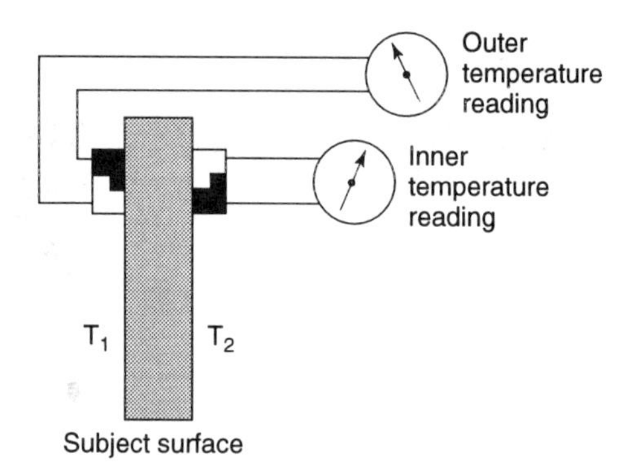

FIGURE 7.49. Heat flux sensor using a pair of thermocouples. This sensor is sold by RdF [34].

7.6.2 Peltier and Thomson Effects

The Peltier effect was discovered by Jean C. A. Peltier in 1834, and relates the heat generated or absorbed at the junction between two dissimilar conductors to the current going through the junction. Specifically, it is described by the following

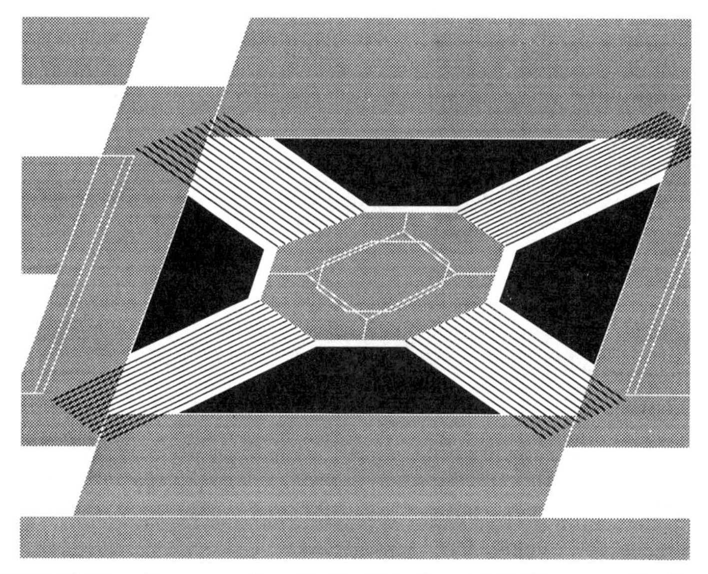

FIGURE 7.50. Microflow sensor described by Robadey et al. [36] incorporating thermopiles.

relation:

$$\dot{Q}_{12} = \Pi_{12} i, \tag{31}$$

where $\dot{Q}_{12}$ is the rate of heat absorption at the junction between materials 1 and 2, Π_{12} is the Peltier coefficient, and i is the current from material 1 to 2. The Peltier coefficient is a function only of the materials at the junction and the junction temperature.

The implications of (31) were not immediately obvious to Peltier and others who studied this phenomenon. It was Lenz, in 1838, who demonstrated that, using the Peltier effect, water could be frozen at a bismuth–antimony junction. He also showed that by reversing the direction of the current, the ice could be melted.

The Peltier effect has been used to create small-scale refrigerators for cooling localized regions. Because the Peltier effect only requires a small junction between two dissimilar conductors, the construction of such refrigerators is a relatively simple matter.

Subsequent to the discovery of the Peltier and Seebeck effects, William Thomson (a.k.a. Lord Kelvin) explained their relationship to each other and postulated the existence of a third thermoelectric effect which now bears his name. The Thomson effect is a reversible transformation of electrical energy and heat due to a finite temperature gradient. It is described by

$$\dot{Q} = N i \Delta T, \tag{32}$$

where N is the Thomson coefficient in a conductor with current flow i and with a temperature difference ΔT in the x-direction. Unlike the Peltier and Seebeck effects, the Thomson effect is a thermoelectric mechanism present in a pure material, and N is a function of that material and the average temperature.

By invoking the first and second laws of thermodynamics, Thomson was able to relate the three thermoelectric effects mentioned in this section. While it is beyond the scope of this book to present his derivations in detail, readers may find the straightforward analysis in books on thermoelectric effects such as by Fuschillo [37] and Jaumot [38]. What is important from the perspective of transduction is simply the resulting relationship between the Peltier and Seebeck coefficients of a material. Thomson's analysis showed that

$$\Pi_A = S_A T, \tag{33}$$

which permits definition of an absolute Peltier coefficient for a material, Π_A in terms of its Seebeck coefficient. The Peltier coefficient in (31), Π_{12}, is the difference in the Peltier coefficients for materials 1 and 2.

7.7 Magnetoresistivity

In the previous two sections of this chapter, we have discussed transducers based on changes in resistance due to variations in mechanical variables—namely, pressure, temperature, and heat flow. In addition to variations with mechanical variables, the resistivity of some materials is affected by the presense of a magnetic field. This effect is referred to as magnetoresistivity, and forms the basis for most magnetic disk heads.

Magnetoresistivity, because it links magnetic and electric energy domains, is not strictly an electromechanical transduction mechanism. However, because it is often used to sense the position of a magnetic field, it can be used to link mechanical and electrical variables in much the same manner that the Hall effect is used. Indeed, the Hall effect is sometimes taken to be one particular example of magnetoresistivity. Here, we will briefly examine magnetoresistivity.

In general there are three galvanomagnetic effects that can occur when a current-carrying conductor is exposed to a magnetic field. First, we can observe changes in the resistivity of some ferromagnetic materials in the presense of a magnetic induction field oriented parallel to the direction of current flow. This is referred to as the longitudinal magnetoresistance effect. Second, if the magnetic induction field and current flow direction are orthogonal, some materials exhibit significant changes in resistivity with field strength as measured in the direction of current flow. This is called the transverse magnetoresistance effect. Third, if the magnetic induction field and current flow are orthogonal, there may be important changes in the resistivity in the third orthogonal direction. This is referred to as the Hall effect and has been described previously.

In general, the longitudinal resistivity of nonmagnetic materials is observed to increase quadratically with the magnetic induction field strength for small fields, and as a linear function of B for very high fields. Note that this distinguishes magnetoresistivity from the Hall effect, in which there is a linear dependence on the magnetic induction field strength. It is the small magnetic induction field strengths that tend to be of greatest interest in sensing based on magnetoresistivity, and for small fields the model used is as follows:

$$\frac{\rho(B) - \rho(0)}{\rho(0)} = \frac{\Delta\rho}{\rho_0} = AB^2, \tag{34}$$

where A is the coefficient of magnetoresistivity, $\rho(0) = \rho_0$ is the resistivity in the absense of a magnetic induction field, and $\rho(B)$ is the resistivity in the presense of a magnetic induction field. This model is in good agreement with free electron theory [40] which says that

$$\frac{\Delta\rho}{\rho(B)} = \frac{AB^2}{1 + CB^2}, \tag{35}$$

where C is the square of the electron mobility in the material. Note that for small field strengths, the right-hand side of (35) reduces to (34).

The most common use of magnetoresistivity today is in the read-back heads in magnetic disks. In this application, a magnetoresistive sensor is moved in close proximity to a disk in which information is embedded in the form of magnetic transitions in the medium. These transitions correspond to changes in the magnetic induction as a function of position, and thus a change in resistivity (and resistance) as the disk head moves. Magnetoresistive sensors are ideal for this application because they can be quite small, have high sensitivity, be low in cost, and be immune to electromagnetic noise, dirt, and light.

In addition to magnetic disk heads, magnetoresistive sensors can be used for virtually any application in which Hall effect sensors are employed. In essense, it is simply a question of changing the direction in which the resistance is sensed. Operating with a magnetoresistive sensor rather than a Hall effect sensor, means trading higher sensitivity for poorer linearity.

7.8 Shape Memory Alloys in Transduction

The previous section describing magnetoresistivity dealt with a transduction mechanism that is electromagnetic and not truly electromechanical, but which has been used so as to link mechanical and electrical behavior. It was presented because the distinction between truly *electromechanical* transduction mechanisms and those which merely act as though they are electromechanical is so trivial that to treat them differently only causes confusion. Another mechanism which is not truly electromechanical but which is coming into vogue in electromechanical transduction involves shape memory alloys. In this section the behavior of shape memory alloys is explained and applications (primarily in actuation) are presented.

Shape memory alloys are metals which undergo a thermally induced phase transition. At low temperatures, the material is in a martensitic phase and is very ductile. At high temperatures, the material is in an austenitic phase which is quite strong and stiff. The materials are said to have memory because the shape of the parent, austenitic phase is recovered as the temperature is raised and the phase

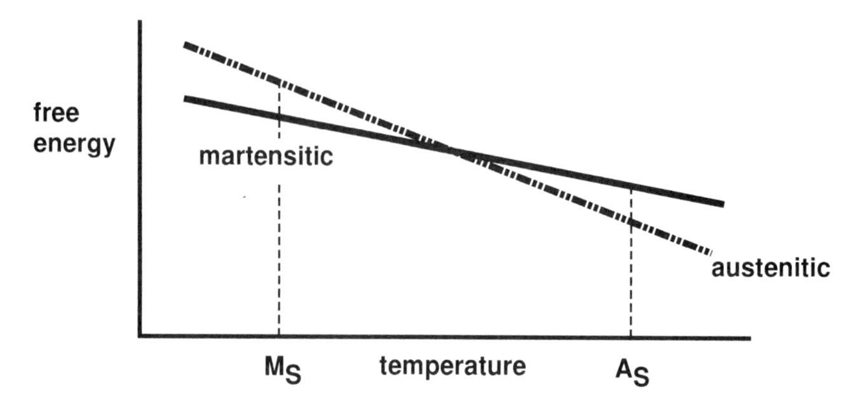

FIGURE 7.51. Free energy of a shape memory alloy in the austenitic and martensitic phases as a function of temperature.

transition takes place. The temperature at which the transition occurs is determined by the free energy of the material in the two phases. Figure 7.51 shows the free energy of the two phases of a shape memory alloy as a function of temperature. Note that the martensitic phase has lower energy at low temperatures while the austenitic phase has lower energy at high temperatures. If there were no hysteresis in the transition process, then the transition from the martensitic to austenitic phase would occur precisely at the crossing point of the two lines, with nature always opting for the lowest energy state. As is normal, the process does have hysteresis, so the transformation from the austenitic to martensitic phase starts at temperature M_S, below the crossing point of the lines, and the transformation from the martensitic to austenitic phase starts at temperature A_S, above the point of equal free energy of the phases.

The martensitic transformation from one phase to another is a lattice transformation involving cooperative atomic movement rather than atomic diffusion. It involves shear transformation from a typically face-centered cubic shape in the austenitic phase to a plate-like shape in the martensitic phase. This shear deformation is virtually always accompanied by macroscopic shape changes of the material. Thus, the transformation is thermomechanical in that temperature changes produce dimensional changes. However, virtually any property of a shape memory alloy changes as a phase transition occurs. For instance, Fig. 7.52 shows a typical plot of the electrical resistivity of a shape memory alloy as a function of temperature. The austenitic phase has a higher electrical resistivity than the martensitic. The shape memory alloys are converted to an electromechanical transducer through the use of ohmic heating to affect the temperature rise. Thus electrical power input causes the phase transition and comcommitant thermomechanical change. Because the heating is resistive, this overall energy transformation is modeled here as a resistive electromechanical mechanism.

Table 7.5, using data from Funakubo [41], shows the main shape memory alloys, the start temperature for a martensitic transformation, and the transformation

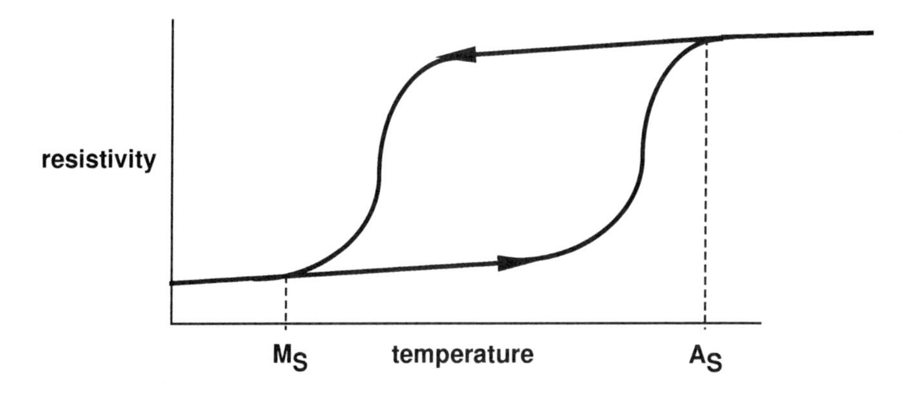

FIGURE 7.52. Electrical resistivity as a function of temperature in a shape memory alloy.

TABLE 7.5. Shape memory alloys. Data taken from Funakubo [41].

Material	M_S(° C)	Hysteresis (°C)
AgCd	−190–−50	15
AuCd	30–100	15
CuAlNi	−140–100	35
CuAuZn	−190–40	6
CuSn	−120–30	
CuZn	−180–−10	10
CuZnX (X=Si, Sn, Al, Ga)	−180–100	10
InTl	60–100	4
NiAl	−180–100	10
TiNi	−50–100	30
FePt	≈ −130	4
FePd	≈ −100	
MnCu	−250–180	25

temperature hysteresis. The earliest of these materials studied was AuCd, which was investigated in the early 1950s. The interest in this material waned with the discovery of the shape memory effect in TiNi in 1963. TiNi is still one of the shape memory alloys of choice, because it is strong, ductile, and corrosion resistant. It is capable of strains of up to 8% and stresses up to 400 MPa with complete reversibility. However, it is roughly ten times the cost of the copper-based alloys, and thus a significant effort is now going into improvements on CuZnAl. CuZnAl is capable of strains of up to 4% and stresses up to 200 MPa with complete reversibility. It is far less corrosion resistant than TiNi and more difficult to process to get a shape memory effect.

There are two types of shape memory effects in shape memory alloys: the one-way and the two-way effect. The one-way effect is that which occurs naturally

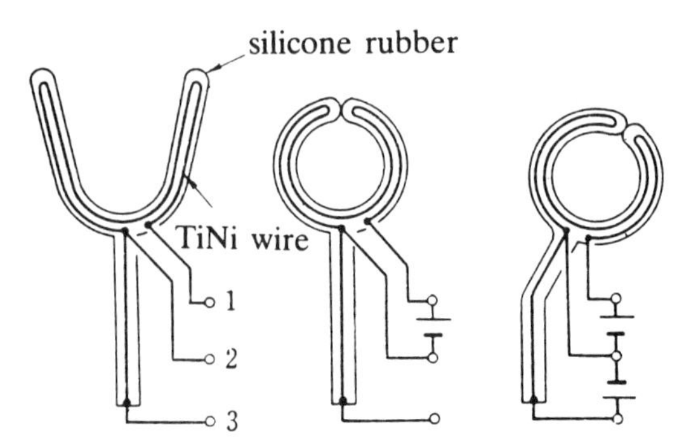

FIGURE 7.53. Shape memory alloy robot hand. Figure taken from Funakubo [41].

and refers to the martensitic to austenitic transformation. This transformation will always return the material to the original (remembered) shape of the austenitic phase as long as the material has not been strained beyond its limit. Under some conditions it is possible to train a material so that it also reverts (partially) to a deformed state upon transformation from the austenitic phase to the martensitic phase, i.e., upon cooling. This produces the two-way effect. The two-way effect is usually used to produce structures such as shape memory alloy springs. Transducers using shape memory alloys use both of these effects. Because the materials are capable of producing large strains and stresses, they are typically used in actuation rather than sensing.

Figures 7.53 and 7.54 show two examples of actuators using shape memory alloys. Both are taken from Funakubo [41]. Figure 7.53 shows a robotic wrist in which application of a current causes a wrist to bend or fingers to close. Figure 7.54 shows a crank engine driven by a shape memory alloy. In both cases, the speed of heating and cooling limits the actuator bandwidth.

Because transducers based on shape memory alloys rely on heating and cooling of the material, there are distinct advantages to applications in which the devices

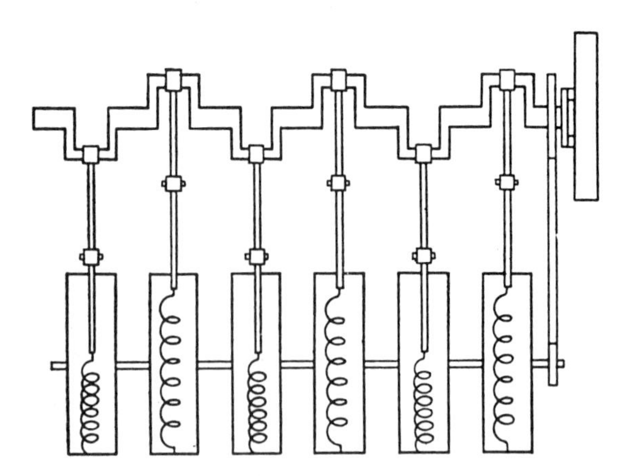

FIGURE 7.54. Shape memory alloy-driven crank engine. Figure taken from Funakubo [41].

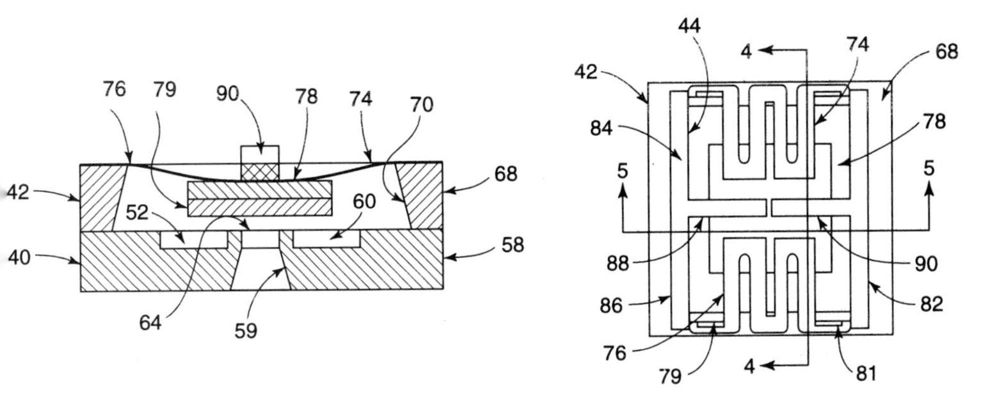

FIGURE 7.55. Shape memory alloy micro-valve described in Johnson and Ray [42].

can be made small as these heat and cool quickly. Figure 7.55 shows one example of a microvalve described in Johnson and Ray [42]. Owing to the small size of this device it is capable of 100 Hz operation. Another shape memory alloy-based actuator which resembles microtweezers, has been described by Krulevitch et al. [43].

For more information on shape memory alloys and their applications, readers are referred to Funakubo [41] and Duerig et al. [44].

7.9 Summary

Resistive transduction mechanisms differ from capacitive and inductive transduction mechanisms in that a single device cannot be used as both a sensor and as an actuator. Generally, resistive transducers are used for sensing purposes. A notable exception are shape memory alloy transducers, in which resistive heating is used to actuate a change of shape.

Several types of resistive transduction mechanisms have been presented. These include conductive switches, potentiometric sensors, piezoresistive sensors, thermoresistive transducers, thermoelectric sensors based on the Seebeck effect, magnetoresistive sensors, and shape memory alloy transducers. Practical examples of these sensors include strain gauges, resistive pressure switches, position sensors, RTDs, thermistors, and thermocouples.

7.10 References

1. Dwyer Instruments Inc., PO Box 373, Michigan City, IN 46360-0373, USA.

2. Micro Pneumatic Logic Inc., 2890 NW 62nd St., Fort Lauderdale, FL 33309-1786, USA.
3. Gems Sensors, Div. IMO Industries Inc., Cowles Rd., Plainville, CT 06062, USA.
4. Qualimetrics Inc., 1165 National Dr., Sacramento, CA 95834, USA.
5. American Electronic Components Inc., 1010 N. Main St., PO Box 280, Elkhart, IN 46515, USA.
6. M. Rossi, *Acoustics and Electroacoustics* (Artech House, Norwood, MA, 1988).
7. C. F. Coombs, Jr. (ed.), *Electronic Instrument Handbook*, 2nd ed. (McGraw-Hill, New York, 1995).
8. J. Fraden, *AIP Handbook of Press Modern Sensors: Physics, Designs and Applications* (AIP Press, New York, 1993).
9. E. O. Doebelin, *Measurement Systems: Application and Design*, 3rd ed. (McGraw-Hill, New York, 1983).
10. SpaceAge Control Inc., 38850 20th Street East, Palmdale, CA 93550, USA.
11. Novotechnik US, Inc., 237 Cedar Hill St., Marlborough, MA 01752, USA.
12. A. Belu-Marian, E. Candet, and A. Devenyi, *Piezoresistive Sensors*, in *Thin Film Resistive Sensors* (Institute of Physics, Bristol, UK 1992).
13. E. E. Simmons, US Patent 2,365,015, Fluid Pressure Gauge, Dec. 12, 1944.
14. R. Grigorovici, A. Devenyi, V. Florea, Gh. Korony, A. Belu, and C. Costachescu, US Patent 3,705,993, Piezoresistive Transducers and Their Devices with Semiconductor Films and their Manufacturing Process, Dec. 12, 1972.
15. J. S. Johanessen, Phys. Status Solidi, A **10**, 569 (1972).
16. C. S. Smith, *Piezoresistance Effect in Germanium and Silicon*, Phys. Rev., **94**, 42 (1954).
17. BLH Electronics, Inc., 75 Shawmut Rd., Canton, MA 02021, USA.
18. Entran® Sensors and Electronics, 10 Washington Ave., Fairfield, NJ 07004-3877, USA.
19. Advanced Mechanical Technology Inc., 176 Waltham St., Watertown, MA 02172-4800, USA.
20. Omega Engineering Inc., PO Box 4047, Stamford, CT 06907, USA.
21. Silicon Microstructures Inc., 48729 Kato Rd., Fremont, CA 94538-7312, USA.
22. J. B. Angell, S. C. Terry, and P. W. Barth, *Silicon Micromechanical Devices*, Sci. Am. **44** (Ap., 1983).
23. C. I. Popescu and M. Popescu, *Thermoresistive Sensors* in *Thin Film Resistive Sensors* (Institute of Physics, Bristol, UK, 1992).
24. Thermometrics, 808 U.S, Highway 1, Edison, NJ 08817-4695, USA.
25. R. S. Figliola and D. E. Beasley, *Theory and Design for Mechanical Measurements*, 2nd ed. (Wiley, New York, 1995).
26. C. I. Popescu and M. Popescu, ICPE Internal Res. Rep., 1989.
27. Hy Cal Sensing Products, 9650 Telstar Ave., El Monte, CA 91731-3004, USA.
28. Dantec Measurement Technology, 777 Corporate Dr., Mahwah, NJ 07430, USA.

29. G. B. Hocker, R. G. Johnson, R. E. Higashi, and P. J. Bohrer, *A Microtransducer for Air Flow and Differential Pressure Sensor Applications*, in C. D. Fung et al. (ed.), *Micromachining and Micropackaging of Transducers* (Elsevier, Amsterdam, 1985).
30. T. Betzner, J. R. Doty, A. M. A. Hamad, H. T. Henderson, and F. G. Berger, *Structural Design and Characteristics of a Thermally-Isolated, Sensitivity-Enhanced, Bulk-Machined, Silicon Flow Sensor*, J. Micromech. Microengng. **6**, 217 (1996).
31. Fluid Components International, 1755 La Costa Meadows Dr., San Marcos, CA 92069-5115, USA.
32. H. Le Chatelier, *Sur la Dissociation du Carbonate de Chaux*, C. R. Acad. Sci. **102**, 1243 (1886).
33. R. J. Moffat, *The Gradient Approach to Thermocouple Circuitry*, Cp. 2 of A. I. Dahl (ed.), *Temperature: Its Measurement and Control in Science and Industry*, Volume 3, Part 2 (Reinhold, New York, 1962).
34. Nanmac Corp., 9–11 Mayhew St., Framingham Center, MA 01701-2439, USA.
35. RdF Corporation, 23 Elm Ave., PO Box 490, Hudson, NH 03051-0490, USA.
36. A. W. van Herwaarden and G. C. M. Meijer, *Thermal Sensors*, in S. M. Sze (ed.), *Semiconductor Sensors* (Wiley, New York, 1994).
37. J. Robadey, O. Paul, and H. Baltes, *Two-Dimensional Integrated Gas Flow Sensors by CMOS IC Technology*, J. Micromech. Microengng., **5**, 243 (1995).
38. N. Fuschillo, *Thermoelectric Phenomena*, in I. B. Cadoff and E. Miller (eds.), *Thermoelectric Materials and Devices* (Reinhold, New York, 1960).
39. F. E. Jaumot, Jr., *Review of Thermoelectric Effects*, Cp. 14 in J. Kaye and J. A. Welsh (eds.), *Direct Conversion of Heat to Electricity* (Wiley, New York, 1960).
40. P. Ciureanu, *Magnetoresistive Sensors*, in *Thin Film Resistive Sensors* (Institute of Physics, Bristol, UK, 1992).
41. H. Funakubo (ed.), J. B. Kennedy (transl.), *Shape Memory Alloys* (Gordon and Breach, New York, 1987).
42. A. D. Johnson and C. A. Ray, *Shape Memory Alloy Film Actuated Microvalve*, US Patent 5,325,880, July 5, 1994.
43. P. Krulevitch, A. P. Lee, P. B. Ramsey, J. C. Trevino, J. Hamilton, and M. A. Northrup, *Thin–Film Shape Memory Alloy Microactuators*, J. MEMS, **5**, 270 (1996).
44. T. W. Duerig, K. N. Melton, D. Stockel, and C. M. Wayman (eds.), *Engineering Aspects of Shape Memory Alloys* (Butterworth-Heinemann, London, 1990).

8
Optomechanical Sensors

In the last decade, one of the most rapidly growing sensor areas has been that of optical sensors. Partly, this trend has been prompted by the rapid decline in optical sensor prices, partly it is the result of a growing need for noncontact sensors with high precision, and partly it stems from the advent of new types of optical devices which are easily used. Optomechanical sensors are devices in which some aspect of light propagation is changed (modulated) by a mechanical variable. As will be demonstrated in this chapter, the precise nature of the modulation can take many different forms. This chapter discusses optomechanical sensors in general and concentrates on fiber optic devices in particular. Please keep in mind that the intent of this chapter is an introduction. It is not possible to do justice to optomechanical sensors in one short chapter, and the interested reader is referred to the many books on the topic. (For instance, see [1]–[3].)

To put optical sensors in perspective, let us begin with a brief history of the physical view of light. The first well-documented consideration of light resulted in a disagreement between Huygens and Newton. Huygens suggested that light is a wave, similar in nature to sound. Newton objected to this view, saying that if light were a wave it should bend around corners the way sound does. Experience told Newton that light does not bend around corners. We now know that both Huygens and Newton were correct and that light does bend around corners. Because light has such a small wavelength compared to sound, we simply lacked knowledge about the bending of light (refraction) until Fresnel conducted careful experiments in 1827.

The view of light as a special type of wave phenomenon persisted until Maxwell presented light as simply one part of a system of electromagnetic waves. While Maxwell's theory was well received, Hertz showed in 1887 that there is a problem with this view of light because it cannot explain the photoelectric effect in metals.

There was a flurry of research on the nature of light following Hertz's arguments, culminating with Planck's incorporating light into his work on quantum theory. Planck suggested, in 1900, that light exists and travels as quanta. The fundamental idea is that light is pure energy (photons). When photons are captured, the energy is converted to another form. In 1905 Einstein used quantum theory to describe the photoelectric effect, and the quantum view of light persists today.

Given the way we view light, it is an open question whether optomechanical sensors are a type of *electro*mechanical transducers, i.e., whether optics is a branch of electromagnetic field theory. Although an interesting question, what is more important than its answer is why the question arises. Clearly light waves obey Maxwell's equations and this could be taken as proof that optics is a branch of electromagnetism. However, optics and optical systems are distinctly different from standard electrical and magnetic systems. In particular, there is a fundamental difference that affects our ability to model optical sensors, namely, that they rely on the continuum nature of matter for operation. Most electromechanical and magnetomechanical transducers generally can be modeled using a finite number of discrete elements to represent the system. By contrast, optomechanical sensors take advantage of wave propagation and would require an infinite number of discrete elements to develop a suitable model. This optical reliance on the continuum nature of matter is shared with other types of transducers. For instance, ultrasonic sensors operate using sound waves which propagate through a medium. Thus, they, too, would require an infinite number of discrete elements in order to develop a model.

In addition to the discrete/continuum distinction between optical systems and electrical and magnetic systems, there is a difference between them in terms of how we choose to describe their behavior. In electrical and magnetic systems it is possible to define power conjugate variables which have clear physical interpretation. This has not been the case for optical systems where we tend to use variables, such as intensity, which are themselves proportional to power. As a result, optical systems are routinely described in terms of scattering formalisms rather than circuit-like models, and it has not been obvious how to translate between these two approaches. While progress in this area is being made by Longoria [4] and Paynter and Busch-Vishniac [5], the way we describe optical systems has certainly emphasized the distinctions between them and their electrical and magnetic cousins. In this chapter, we will avoid the modeling issue to prevent confusion, but it is important to note that we are not simply forgetting to include bond graph models of optomechanical sensors. Such bond graph models simply have not been developed.

Because optomechanical transducers rely on wave propagation, the wavelength of light is an important descriptor. Optical wavelengths are generally taken to span 0.3 μm to 3.0 μm. Visible light is confined to 0.4 μm to 0.7 μm. Hence a significant portion of the long wavelength region is not visible to humans.

Optomechanical sensors generally consist of four elements: a source of light, an optical guide, a modulating agent, and a light detector. Typical light sources include incandescent filament lamps, arc lamps, infrared sources, and lasers. Incandescent filament lamps are like ordinary light bulbs and produce incoherent light. Arc lamps produce light by passing an electrical discharge through an ionized gas. Infrared sources produce light through heating of certain chemical mixtures. Lasers produce light through the coherent stimulation of certain materials which are able to maintain a metastable energy inversion state. Of these four common sources, lasers have been changing the most rapidly over the last decade, with the advent of semiconductor lasers of low cost and high controllability (See Parker [1], for example).

Optical guide components vary from very elementary to quite complex. Some systems use no guide component, relying on the propagation of light through air. These systems either compensate for dispersion of the light beam or use sensors which minimize dispersion. Other systems use sophisticated lenses and other optical components. Fiber optic systems use glass or polymer fibers as waveguides for the light.

The modulating component of an optomechanical sensor is the part which uses the variable we desire to measure to modulate the optical signal in some specified manner. There are a wide range of options available in optical sensors. For example, fiber optic sensors can monitor any variable which causes a change in the optical path length. However, sensing is not limited to path length changes. Changes in the index of refraction of a medium, the direction of light propagation, and many other properties can be detected.

Optical detectors convert light energy to electrical energy and thus represent the optoelectric transduction part of an optomechanical sensor. These detectors generally fall into two categories: quantum detectors, and thermal detectors. In this chapter we focus on the quantum detectors, as thermal detectors have been discussed previously (See Chapter 7).

8.1 Quantum Detectors

Quantum optical detectors are sensors based on the photoelectric effect first described by Einstein. The fundamental idea of quantum detectors is that photons create charge carriers which are observed as the indirect measure of optical activity. The energy of a photon, $\mathcal{E}_{\text{photon}}$ is

$$\mathcal{E}_{\text{photon}} = h\nu, \tag{1}$$

where h is Planck's constant (6.63×10^{-34} J s or 4.13×10^{-15} eV s), and ν is the frequency. When a photon strikes a conducting surface the energy can generate a free electron. This process is described by the following:

$$h\nu = \Phi + \text{kinetic energy}, \tag{2}$$

where Φ is the energy required to free an electron (known as the work function). The value of Φ is determined by the material. Equation (2) merely states that an electron is freed if the energy of a photon striking a conductor exceeds the work function. The additional energy supplies kinetic energy to the freed electron.

In crystalline materials, the electrical conductivity depends on the number of charge carriers present. Thus, the conductivity may be expressed as

$$\sigma = q_e(\mu_e c_n + \mu_h c_h), \tag{3}$$

where σ is the conductivity, q_e is the charge of an electron, μ_e is the electron mobility, μ_p is the hole mobility, c_n is the concentration of electrons (negative charge carriers), and c_h is the concentration of holes (positive charge carriers). The mobilities, μ_e and μ_h are functions of the material.

Equations (2) and (3) may be combined to explain how many optomechanical sensors work. Light incident on a material creates charge carriers and this affects the conductivity of that material. The change in conductivity is monitored to provide a measure of the light striking the surface. Detectors that operate in this manner are referred to as photoconductive or photoresistive sensors. In addition to photoconductive sensors, there are photovoltaic and photoemissive optical transducers. Photovoltaic devices monitor the voltage induced by light shone on the surface of a material. Typical devices now use silicon or selenium and are roughly 10–20% efficient at converting light power to electrical power. Photoemissive devices, such as light emitting diodes (LEDs), use current to stimulate photon emission in the effective inverse of the photoelectric effect described above. Below we focus on photoconductive detectors, as these are by far the most common optomechanical sensors in use today.

8.1.1 Photoconductive Sensors

Photoconductive sensors were among the first light sensors made and are still the most common devices in use. The first material discovered to be a photoconductor was selenium. Selenium is now rarely used in photodetectors because of its slow response time and poor stability. Instead a number of materials have replaced selenium in various applications. Table 8.1 shows some of these materials, their band gap (i.e., their work function), and the corresponding longest wavelength at which these materials can operate. The materials with the longest wavelengths of over 3 μm are infrared detectors. The most popular of the materials listed is silicon, although cadmium sulfide (CdS) and cadmium selenide (CdSe) are popular for sensors of light in the visible spectrum. Light meters today tend to use cadmium sulfide; smoke detectors often use cadmium selenide; intruder alarms exploit lead sulfide (PbS); weapons use lead sulfide, lead selenide (PbSe), or indium antimonide (InSb). New work on a number of polymer gels is promising significant advances in this technology [6].

Figure 8.1, taken from Fraden [7], shows how a photoconductive sensor works. A voltage is applied across two electrodes with a photoconductive material between them. The current flow resulting from the voltage is monitored. Light incident on the photoconductor increases the conductivity, which causes a corresponding increase in the observed current. Because of our ability to detect small currents,

TABLE 8.1. Photoconductive materials in common usage in sensors.

Material	Band Gap (eV)	Longest Wavelength (μm)
ZnS	3.6	0.345
CdS	2.41	0.52
CdSe	1.8	0.69
CdTe	1.5	0.83
Si	1.12	1.10
Ge	0.67	1.85
PbS	0.37	3.35
InAs	0.35	3.54
Te	0.33	3.75
PbTe	0.3	4.13
PbSe	0.27	4.58
InSb	0.18	6.90

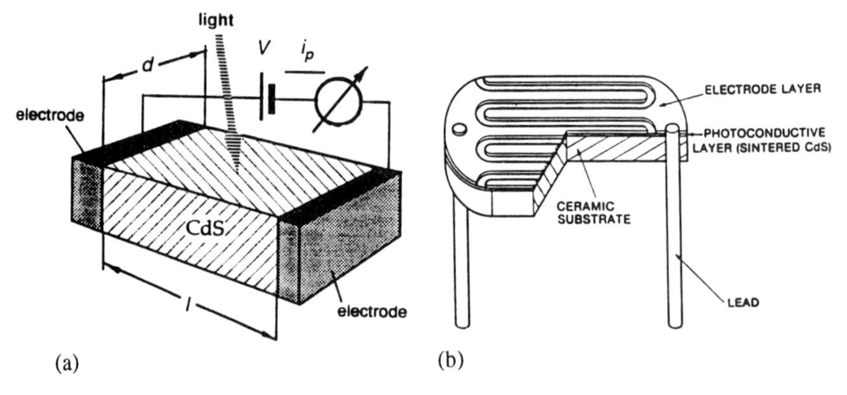

FIGURE 8.1. Operation of a photoconductor: (a) schematic of operation, and (b) typical sensor. This figure appeared in Fraden [7].

it is possible to make photoconductive sensors quite small. This is demonstrated in Fig. 8.1 which shows a typical photoconductor with a serpentine layer of cadmium sulfide deposited onto a ceramic substrate and packaged using standard microelectronic techniques.

One step up in sophistication from the straightforward photoconductor shown in Fig. 8.1 is a photodiode. Photodiodes are the most common optical transducers in use today. In their simplest form, photons switch on a transistor so that current flows. The current is observed at one or more electrodes.

Suppose that η is the probability that a photon at frequency ν will produce a free electron in the photodiode material. Then the rate of electron production, r, is given by

$$r = \frac{\eta \mathcal{P}}{h\nu}, \tag{4}$$

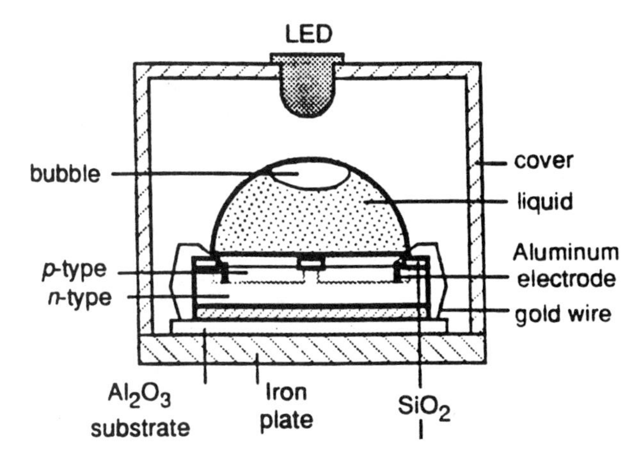

FIGURE 8.2. Optomechanical inclination sensor. Figure taken from Fraden [7].

where $\mathcal{P}$ is the optical power striking the photodiode. The current is simply the product of this rate of electron production and the charge of an electron, $i = rq_e$. Thus, the induced photocurrent is proportional to the light power striking the photodiode surface.

Figure 8.2, taken from Fraden [7], shows an example of an optomechanical sensor which uses photodiodes. In this sensor, a bubble in a liquid casts a shadow on an array of four photodiodes. The position of the shadow, and thus of the bubble, is determined from the relative photocurrents seen by the four photodiodes.

Figures 8.3 and 8.4 show optomechanical sensors which use photodiodes. In both cases, the modulating agent is a mechanical structure which moves in response to a load. This motion causes some of the light from a source to be blocked, so that the optical power reaching the photodiode, and thus the photocurrent, is diminished as load increases. Figure 8.3, taken from Webster [8], shows a tactile sensor using a U-shaped mechanical structure which bends. Figure 8.4, taken from McAlpine [9], uses a translating structure.

Some of the most common photodiodes available today use the lateral photoeffect. These photodiodes are *p–n* junctions with electrodes typically on a p-surface. Six geometries are shown in Fig. 8.5, taken from Fraden [7]. The photocurrent induced by light travels to the electrodes. The amount of photocurrent received by each electrode is directly related to its proximity to the centroid of the light. Thus, if two electrodes are used, we may determine the position of the light centroid between the electrodes by differencing the photocurrents they receive, and dividing by the sum of the photocurrents. The result is a position measurement that is normalized and thus independent of the light power amplitude. By using two pairs of electrodes, with a 90° rotation between the parallel electrodes pairs, we can locate the light spot centroid in two dimensions.

Figure 8.6 shows an example of a position sensor which exploits lateral effect photodiodes. In this device, described by Pavuluri et al. [10], light sources and

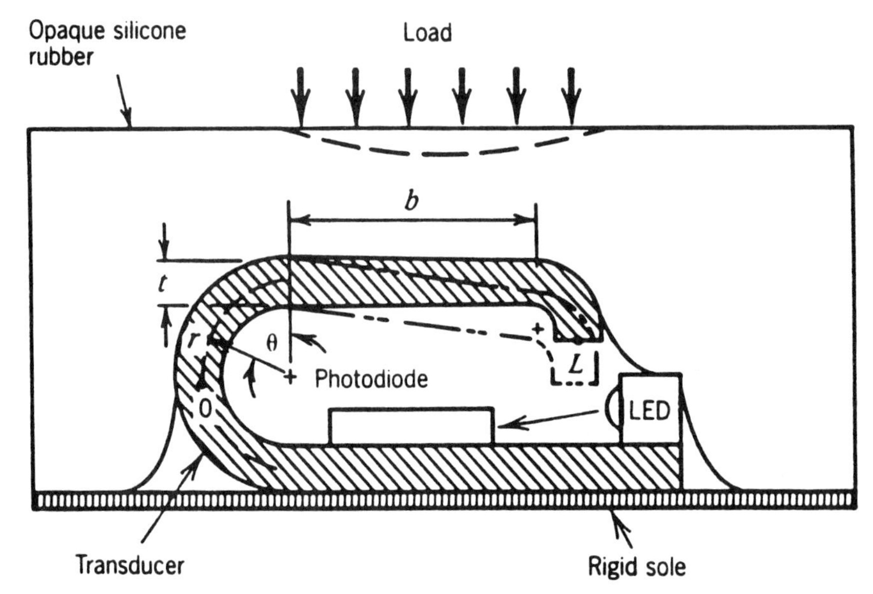

FIGURE 8.3. Tactile sensor incorporating a photodiode described in J. G. Webster (ed), **Tactile Sensors for Robotics and Medicine**, (Wiley, NY, 1988). Reprinted with permission of John Wiley and Sons, Inc.

photodiodes are located on a plane below a target. The full six coordinates of the target are determined from the positions of the reflections on the two-dimensional lateral effect photodiodes. Similar sorts of position detection have been exploited by Ni and Wu [11] and Trethewey et al. [12].

8.2 Fiber Optic Waveguide Fundamentals

The previous section of this chapter described the most common optical detectors and presented a few examples of optomechanical sensors which use them and have an optical guide of air. The second common kind of optical guide in sensors is fiber. To understand the operation of fiber optic sensors, we must consider reflection and transmission at the interface between media with different optical indices of refraction. It is this topic which is presented below.

Regardless of the precise wave which encounters an interface, we may determine the amplitude of the reflected and transmitted waves if the material properties of the two media at the interface are known. For optical waves, the property that characterizes a medium is the index of refraction. The index of refraction, n, is defined as the ratio of the speed of light in vacuum to the speed of light in that material. Table 8.2 lists the index of refraction for materials of interest in optical sensors. Note that the index of refraction of air is essentially 1. Of the materials shown in Table 8.2, fused silica SiO_2 and Pyrex are the most common for fibers.

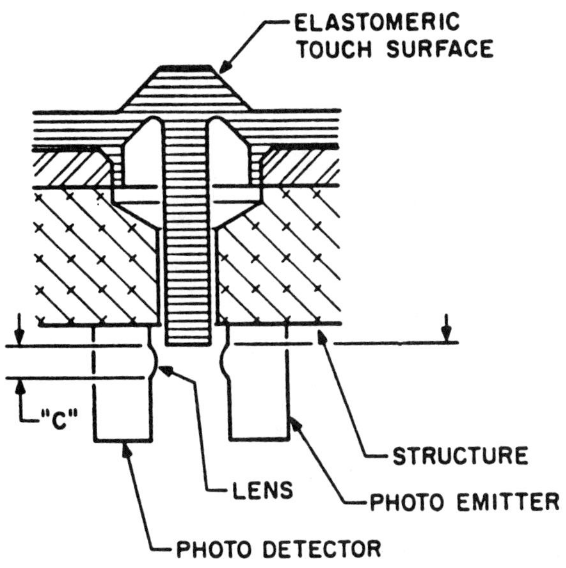

FIGURE 8.4. Tactile sensor incorporating a photodiode described by McAlpine [9].

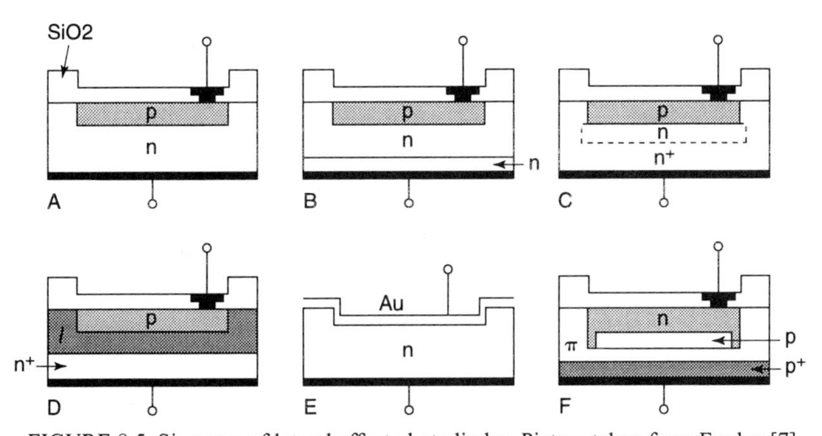

FIGURE 8.5. Six types of lateral effect photodiodes. Picture taken from Fraden [7].

TABLE 8.2. Index of refraction for a variety of materials.

Material	Index of refraction
Air	1.00029
Fused silica SiO_2	1.46
Pyrex 7740	1.47
Quartz	1.54
Diamond	2.42
GaAs	3.13
Si	3.42
Ge	4.00

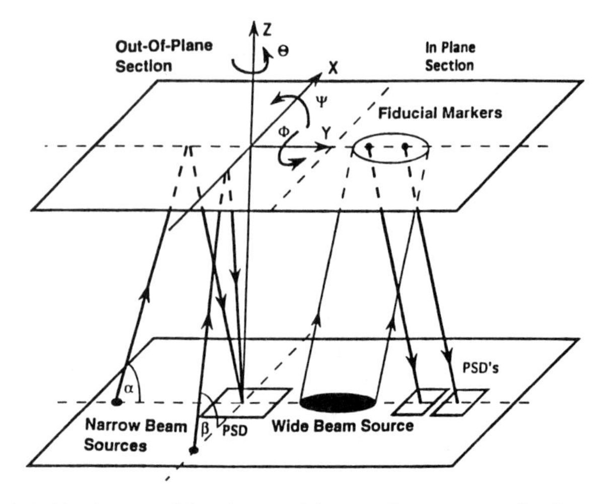

FIGURE 8.6. Six degree-of-freedom position sensing system using lateral effect photodiodes as the optical detectors. This system was described by Pavuluri et al. [10].

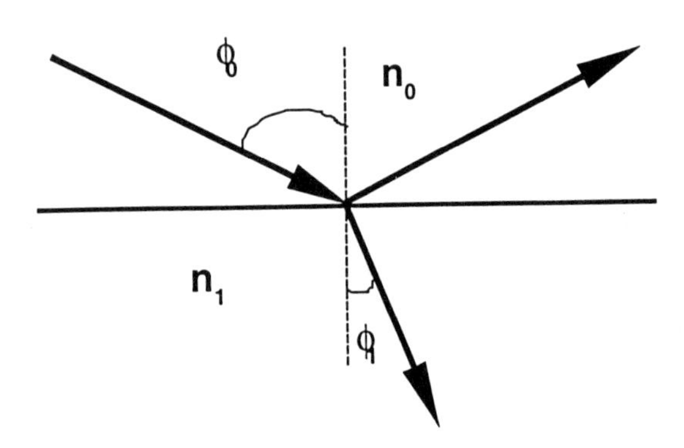

FIGURE 8.7. Optical transmission at an interface.

Consider the system shown in Fig. 8.7. At the interface the trace wavelengths must match. In other words, the projection of the wave onto the interface in medium 0 must match the projection in medium 1. This results in the well-known Snell law, which for optical waves relates the angles of incidence and transmission to the indices of refraction of the two media:

$$n_0 \sin(\phi_0) = n_1 \sin(\phi_1), \tag{5}$$

where n_0 and n_1 are the two indices of refraction, ϕ_0 is the angle of incidence measured from normal to the interface, and ϕ_1 is the transmission angle.

We may solve Snell's law for the transmission angle:

$$\phi_1 = \sin^{-1}\left[\frac{n_0}{n_1}\sin(\phi_0)\right]. \tag{6}$$

Note that this equation indicates that if the light is normal at the interface between two media, then the transmitted light also propagates normal to the interface. As the angle of incidence approaches 90°, ϕ_1 is approximately given by $\sin^{-1}(n_0/n_1)$. Note that if the transmitting medium has a smaller index of refraction than the incident medium, then the above equation shows that there is no angle at which light can be transmitted into the medium. We broaden this concept to define the critical angle, ϕ_c, as that angle beyond which there is no transmission into the medium:

$$\phi_c = \sin^{-1}(n_0/n_1). \tag{7}$$

Now let us apply the above analysis to an optical fiber. Most common fibers used in optical sensors are polished glass with cladding that is designed to ensure total internal reflection of the light. Figure 8.8 shows a view of the fiber end. Imagine that light in the medium with index of refraction n_0, is incident on the fiber at angle ϕ_0. Then the fiber, with index of refraction n_1, has light transmitted at initial angle ϕ_1, described by (5). Defining θ as the complementary angle, we may rewrite Snell's law for this case as

$$\begin{aligned} n_0 \sin(\phi_0) &= n_1 \cos(\theta) \\ &= n_1\sqrt{1-\sin^2(\theta)}. \end{aligned} \tag{8}$$

If θ is roughly equal to the critical angle for tranmission from the fiber to the cladding, then from (7) and (8) we have

$$\begin{aligned} n_0 \sin(\phi_0) &= n_1\sqrt{1-\left(\frac{n'}{n_1}\right)^2} \\ &= \sqrt{n_1^2 - n'^2}. \end{aligned} \tag{9}$$

We define the right-hand side of this equation as the numerical aperture (NA) of the fiber. It is a measure of the light acceptance capability of the fiber. As the NA increases, the ability to couple light into the fiber increases. This is shown graphically in Fig. 8.9 which shows the light cone which couples into the fiber. Note that the coupling efficiency also increases as the fiber diameter increases since the fiber captures more light (assuming uniform illumination).

Now consider-propagation of light through the fiber. While some rays propagate along the axis without encountering the interface between the fiber and the cladding, most of the light rays do encounter the interface. These rays which bounce off the interface are known as meridional rays and they must travel a longer path than the axial or nonmeridional rays in order to cover the same length of fiber. Consider a meridional ray as shown in Fig. 8.10. Assuming that the light is incident

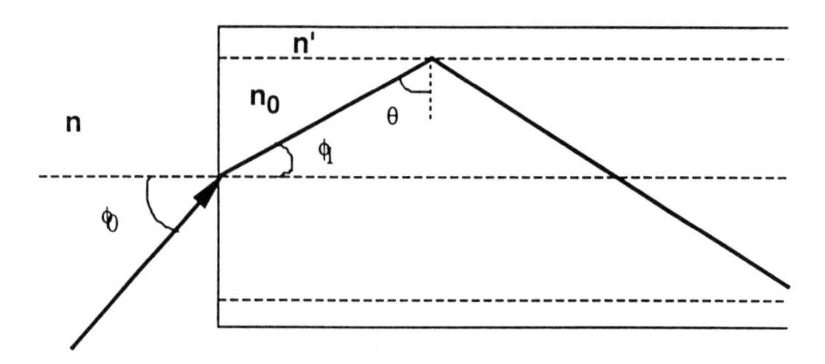

FIGURE 8.8. End of a fiber in an optical sensor.

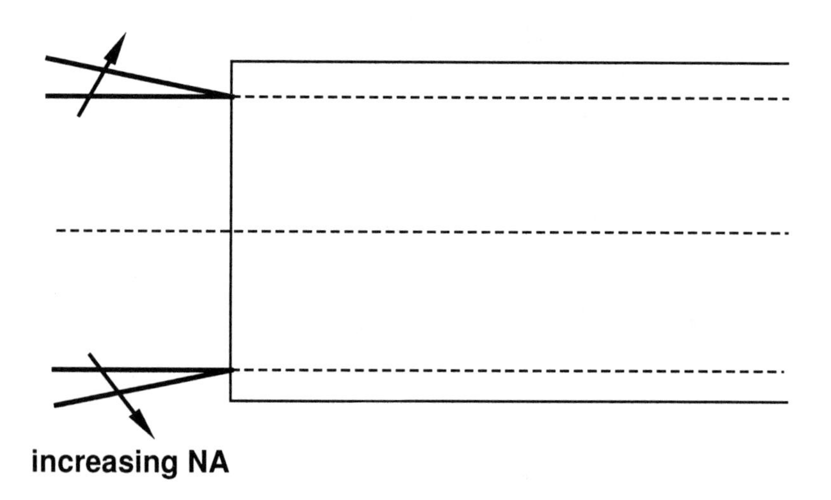

FIGURE 8.9. Acceptance light cone for coupling of light into a fiber.

from air ($n_0 = 1$) into a fiber, we have

$$l \cos(\phi_1) = L, \tag{10}$$

$$n_1 \sin(\phi_1) = \sin(\phi_0). \tag{11}$$

From (11) we have

$$\begin{aligned} \cos(\phi_1) &= \sqrt{1 - \sin^2(\phi_1)} \\ &= \sqrt{1 - \frac{\sin^2(\phi_0)}{n_1^2}} \end{aligned} \tag{12}$$

Substituting this into (10) yields

$$l = L \sec(\phi_1)$$

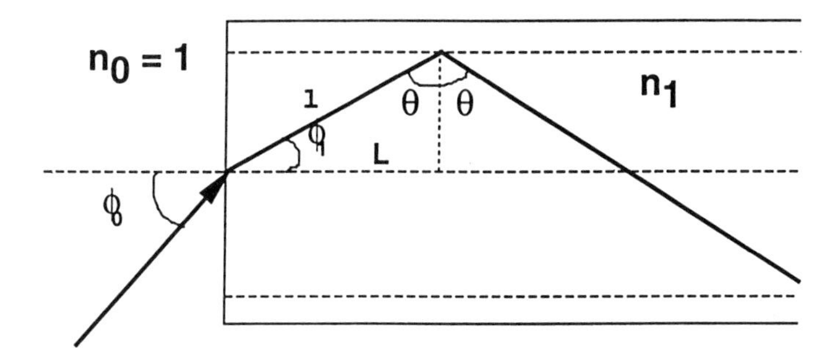

FIGURE 8.10. Path traveled by a meridional ray.

$$= \frac{L}{\sqrt{1 - \frac{\sin^2(\phi_0)}{n_1^2}}}. \tag{13}$$

Note that from (13) the length of a light ray is independent of the diameter of the fiber and depends only on the angle of incidence, fiber length, and fiber refractive index. However, the number of reflections of the ray as it moves along the path *is* dependent on the fiber diameter. The number of reflections N is given by

$$N = \frac{L}{d} \tan(\phi_1), \tag{14}$$

where d is the fiber diameter.

The effect of having light rays of nonuniform path length can be significant. To understand this, consider a source that is turned on suddenly at time $t = 0$. All of the light rays are emitted instantaneously, but the first to arrive at the opposite end of the fiber will be those with the shortest path length, i.e., the axial mode rays. The next rays to arrive will be those with the fewest reflections off the fiber-cladding interface. The last to arrive will be those with the greatest number of reflections. The net effect of the nonuniform path lengths is to broaden the light pulse. This effect is referred to as modal dispersion.

One way to deal with modal dispersion is to make the fiber sufficiently small in diameter that the light acceptance cone is narrow so that there are few reflections per unit length of fiber. Such fibers are referred to as single mode fibers and trade reduced modal dispersion for the lower efficiency of coupling light into the fiber. Multimode fibers accept modal dispersion in return for greater coupling efficiency from the source to the waveguide. In addition to uniform single mode or multimode fibers with step refractive changes at the cladding, there are multimode graded fibers in which the refractive index varies continuously between the fiber center and the cladding.

The above discussion has completely ignored losses of energy as the light propagates in the fiber. One of the great advantages of fiber optic devices is that the

TABLE 8.3. Characteristics of typical fibers.

Type	Loss (dB/km)	NA	Outer diam. (μm)	Core/ Cladding	Bandwidth (MHz-km)
Multimode step:					
Glass-clad glass bundle	400–600	0.4–0.6	50–70	0.90–0.95	20
Glass-clad glass low loss	2-6	0.2-0.3	50-200	0.4-0.8	20
Plastic-clad silica	3-10	0.3-0.4	200–600	0.7	20
Single mode	2–6	0.15	5-8	0.04	1000

losses are quite small. Attenuation is defined by

$$A = -10 \log_{10} \left(\frac{\mathcal{P}_{\mathrm{i}}}{\mathcal{P}_{\mathrm{out}}} \right), \tag{15}$$

where $\mathcal{P}_{\mathrm{i}}$ is the input power coupled into the fiber and $\mathcal{P}_{\mathrm{out}}$ is the output power. The attenuation for fibers varies between about 1–1000 dB/km. Typical telecommunications fibers have losses of about 5 dB/km. Causes of energy loss are absorption, scattering, end loss, and loss into the cladding due to bending. Multimode fibers generally have less attenuation at a given wavelength than a single mode fiber. Table 8.3 summarizes the physical properties of typical fibers including the geometrical dimensions, the numerical aperture, and the attenuation.

8.3 Intensity-Modulated Fiber Optic Sensors

As described in the second section of this chapter, the most important part of an optical sensor is that area in which the process that we desire to monitor modulates the light in some fashion. In fiber optic devices the modulation can take two fundamental forms: intensity modulation and phase modulation. In intensity modulation the magnitude of the monitored light is directly related to a process variable we desire to obtain. In phase modulation, such as interferometry, the signal magnitude is rendered effectively unimportant and the phase difference in two light signals is monitored. In this section we discuss the intensity modulation approaches.

There are a number of ways that we can construct intensity-modulated fiber optic sensors. First consider a transmissive sensor in which the light leaves one fiber and couples into a second, sensing fiber. Figure 8.11 shows two distinct ways in which such a sensor could be used. In Fig. 8.11(a) the light leaves the source fiber and couples into the sensing fiber which is able to move along the axis connecting the two fibers. The larger the gap between the two fiber ends, the less intense the light carried in the sensing fiber. Figure 8.11(b) shows a similar arrangement but with the sensing fiber moving normal to the axis connecting the two fibers. In this sensor, the maximum light intensity in the sensing fiber occurs when the two fibers are aligned. Motion of the sensing fiber normal to the axis results in a

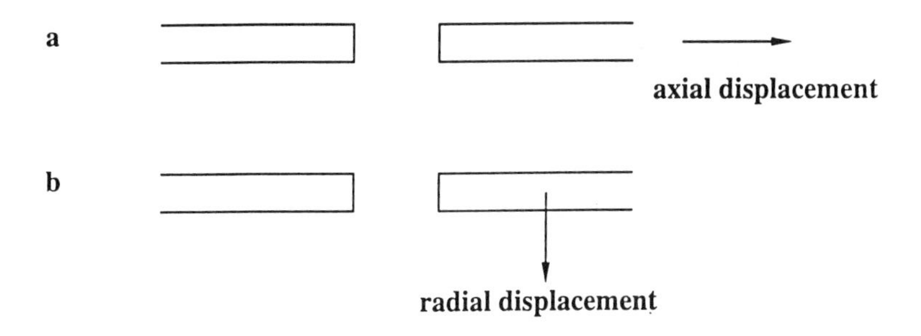

FIGURE 8.11. Typical transmissive intensity-modulated sensor geometries.

rapid drop in the sensing light intensity. Figure 8.12 shows the typical results for sensing light intensity as a function of distance in these transmissive, intensity-modulated sensors. Note that the off-axis sensor is more sensitive to motions but has a narrower range than the axial motion sensor.

Figure 8.13 shows an accelerometer described by Carome et al. [13]. In this device, one multimode fiber goes through a cantilever beam which responds to vibration. A second multimode fiber is located on the anvil opposite the beam. Thus as the beam vibrates, the vibrating fiber moves radially with respect to the receiving fiber.

Another approach to transmissive intensity-modulated sensors is to take advantage of the internal reflection of light at the end of a fiber. Consider the two fibers shown in Fig. 8.14. The angular cut at the endpoint of the source fiber causes total internal reflection of the light provided that there is a significant gap between the source and the sensing fiber. When that gap becomes comparable to or smaller than an optical wavelength, some of the energy leaves the source fiber and is received by the sensing fiber. The amount of light coupling into the sensing fiber is a function of the gap between the two fibers, as shown in Fig. 8.15. Compared to the two transmissive methods shown in Fig. 8.11, the frustrated total internal reflection (FTIR) approach is higher sensitivity and smaller range.

Figure 8.16, taken from Krohn [14], shows a pressure sensor which is based on an FTIR sensor. In this device, pressure on a diaphragm causes motion of a ferrule mounted onto a cantilever spring. This motion causes a variation in the separation between two ends of a cut fiber.

Another general approach to intensity modulation is to rely on reflection rather than transmission. In reflection approaches, the light in a source fiber leaves the

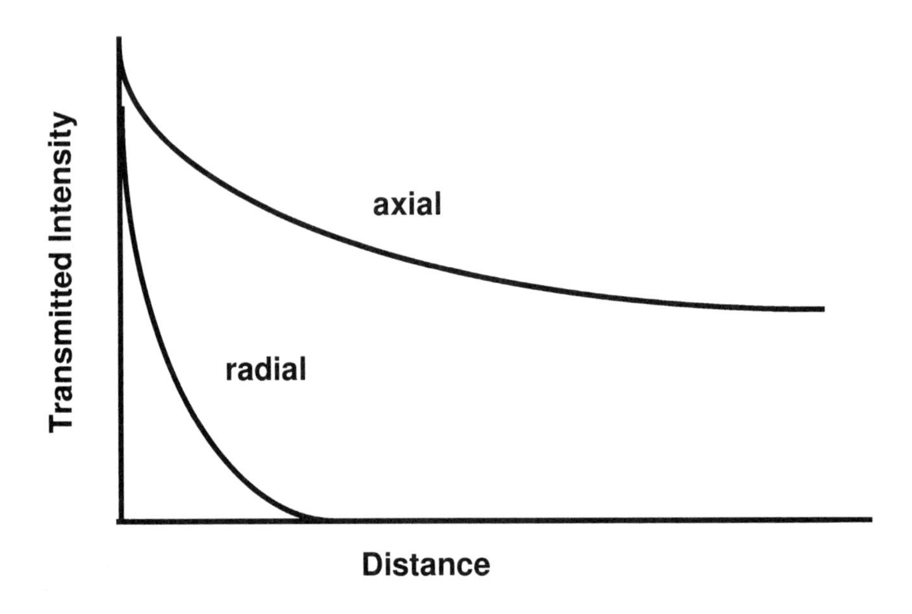

FIGURE 8.12. Intensity versus distance for the axial motion and radial motion transmissive sensors.

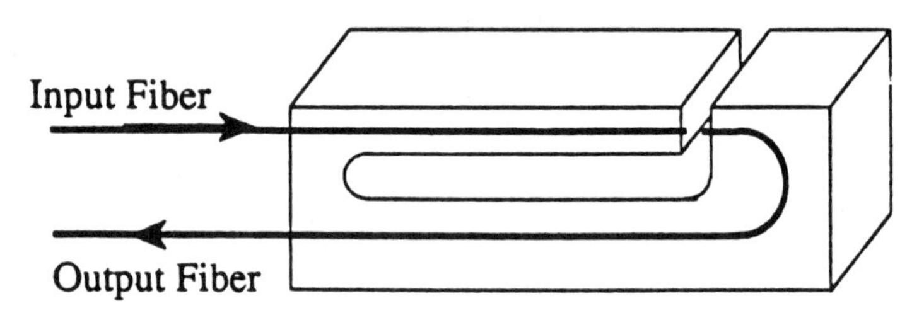

FIGURE 8.13. Fiber optic accelerometer described by Carome et al. [13].

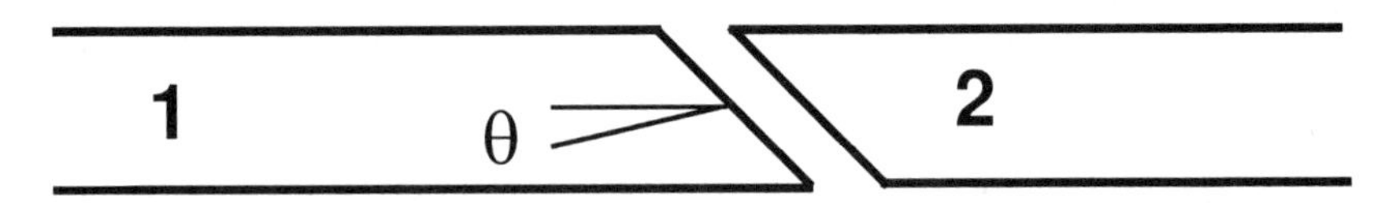

FIGURE 8.14. Intensity modulation based on frustrated total internal reflection.

fiber, reflects off a target, and is then received by a sensing fiber. The geometry is as shown in Fig. 8.17. The assumption inherent in this approach is that the light reflects specularly off the smooth target. Beyond a certain gap between the fibers and the target, the intensity of the received light decreases as $1/r^2$, where r is the gap length. This decay is caused by spreading of the light in a conical shape. For very short gap lengths, the received light intensity actually increases as the gap length increases. We can understand this result when we consider the

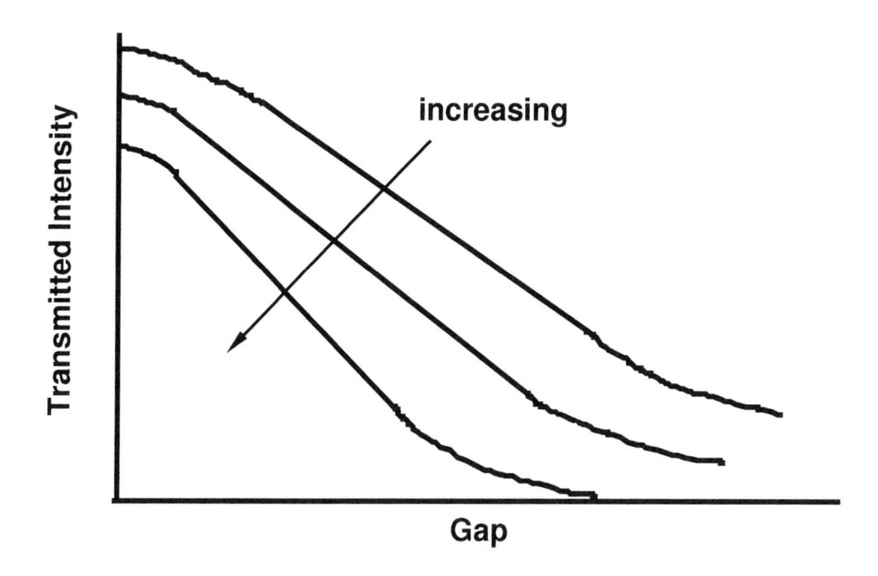

FIGURE 8.15. Sensing light intensity as a function of distance for the sensor using frustrated total internal reflection.

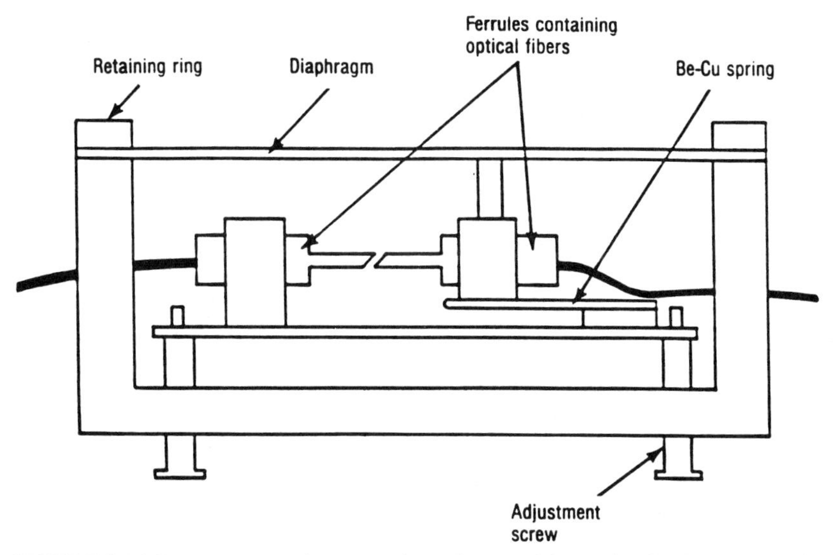

FIGURE 8.16. Pressure sensor incorporating a frustrated internal reflection sensor. Figure taken from Krohn [14].

limit of the target pressed up against the end of the fibers. In this condition the light leaving the source reflects directly back into the source fiber rather than the sensing fiber. Hence no reflected light is received. As the target moves slightly away, the spreading of the light exiting the source fiber permits some of the light to

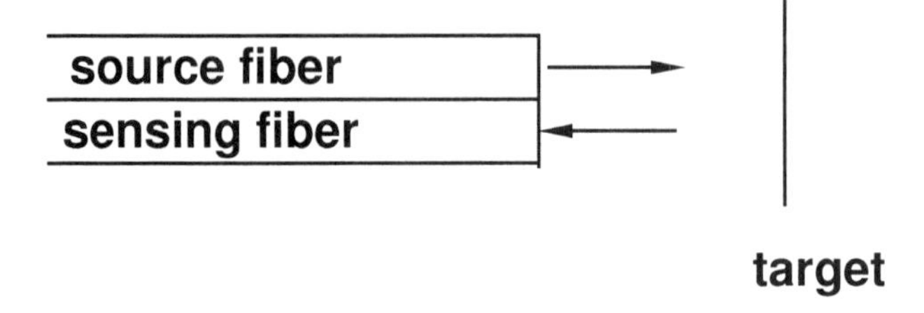

FIGURE 8.17. Reflective intensity-modulated sensor geometry.

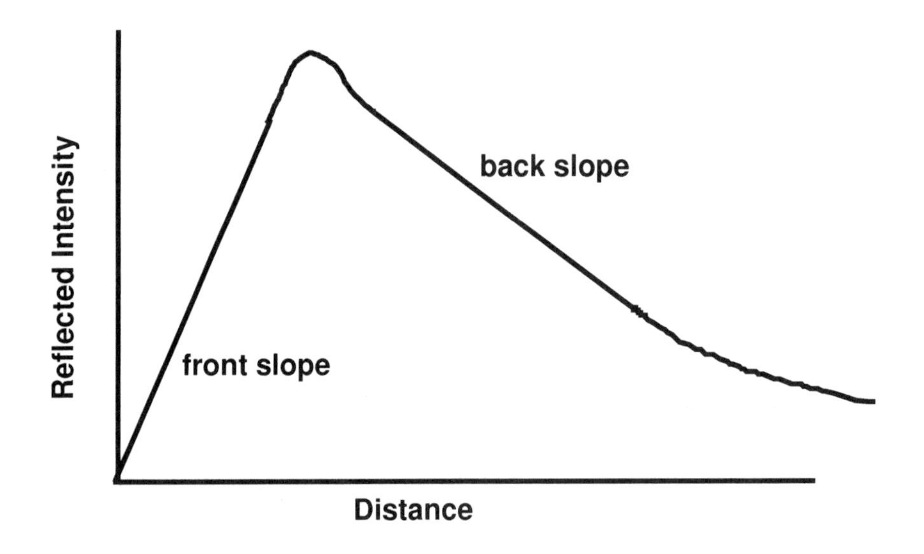

FIGURE 8.18. Sensed light intensity as a function of target/fibers separation.

reflect into the sensing fiber. The net result for sensing fiber light intensity versus target/fiber separation is shown in Fig. 8.18.

As a practical matter, most commercial devices use a fiber optic probe with many fibers emitting light and many receiving light. The details of the sensing response as a function of target/fiber probe separation depend on how the sensing and receiving fibers are arranged within the probe: randomly, top half-sensing and bottom half-receiving, or strictly alternating (See product literature from Mechanical Technology Inc. [15]). These options are shown in Fig. 8.19, taken from Krohn [14]. While reflecting intensity modulated sensors are accurate, simple, and relatively low in cost, it should be noted that they rely on well-aligned fiber ends and targets. A slight angular skewing of the target can cause wildly wrong results.

There are many examples of optomechanical fiber optic sensors based on the reflection of light. For instance, Fig. 8.20, taken from Coombs [16], shows a fiber optic pressure sensor. In this device, pressure causes deflection of a mirrored diaphragm. This effectively changes the distance from the sensing and receiving

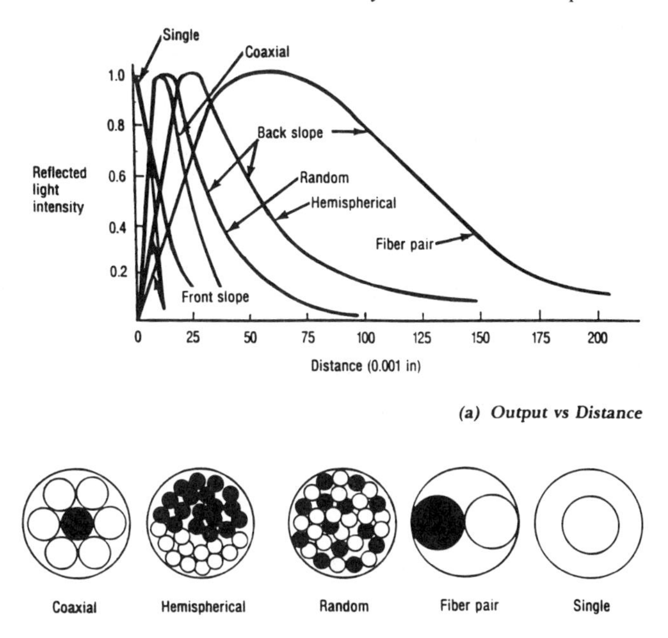

FIGURE 8.19. Fiber optic displacement sensor operating on a reflection principle. Figure taken from Krohn [14].

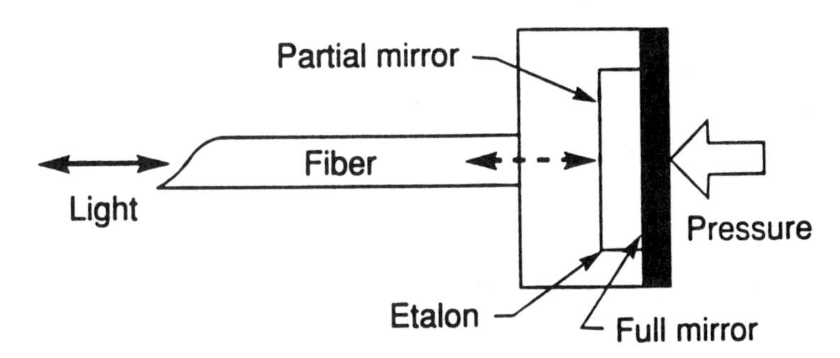

FIGURE 8.20. Pressure sensor based on reflection of light off a membrane reflecting light into a fiber. Figure taken from C. F. Coombs, Jr. (ed), **Electronic Instrument Handbook**, 2nd ed., (McGraw-Hill, NY, 1995). Reproduced with permission by The McGraw-Hill Companies.

fibers to the reflecting target. The same approach can be used to create a temperature sensor, in which case a material with a high thermal coefficient of expansion is used to separate the reflecting surface from the fibers. Temperature changes thus produce changes in the distance to the reflecting target.

Another example of an optomechanical sensor operating via reflection of light off a target and into a fiber is the torque sensor shown in Fig. 8.21 and taken from Coombs [16]. In this device, an eccentric cam serves as the reflecting target.

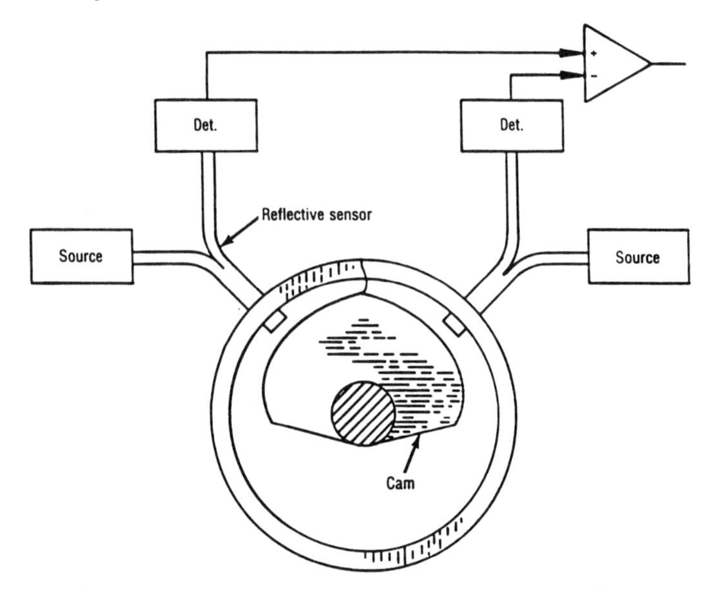

FIGURE 8.21. Torque sensor based of fiber optic reflection sensor. Figure taken from C. F. Coombs, Jr. (ed), **Electronic Instrument Handbook**, 2nd ed., (McGraw-Hill, NY, 1995). Reproduced with permission by The McGraw-Hill Companies.

Distance to the cam is sensed at two locations in order to compensate for slight misalignments.

One last example of a reflective fiber optic sensor is shown in Fig. 8.22, taken from Krohn [14]. In this device, fluid flow turns a turbine and a stationary optical sensor is used to count the blade passages. This measures the turbine speed and thus the flow rate.

The transmissive and reflective intensity modulating sensors discussed so far rely on light leaving one fiber and entering another. Sensors of this sort are referred to as extrinsic sensors because the light leaves the waveguide. Another class of sensors is intrinsic sensors in which the light remains in a waveguide for the entire path. There are two basic types of intensity-modulated intrinsic sensors. In one type, the light intensity is modulated as it propagates down a waveguide through changes in the properties of the core and cladding. The physical properties that can be affected are absorption, scattering, fluorescence, index of refraction, and polarization. Any process which causes changes in these physical properties can be sensed by simply monitoring shifts in the light intensity observed at the two ends of a single fiber. For example, many of the fiber properties are temperature sensitive. Hence a thermal probe can be made by simply monitoring the changes in light intensity along a fiber of known length and composition.

A second type of intrinsic intensity-modulated sensor relies on bending of the fiber as shown in Fig. 8.23, taken from Coombs [16]. A force on a mechanical piece with etched grooves causes bending in the fiber caught between the moving piece and its fixed mate. As a result of the bending, there is light lost into the cladding

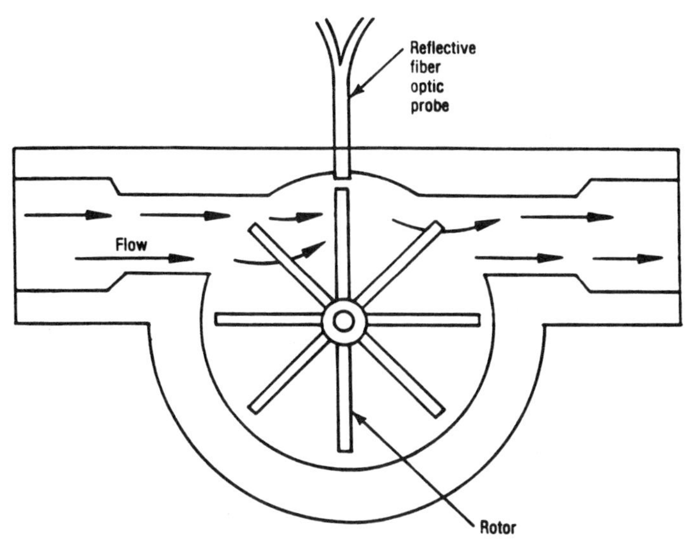

FIGURE 8.22. Fiber optic flow rate sensor operating on a reflection principle. Figure taken from Krohn [14].

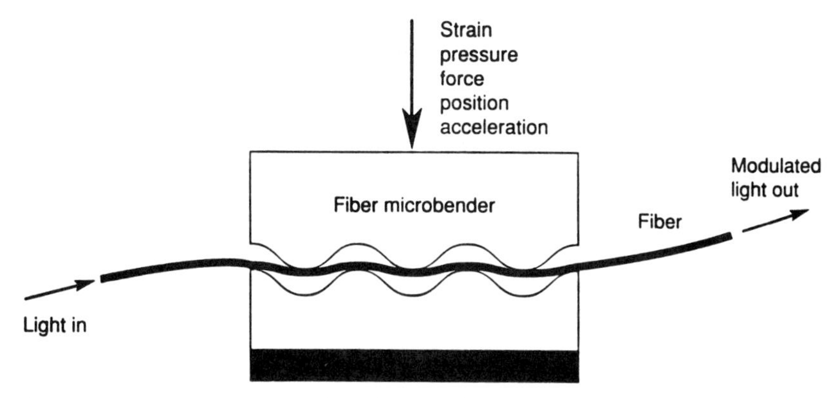

FIGURE 8.23. A microbend sensor in which a moving piece presses the fibers into a bending configuration. Figure taken from C. F. Coombs, Jr. (ed), **Electronic Instrument Handbook**, 2nd ed., (McGraw-Hill, NY, 1995). Reproduced with permission by The McGraw-Hill Companies.

because the critical angle for total internal reflection is no longer exceeded at the bends. The amount of light lost to the cladding increases with an increasing amount of bending. Figure 8.24 shows a typical graph of the sensed light intensity beyond the bends as a function of the displacement of the moving press.

Figure 8.25, from Krohn [14], shows a torque sensor which exploits the bending principle to create an intrinsic fiber optic sensor. Here motion of a cam affects the bending and thus the amount of light reaching the optical sensor.

Another example of an intrinsic fiber optic sensor based on the bending mechanism is shown in Fig. 8.26, taken from Fraden [7]. Here a U-shape is given to a

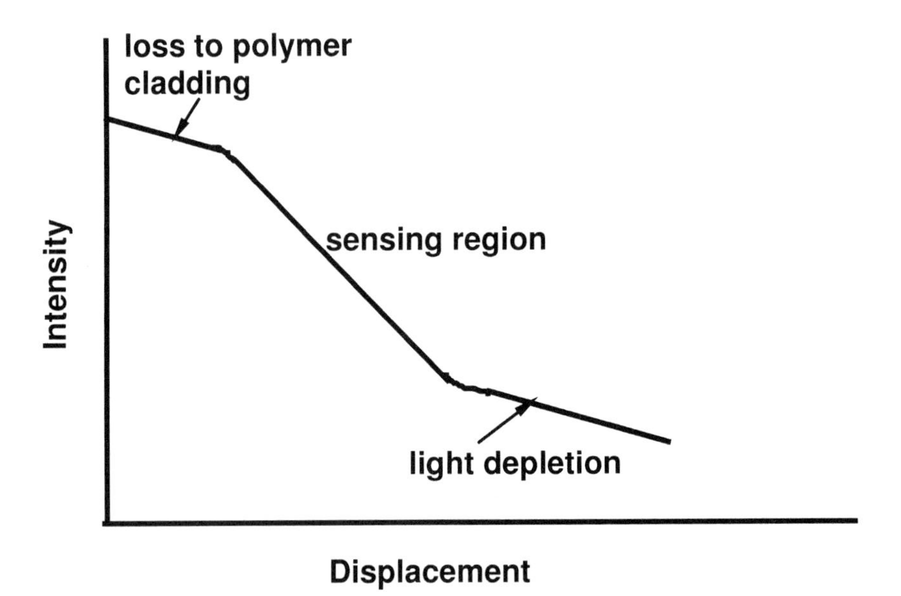

FIGURE 8.24. Intensity versus displacement of the press for the microbend sensor.

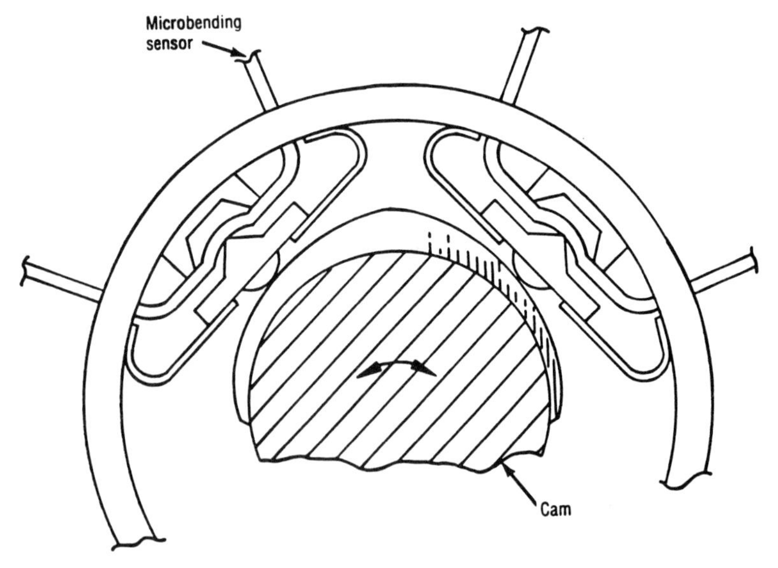

FIGURE 8.25. Fiber optic torque sensor exploiting the intrinsic bending mechanism. Figure taken from Krohn [14].

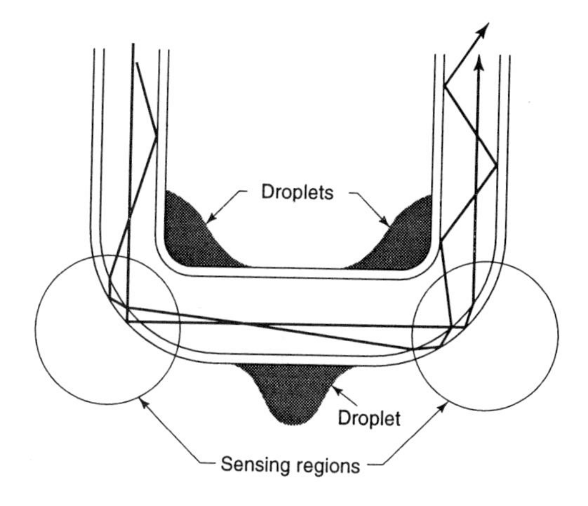

FIGURE 8.26. Liquid level sensor using bent fiber. Figure taken from Fraden [7].

fiber which is then used to detect the height of a liquid. When the height reaches the curved edges of the U, the liquid serves as cladding and, if the sensor is properly designed, total internal reflection is replaced by significant light leakage. Thus, the intensity of light reaching the sensor can be used to determine whether a threshold liquid level has been reached.

8.4 Phase-Modulated Sensors

While the intensity-modulated sensors discussed in the previous section are simpler and generally less expensive than phase-modulated sensors, phase-modulated fiber optic sensors usually are more sensitive. The typical operation of a phase-modulated sensor requires a coherent source and two single mode fibers. Figure 8.27 shows the simplest geometry. The light source is used to supply light to two fibers, one of which serves as a reference and one of which is the sensing path. The process being monitored modulates the signal in the sensing path by causing a change in some physical characteristic such as the length or index of refraction. The detector compares the received light phase from the two fibers.

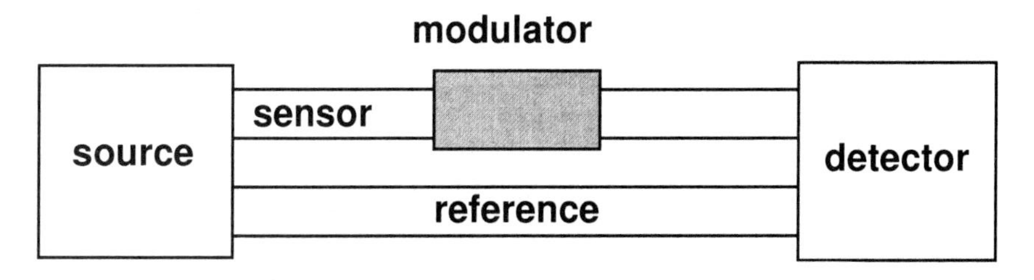

FIGURE 8.27. Simple diagram of a phase-modulated sensor.

An example of a transducer based on such a simple phase-modulated fiber optic sensor is shown in Fig. 8.28, and described in Brown et al. [17]. Here a piece of rubber is wrapped N times by a fiber. A pressure incident on the face of the rubber compresses it. Coupling described through Poisson's ratio causes simultaneous expansion in the radial direction. Hence the fiber path length changes from $2\pi N r_0$ to $2\pi N(r_0 + \Delta r)$. The amplitude of light in the fiber is given by

$$I = A\cos(c/\omega x), \tag{16}$$

where c is the speed of light propagation in the medium and ω is the frequency of the light. Based on this, the change in phase which results from compression of the rubber is given by

$$\Delta\phi = \frac{2\pi\omega N\Delta r}{c}. \tag{17}$$

It remains simply to relate the incident pressure to Δr in order to have a phase change measurement which is directly related to the pressure. The relationship between pressure and Δr is entirely a function of the material properties of the rubber and is easily expressed.

An alternative to the simple phase modulation technique described above is the use of interferometry. In interferometry we add two waves and examine the resulting patterns in detail. Suppose the first fiber is carrying light described by $a_1 \sin(\omega t + \phi_1)$ at its end, and the second fiber is carrying light described by $a_2 \sin(\omega t + \phi_2)$ at its end. The addition of these two signals yields

$$S = a\sin(\omega t + \theta), \tag{18}$$

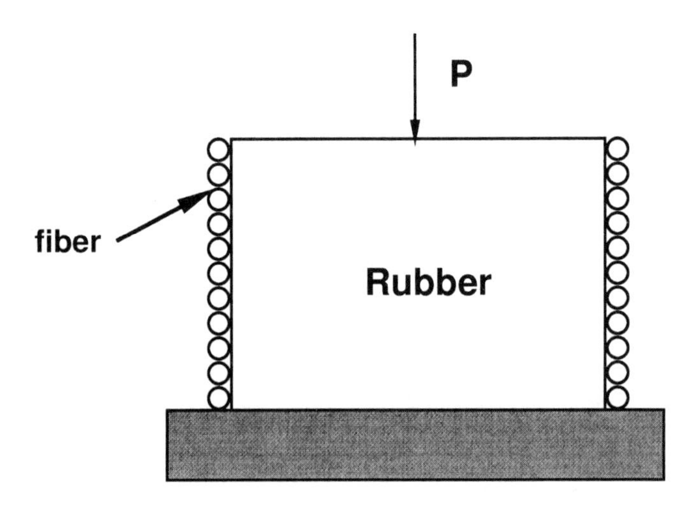

FIGURE 8.28. Fiber optic geophone, as described in Brown et al. [17], based on phase-modulated fiber optic sensor.

where

$$a = \sqrt{a_1^2 + a_2^2 + 2a_1a_2 \cos(\phi_1 - \phi_2)} \quad (19)$$

and

$$\theta = \tan^{-1}\left(\frac{a_1 \sin\phi_1 + a_2 \sin\phi_2}{a_1 \cos\phi_1 + a_2 \cos\phi_2}\right). \quad (20)$$

By monitoring the magnitude and phase of the summed signal we can determine ϕ_1 and ϕ_2. The difference between these two phases is then related to the process causing the difference in light propagation in the two fibers.

From (19) we see that the summed intensity is maximum if $\phi_1 - \phi_2 = 2\pi m$ and minimum when $\phi_1 - \phi_2 = (2m + 1)\pi$, where m is an integer. At these extremes we have, from (19),

$$I_{max} = \sqrt{a_1^2 + a_2^2 + 2a_1a_2}, \quad (21)$$

$$I_{min} = \sqrt{a_1^2 + a_2^2 - 2a_1a_2}. \quad (22)$$

We define the fringe visibility or contrast as

$$\text{fringe visibility} = \frac{I_{max} - I_{min}}{I_{max} + I_{min}}. \quad (23)$$

For a given amplitude of light at the end of the first fiber (a_1), we may use (21)–(23) to find the best choice of a_2 by differentiating (23) and setting the result to zero. After several lines of algebra, the result we obtain is that the best fringe visibility is obtained when $a_1 = a_2$. This is an advantageous result because it means that rather than using two sources or one source and an attenuator, we can simply use a single source and a beam splitter in an interferometer. The result for $a_1 = a_2$ is that $I_{\max} = 2a_1$ while $I_{\min} = 0$.

Fiber optic interferometry is generally quite expensive due to the need for a coherent source and good signal processing for fringe detection. However, it produces sensors with extremely high sensitivity to small motions or changes in physical properties of the fiber. There are a number of specific configurations for interferometers, but it is beyond the scope of this chapter to consider the advantages and disadvantages of each of these geometries. For further information on interferometry, see Udd [18].

8.5 Photostriction

The previous sections of this chapter have concentrated on optomechanical sensors, and until recently it was not clear that we could devise optomechanical actuators as well. In particular, there has been a recent flurry of work on photostrictive materials. Photostrictive materials are materials which produce a strain in response to incident light. It is thought that this is the result of two effects combining:

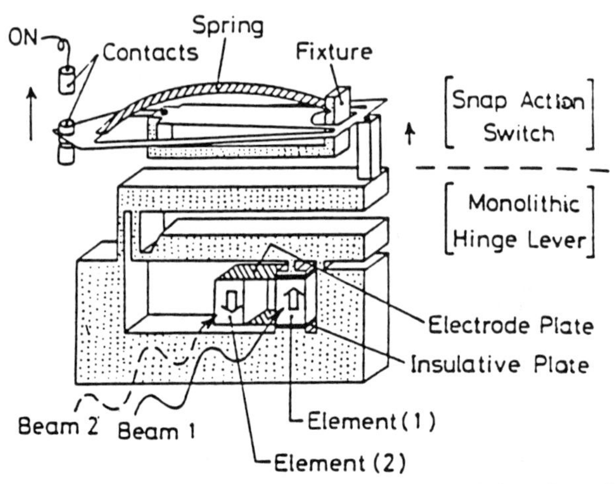

FIGURE 8.29. Photodriven relay of Uchino and Aizawa [19], based on photostriction in PLZT.

the photovoltaic effect and piezoelectricity. Hence, incident light gives rise to a potential in the material due to photovoltaic action. The potential then causes a strain due to piezoelectricity. The figure of merit for conversion of light to strain is the product of the photovoltaic electric field and a piezoelectric constant.

The photostrictive material receiving the vast majority of attention is PLZT, lead lanthanum zirconate titanate. Using this material Uchino and Aizawa [19] have developed a photodriven relay shown in Fig. 8.29. This relay uses two light beams to drive a relay arm which is connected to a mechanical amplifier and a snap switch and which contains two photostrictive elements. The two light beams are connected in parallel but exactly out of phase. Thus when light is shone on both photostrictive elements, the induced voltages cancel. Operation requires closing the relay by shining light on only one element, and then opening the relay by shining light on both photostrictors. This apparently produces a faster change than simply relying on photoconductivity in the material.

While photostriction makes it possible to conceive of optomechanical actuators for the first time, there is a long way to go before such actuators will become commercially available. At present, they are not capable of supporting much load, and their characteristics are not well documented.

8.6 Summary

Optical sensors are capturing a steadily increasing share of the sensor market. They offer the advantages of low mass, high sensitivity, and an ability to operate in a noncontact fashion. However, they tend to be expensive and to require special signal processing.

This chapter has presented popular optical sensors of the quantum detector class, focusing on photoconductive sensors. Additionally, fiber optic sensors were presented. These may be divided into two main types: intensity modulated and phase modulated. The intensity-modulated sensors are generally less expensive but less sensitive than their phase modulation counterparts.

8.7 References

1. S. P. Parker (ed.), *Optics Source Book* (McGraw-Hill, New York, 1988).
2. S. M. Juds, *Photoelectric Sensors and Controls, Selection and Applications* (Marcel Dekker, New York, 1988).
3. W. O. Grant, *Understanding Lightwave Transmission: Applications of Fiber Optics* (Harcourt Brace Jovanovich, San Diego, CA, 1988).
4. R. G. Longoria, *Wave-Scattering Formalisms for Multiport Energetic Systems*, J. Franklin Inst., **333(B)**, 539 (1996).
5. H. M. Paynter and I. J. Busch-Vishniac, *Wave-Scattering Approaches to Conservation and Causality*, J. Franklin Inst., **325**, 295 (1988).
6. A. Mamada, T. Tanaka, D. Kungwatchakun, and M. Irie, *Photoinduced Phase Transition of Gels*, Macromolecules, Feb. 5, 1990.
7. J. Fraden, *AIP Handbook of Modern Sensors* (AIP Press, New York, 1993).
8. J. G. Webster (ed.), *Tactile Sensors for Robotics and Medicine* (Wiley, New York, 1988).
9. G. A. McAlpine, *Tactile Sensing*, Sensors, April 1986.
10. J. K. Pavuluri, V. Mancevski, I. J. Busch-Vishniac, and A. B. Buckman, *Design and Performance of a Six Degree-of-Freedom, Single-Sided, Noncontact Position Sensor*, IEEE Trans. Instrumentation and Meas., under review.
11. J. Ni and S. M. Wu, *A New On-Line Measurement System for Machine Tool Geometric Errors*, Proc. 15th North Am. Manuf. Research Conf., May 1987.
12. M. W. Trethewey, H. J. Sommer, and J. A. Cafeo, *A Dual-Beam Laser Vibrometer for Measurement of Dynamic Structural Rotations and Displacements*, J. Sound Vibration, **164**, 67 (1993).
13. E. F. Carome, V. E. Kubulins, and R. L. Flanagan, *Los Cost Fiberoptic Vibration Sensors*, SPIE, vol. 1589, Specialty Fiber Optic Systems for Mobile Platforms, 133 (1991).
14. D. A. Krohn, *Fiber Optic Sensors: Fundamentals and Applications* (Instrument Society of America, Research Triangle Park, NC, 1988).
15. Mechanical Technology, Inc., Advanced Products Division, 968 Albany-Shaker Rd., Latham, NY 12110-1401, USA.
16. C. F. Coombs, Jr. (ed.), *Elecronic Instrument Handbook*, 2nd ed. (McGraw-Hill, New York, 1995).
17. D. A. Brown, D. L. Gardner, and S. L. Garrett, *Fiber-Optic Interferometric Push-Pull Hydrophones*, Naval Res. Rev., **44**, 17 (1992).
18. E. Udd (ed.), *Fiber Optic Sensors* (Wiley, New York, 1991).
19. K. Uchino and M. Aizawa, *Photostrictive Actuator Using PLZT Ceramics*, Jpn. J. Appl. Phys. **24(S3)**, 139 (1985).

Part III

Analysis of Transducers

9

2-port Theory

In the previous parts of this book we have developed tools which are useful in the modeling and analysis of transducers, and we have discussed various mechanisms through which electrical signals can be generated from mechanical signals and vice versa. In this section of the book we abandon our detailed concentration on transduction mechanisms, in favor of a discussion of general transducer characteristics. We begin this discussion with a presentation of the classical view of transducers as 2-port elements. Using 2-port models, we will develop reciprocity relations and discuss arrays of transducers which are connected. Following this development, the next chapter will discuss transducer measurement specifications, concentrating on the precise definitions (when they exist) for the various measures of performance.

9.1 Basic 2-Port Equations

Regardless of the type of transducer or its function, it can be viewed as a device which has input power and output power, i.e., in very much the way viewed in Fig. 1.4. Each port is defined by two power conjugate variables, making for a total of four variables for the transducer. Two of these variables are designated as independent, and the remaining two are determined by the details of the transducer, so they are dependent variables. The goal of 2-port theory is to develop the links between the independent and dependent variables. This is tantamount to developing a model of the performance of a transducer, and provides more or less the same information as a bond graph model, but not in the state equation formulation.

For instance, consider the piezoelectric compression accelerometer whose bond graph model is shown in Fig 5.20. We can simplify the bond graph model by noting that the case inertance is a *dependent* energy storage element (as its velocity is entirely determined by the structure to which it is attached). This means that we can eliminate the case inertance from the model, producing the bond graph shown in Fig. 9.1. The dashed box around the majority of the bond graph model permits the view of this device as a 2-port. The input port is associated with power conjugate variables V and F, while the output port is associated with variables e and i. Were we to use the analysis techniques presented in Chapter 2, we would proceed to determine state equations for the accelerometer and the output equations in terms of the state variables. In 2-port theory, essentially the same information is developed, but then presented in a manner that is much more advantageous for the evaluation of input/output relations. However, as will be discussed below, conventional 2-port theory is severely restricted in application. It is generally regarded as a purely frequency domain approach (applicable to steady-state performance analysis only) that applies to 2-ports whose elements are entirely linear. Thus while 2-port theory is useful in many cases, it is much more narrowly applicable than the modeling approaches presented previously.

There are six possible sets of choices for the independent variables used to represent any electromechanical transducer. For each of these we may generate the functional representation linking the dependent and independent variables.

For example, suppose we choose the flows f_{in} and f_{out} as the independent variables. Then, in general, we may write the efforts as

$$\begin{aligned} e_{\text{in}} &= g(f_{\text{in}}, f_{\text{out}}) \\ e_{\text{out}} &= h(f_{\text{in}}, f_{\text{out}}), \end{aligned} \tag{1}$$

where g and h are functions defined by the details of the transduction. If g and h are linear functions, then we may rewrite (1) as

$$\begin{aligned} e_{\text{in}} &= af_{\text{in}} + bf_{\text{out}} \\ e_{\text{out}} &= cf_{\text{in}} + df_{\text{out}}. \end{aligned} \tag{2}$$

(3)

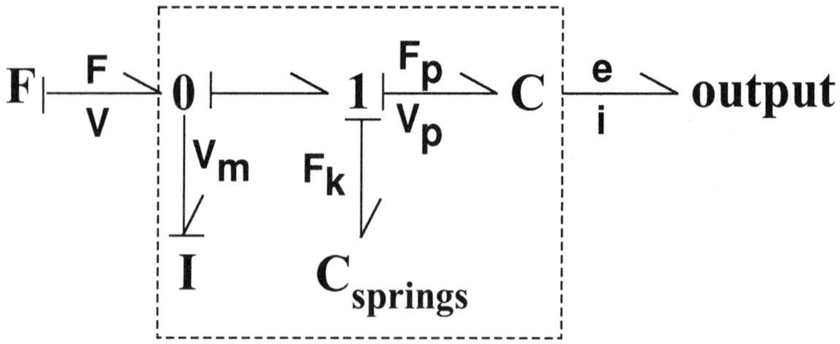

FIGURE 9.1. Bond graph and 2-port view of a piezoelectric compression accelerometer.

Here the coefficients $a, b, c,$ and d have units of impedance. This representation is called, therefore, the impedance representation and the matrix $[\mathbf{Z}]$ is the coupler between the flows and the efforts.

$$\mathbf{e} = [\mathbf{Z}]\mathbf{f}. \tag{4}$$

The $1, 1$ element of the impedance matrix is referred to as the input impedance since it is the ratio of the input effort to input flow with the output flow set to zero. Similarly the $2, 2$ element, which is defined as the ratio of the output effort to the output flow when the input flow is zero, is referred to as the output impedance. The $1, 2$ and $2, 1$ elements are transfer impedances, relating input effort to output flow, and output effort to input flow.

A second representation for the transducer uses the efforts as the independent variables and the flows as the dependent variables. Assuming linearity, this generally is written as

$$\mathbf{f} = [\mathbf{Y}]\mathbf{e}, \tag{5}$$

where $[\mathbf{Y}]$ is the mobility or admittance matrix. Y_{11} is the ratio of the input flow to the input effort with the output effort set to zero and is referred to as the input mobility. Y_{22} is the ratio of the output flow to the output effort with the input effort set to zero and is called the output mobility. Y_{12} and Y_{21} are the transfer mobilities. Note, that although the inverse of an impedance is a mobility, the definitions above make it clear that $Y_{11} \neq 1/Z_{11}$. The impedance and admittance matrices are inverses of each other however.

A third possible representation for the transducer treats the input effort and output flow as the independent variables. This representation is referred to as the conductance representation and can be written (assuming linearity) as

$$\begin{aligned} f_{\text{in}} &= g_{11}e_{\text{in}} + g_{12}f_{\text{out}}, \\ e_{\text{out}} &= g_{21}e_{\text{in}} + g_{22}f_{\text{out}}. \end{aligned} \tag{6}$$

The elements of $[\mathbf{g}]$ do not all have the same units. For example, g_{11} is the ratio of the input flow to input effort with the output effort set to zero. It is thus the inverse of the input impedance Z_{11}. Similarly g_{22} is the ratio of the output effort to output flow with the input effort set to zero. It is the inverse of the output mobility Y_{22}. The element g_{12} is the ratio of input flow to output flow with the input effort set to zero, and g_{21} is the ratio of the output effort to input flow with the output flow set to zero.

A fourth representation for the transducer is the inverse of the conductance representation and is referred to as the transconductance representation. For a linear system, the transconductance representation can be written as

$$\begin{aligned} e_{\text{in}} &= h_{11}f_{\text{in}} + h_{12}e_{\text{out}} \\ f_{\text{out}} &= h_{21}f_{\text{in}} + h_{22}e_{\text{out}}. \end{aligned} \tag{7}$$

The element h_{11} is the inverse of the input admittance Y_{11}. The element h_{22} is the inverse of the output impedance Z_{22}. Element h_{12} is the ratio of the input effort to output effort with the input flow set to zero. Note that since the constraint conditions differ, this is not the same as the inverse of g_{21}. Element h_{21} is the ratio of the output flow to input flow when the output effort is set to zero. Note that it is not the inverse of g_{12} because of different constraint conditions.

A fifth representation for the linear 2-port transducer is to use the two output port variables as the independent variables. Then we have one of the two representations which generally are referred to as transfer representations

$$\begin{aligned} e_{\text{in}} &= a_{11}e_{\text{out}} - a_{12}f_{\text{out}}, \\ f_{\text{in}} &= a_{21}e_{\text{out}} - a_{22}f_{\text{out}}. \end{aligned} \tag{8}$$

The minus signs in (8) are used to simplify the analysis when elements are cascaded. This will be discussed in Section 8.3. Note that a_{11} is the inverse of g_{21}, a_{12} is the negative inverse of Y_{21}, a_{21} is the inverse of Z_{21}, and a_{22} is the negative inverse of h_{21}.

The last remaining representation for the linear 2-port transducer is to invert the transfer representation just given to produce

$$\begin{aligned} e_{\text{out}} &= b_{11}e_{\text{in}} - b_{12}f_{\text{in}}, \\ f_{\text{out}} &= b_{21}e_{\text{in}} - b_{22}f_{\text{in}}. \end{aligned} \tag{9}$$

Element b_{11} is the inverse of h_{12}, b_{12} is the negative inverse of Y_{12}, b_{21} is the inverse of Z_{12}, and b_{22} is the negative inverse of g_{12}.

Let us return to the piezoelectric compression accelerometer whose model is shown in Fig. 9.1, and demonstrate how one develops the 2-port equations in one of the six forms just presented. First, consider the piezoelectric element represented by the 2-port capacitance. Assuming the element (or stack of elements) to be oriented in the 3-direction and strained in the 3-direction, we may write the equations describing this piezoelectric material directly from Eq. (5.30). The result is

$$\begin{aligned} F_p &= \frac{Ak_p}{l}x - h_{33}q, \\ e &= -h_{33}x + \frac{l}{\epsilon A}q, \end{aligned} \tag{10}$$

where A is the cross-sectional area of the piezoelectric element and is also assumed to be the area of the electrodes, l is the thickness of the element and the separation distance between the electrodes, and k_p is the mechanical stiffness of the element. While these equations are appropriate for a state equation formulation, in conventional 2-port theory we use power conjugate variables exclusively. Hence we must next rewrite the equations in terms of V_p (which we define as dx/dt) and i. If we assume steady-state excitation at frequency ω, then differentiation and integration simply result in multiplication and division by factors of $j\omega$. Hence we have

$$\begin{aligned} F_p &= \frac{Ak_p}{lj\omega}V_p - \frac{h_{33}}{j\omega}i, \\ e &= -\frac{h_{33}}{j\omega}V_p + \frac{l}{\epsilon Aj\omega}i. \end{aligned} \tag{11}$$

Since we seek to express the transducer behavior by linking output and input port variables, we need to eliminate F_p and V_p, so we solve the first of the pair for i in terms of F_p and V_p and then substitute the result into the second equation. This permits us to rewrite the equations as follows:

$$
\begin{aligned}
e &= -\frac{l}{\epsilon A h_{33}} F_p + \left(\frac{k_p}{\epsilon h_{33}} - \frac{h_{33}}{j\omega} \right) V_p, \\
i &= -\frac{j\omega}{h_{33}} F_p + \frac{A k_p}{l h_{33}} V_p.
\end{aligned} \tag{12}
$$

Now we can examine the remainder of the accelerometer elements in order to eliminate the variables F_p and V_p in favor of F and V. We do this by noting, from the bond graph model, the following junction relations:

$$
\begin{aligned}
F_p &= F - F_k, \\
V &= V_m + V_p, \\
V_k &= V_p,
\end{aligned} \tag{13}
$$

and the following constitutive relations:

$$
\begin{aligned}
F_k &= k x_k = \frac{k}{j\omega} V_p, \\
p_m &= m V_m = \frac{F}{j\omega},
\end{aligned} \tag{14}
$$

where k is the stiffness of the retaining springs attached to the lead mass, x_k is the compression of those springs, and p_m is the linear momentum of the mass. These permit us to write the following:

$$
\begin{aligned}
V_p &= -\frac{1}{j\omega m} F + V, \\
F_p &= \left(1 - \frac{k}{\omega^2 m} \right) F - \frac{k}{j\omega} V.
\end{aligned} \tag{15}
$$

Substituting these into (12) yields the following equations in 2-port standard form:

$$
\begin{aligned}
e &= \left[\frac{l}{\epsilon A h_{33}} \left(\frac{k}{\omega^2 m} - 1 \right) - \frac{k_p}{j\omega m \epsilon h_{33}} - \frac{h_{33}}{\omega^2 m} \right] F \\
&\quad + \left[\frac{kl}{j\omega \epsilon A h_{33}} + \frac{k_p}{\epsilon h_{33}} - \frac{h_{33}}{j\omega} \right] V, \\
i &= \left[-\frac{j\omega}{h_{33}} - \frac{k}{j\omega m h_{33}} - \frac{A k_p}{l h_{33} j\omega m} \right] F + \left[\frac{k}{h_{33}} + \frac{A k_p}{l h_{33}} \right] V.
\end{aligned} \tag{16}
$$

As written in (16), we have derived the [**b**] representation for the piezoelectric compression accelerometer. While the expressions are long, it is a simple matter to identify the elements of the matrix, all of which are functions only of frequency and of the system parameters. We might well choose to simplify these expressions by recalling that piezoelectric accelerometers are usually operated below their resonance frequencies. Thus, one could take the low-frequency limit of (16) to eliminate some terms. Similarly, we could take the high-frequency limit to examine the accelerometer behavior in the range not normally used in sensing. This is a particularly good exercise for the curious reader because it provides the justification for the choice to exploit the accelerometer only in the low-frequency regime. Finally, we note that having derived the **[b]** representation for the accelerometer, it is a simple matter to rearrange terms so as to find the **[Z]**, **[Y]**, **[g]**, **[h]**, and **[a]** representations.

In developing the six general representations of the 2-port behavior of a transducer, a linear representation was chosen in each case. It is reasonable to question whether a similar analysis could be applied to transducers with nonlinear behavior. Indeed, given the four port variables associated with a 2-port device, the six representations merely define causal structure in the sense that they establish one pair of variables as the independent variables and the other pair as the dependent variables. Clearly this concept does not rely on the linearity of the relationship. What does rely on the linearity is the conventional use of matrices to represent the functional relationship between the dependent and independent variables. For linear systems, the matrices consist of elements whose values are set by the system parameters (such as mass, nominal capacitance, etc.). The elements are thus not functions of the independent port variables. Were the system to be nonlinear, then the elements of any particular matrix could be functions of the independent port variables. This certainly creates an added sophistication, but not an insurmountable one. However, it should be noted that there is an important distinction between nonlinear and linear transduction representations using conventional 2-port theory. In linear systems we could clearly derive any of the six 2-port representations from any one of the others. In nonlinear systems the matrix elements are permitted to be functions of the independent parameters and it is not always true that we can invert these relations to derive functional forms appropriate for a different choice of independent parameters. Thus, the 2-port representations presented above should be viewed as applying generally to linear or nonlinear transduction relationship, but with the caveat that in the case of nonlinear transducers, it might not be possible to uniquely define all six representations.

An additional point to note is that the traditional view of the various 2-port representations uses a frequency domain approach. For instance, the impedance matrix elements are impedances which are typically defined as a function of frequency. Again, this perspective arises more from history than from necessity. There is nothing in the development of 2-port representations that necessitates using a frequency domain approach. It simply makes it easy to represent differentiation and integration in time. However, the same approach can be used in the time domain, with a differentiation or integration operator replacing multiplication or division by $j\omega$.

With the six possible representations for a transducer given above, it is reasonable to ask which is more appropriate for any given situation. There are generally three issues to consider. First, which sets of coefficients are easiest to measure? When there is a clear answer to this question, it usually is best to choose the most easily measured parameters, as this usually produces the smallest errors. Second, What is the causal structure of the system? In some cases we cannot convert easily from one form to another without significant degradation due to noise. Third, are pieces of the transducer connected in a way which is best described by a particular representation? These issues will be discussed in the following sections of this chapter.

9.2 Reciprocity

Suppose we have a passive transducer where energy can flow in either direction, i.e., which can function as either a sensor or an actuator. (Of course, the transducer is presumably designed to function primarily as *either* a sensor *or* an actuator because it cannot simultaneously perform both tasks well.) We will refer to such a transducer as a reciprocal transducer. For reciprocal transducers, it is possible to derive relations which must hold true and which reduce some of the problems associated with measuring the elements of the various 2-port matrices.

Then it is immaterial which port we label the input port and which is labeled the output port. Let us choose the transfer representations of the transducer. Then from (8) and (9), both of which we assume can be uniquely defined, we have the following:

$$\begin{aligned} \begin{bmatrix} e_{in} \\ f_{in} \end{bmatrix} &= \begin{bmatrix} a_{11} & -a_{12} \\ a_{21} & -a_{22} \end{bmatrix} \begin{bmatrix} e_{out} \\ f_{out} \end{bmatrix} \\ &= \begin{bmatrix} a_{11} & -a_{12} \\ a_{21} & -a_{22} \end{bmatrix} \begin{bmatrix} b_{11} & -b_{12} \\ b_{21} & -b_{22} \end{bmatrix} \begin{bmatrix} e_{in} \\ f_{in} \end{bmatrix} \\ &= \begin{bmatrix} a_{11}b_{11} - a_{12}b_{12} & -a_{11}b_{12} + a_{12}b_{22} \\ a_{21}b_{11} - a_{22}b_{21} & -a_{21}b_{12} + a_{22}b_{22} \end{bmatrix} \begin{bmatrix} e_{in} \\ f_{in} \end{bmatrix}. \end{aligned} \tag{17}$$

We identify the last matrix in (17) as the identity matrix. Hence the following relations must hold true:

$$\begin{aligned} a_{11}b_{11} - a_{12}b_{21} &= 1, \\ -a_{11}b_{12} + a_{12}b_{22} &= 0, \\ a_{21}b_{11} - a_{22}b_{21} &= 0, \\ -a_{21}b_{12} + a_{22}b_{22} &= 1. \end{aligned} \tag{18}$$

The relations of (18) might seem difficult to solve, being a set of four nonlinear algebraic equations in eight unknowns. However, there is a unique, easily verified solution, which is given below:

$$\begin{aligned} a_{11} &= b_{22} = A, \\ a_{12} &= b_{12} = B, \\ a_{21} &= b_{21} = C, \end{aligned}$$

$$a_{22} = b_{11} = D,$$
$$AD - BC = 1. \tag{19}$$

Here A, B, C, and D are the four transfer matrix components that define the transducer. Equations (19) are called the reciprocity relations.

Now let us examine the relations of (19). The first equation tells us

$$\left.\frac{e_{\text{in}}}{e_{\text{out}}}\right|_{f_{\text{out}}=0} = -\left.\frac{f_{\text{out}}}{f_{\text{in}}}\right|_{e_{\text{in}}=0}, \tag{20}$$

and that $g_{12} = g_{21}$. We may interpret this by imagining that we have a bar of solid material. One end of the bar we identify as the input end and the other as the output end, although this assignment is clearly arbitrary. Then (20) says that measurement of the force at the input end, relative to that at the output end when the output end is clamped, is the same as the negative of the output end velocity relative to the input end velocity when the input end is free (i.e., unforced). Of these two measurements, the former is very difficult to make because it is nearly impossible to measure a force at a clamped location. Hence, we could make the latter measurement instead and equate the two in a reciprocal system.

The second equation in (19) says that

$$-\left.\frac{e_{\text{in}}}{f_{\text{out}}}\right|_{e_{\text{out}}=0} = -\left.\frac{e_{\text{out}}}{f_{\text{in}}}\right|_{e_{\text{in}}=0}, \tag{21}$$

and that $Y_{21} = Y_{12}$ in a reciprocal transducer. Equation (21) may be easily interpreted as simply stating that if energy can flow in either direction, then it is clearly arbitrary which end of the transducer is identified as the input end and which as the output end. Simply substituting *out* for *in* and *in* for *out* on the left-hand side of the equation produces the right-hand side of the equation.

The third equation in (19) is similar to the second in that it simply expresses the arbitrariness of assignment of one port as the input port and one as the output port. This equation yields

$$\left.\frac{f_{\text{in}}}{e_{\text{out}}}\right|_{f_{\text{out}}=0} = \left.\frac{f_{\text{out}}}{e_{\text{in}}}\right|_{f_{\text{in}}=0}, \tag{22}$$

and the fact that in a reciprocal transducer $Z_{12} = Z_{21}$.

The remaining identity in Eq. (19) says that

$$-\left.\frac{f_{\text{in}}}{f_{\text{out}}}\right|_{f_{\text{out}}=0} = \left.\frac{e_{\text{out}}}{e_{\text{in}}}\right|_{f_{\text{in}}=0}, \tag{23}$$

and $h_{12} = h_{21}$. Again we may interpret this relation in terms of the bar. Reciprocity guarantees that the negative ratio of the output velocity to the input velocity when the output end is free equals the ratio of the output force to the input force when the input end is clamped. Since it is almost impossible to measure force at a clamped end, the former measurement is the preferred one in a reciprocal system.

It should be noted that the reciprocity relations derived above solve some but *not all* of the potential measurement problems that could arise in the characterization of transducers. For example, consider parameter C and the example of the solid bar. While reciprocity provides two alternatives for the measurement, both require measurement of the force at a clamped end at a mechanical port. Historically, we solve this problem by using something other than reciprocity relations. For example, while it is *not true* that $Y_{ij} = 1/Z_{ij}$, it is true that $[\mathbf{Y}] = [\mathbf{Z}]^{-1}$. Hence, if the mobility representation can be found, than the impedance representation can be determined. For a mechanical system, the mobility representation requires measurement of velocities at the ends which are free.

The problem in determining one representation for the transducer from another is related to causality. The tacit assumption made in using one representation to obtain an alternative representation is that either is causally consistent. This is rarely the case. Consider an element of a system, El, which we assume for now to be a 1-port element. Then there are two causal structures for the element. If the causality is as indicated in Fig. 9.2(a) then the element specifies the effort and the associated flow may be found by multiplying this effort and the element mobility. If the causal structure is an indicated in Fig. 9.2(b), then the element specifies that the flow and the associated effort may be found by multiplying this flow by the element impedance. Hence we may think of the representation shown in Fig. 9.2(a) as the mobility representation and that in Fig. 9.2(b) as the impedance representation. Note that the representation indicates which measurement should be made to characterize the element. The problem with choosing to make an alternative, easier measurement, and somehow to convert the information into the appropriate form is that in so doing causality is violated. This is not a problem in an ideal system which is without noise and significant nonlinearities, and which can be described by constitutive relations that are uniquely invertable. In other systems, it is possible that the conversion process from one representation to another will lead to errors.

For example, consider the mobility versus frequency curve shown in Fig. 9.3. This figure is typical of a mechanical system which displays many resonances. Note that there is a noise floor which obscures the dips in mobility. As a result,

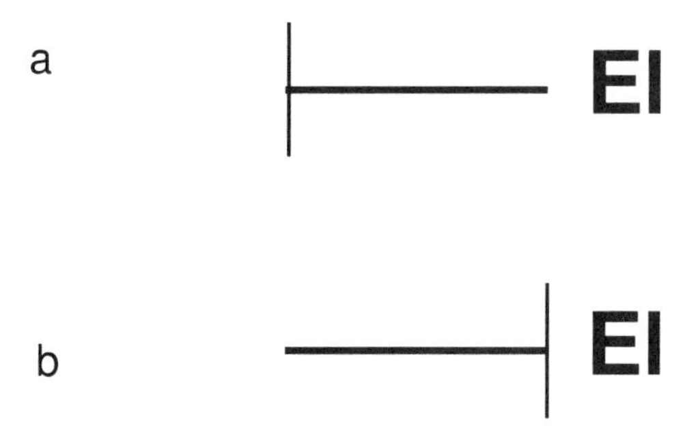

FIGURE 9.2. Element in a vibration isolation system: (a) mobility representation and (b) impedance representation.

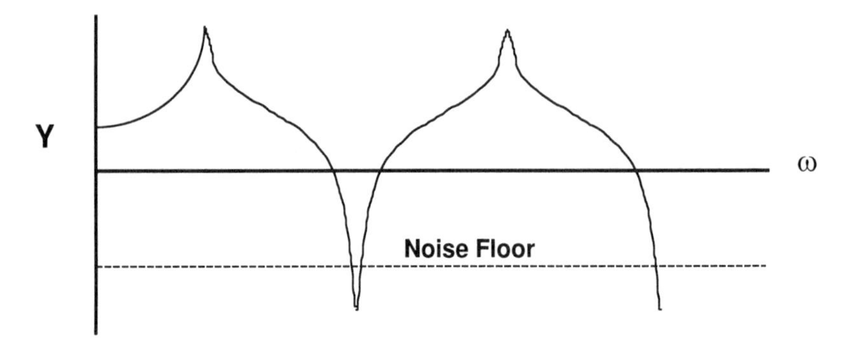

FIGURE 9.3. Mobility versus frequency for a mechanical structure.

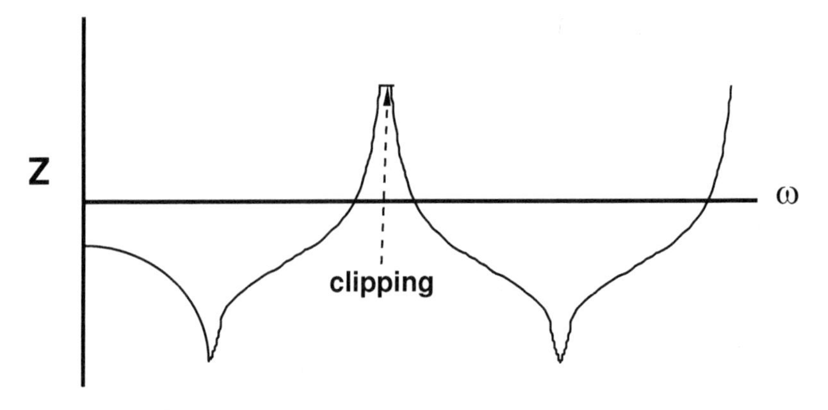

FIGURE 9.4. Impedance versus frequency for a mechanical structure, as obtained by inverting a mobility measurement.

the impedance we would obtain from inversion of the mobility would have peaks that are clipped, as shown in Fig. 9.4. For further information on the effect of causality on measurement, see Busch-Vishniac [1]. For additional information on reciprocity, particularly as it applies to transducer calibration, see References [2] – [4].

The bottom line is that reciprocity is a useful concept that can be employed to solve some thorny measurement problems. However, in real transducers with noise and nonlinearities, reciprocity is only a partial solution.

9.3 Connected 2-Ports

As mentioned previously, we might choose to use a particular representation for a transducer because it makes it possible to measure the 2-port characteristics. Another reason we might choose a particular representation for a transducer is the desire to connect multiple 2-ports in some particular fashion. For example, suppose that we have two 2-port transducers which we wish to connect so that their flows are the same at the input and output ports. Then Fig. 9.5 shows a bond graph model of the combined pair. The input effort is the sum of the input efforts to the two 2-port elements. Similarly the combined output effort is the sum of the output efforts associated with the two 2-ports. Hence we may write:

$$\begin{aligned}\begin{bmatrix} e_{\text{in}} \\ e_{\text{out}} \end{bmatrix} &= \begin{bmatrix} e^1_{\text{in}} + e^2_{\text{in}} \\ e^1_{\text{out}} + e^2_{\text{out}} \end{bmatrix} \\ &= [\mathbf{Z}^1]\begin{bmatrix} f_{\text{in}} \\ f_{\text{out}} \end{bmatrix} + [\mathbf{Z}^2]\begin{bmatrix} f_{\text{in}} \\ f_{\text{out}} \end{bmatrix} \\ &= ([\mathbf{Z}^1] + [\mathbf{Z}^2])\begin{bmatrix} f_{\text{in}} \\ f_{\text{out}} \end{bmatrix},\end{aligned} \tag{24}$$

where the superscripts indicate the number of the element. The result is that the impedance matrices simply add in 2-ports connected so that their input and output flows are common. Hence, the impedance matrix representation is a convenient choice for such a system.

Suppose now that we consider two 2-ports connected as shown in Fig. 9.6. Then the input efforts and the output efforts are common for the 2-ports, while the input flow is the sum of the input flows of each of the 2-ports and the output flow is the sum of the individual output flows. Hence we may write:

$$\begin{aligned}\begin{bmatrix} f_{\text{in}} \\ f_{\text{out}} \end{bmatrix} &= \begin{bmatrix} f^1_{\text{in}} + f^2_{\text{in}} \\ f^1_{\text{out}} + f^2_{\text{out}} \end{bmatrix} \\ &= [\mathbf{Y}^1]\begin{bmatrix} e_{\text{in}} \\ e_{\text{out}} \end{bmatrix} + [\mathbf{Y}^2]\begin{bmatrix} e_{\text{in}} \\ e_{\text{out}} \end{bmatrix} \\ &= ([\mathbf{Y}^1] + [\mathbf{Y}^2])\begin{bmatrix} e_{\text{in}} \\ e_{\text{out}} \end{bmatrix}.\end{aligned} \tag{25}$$

The result is that the admittance matrices add so we naturally tend to choose an admittance representation.

In a similar fashion we could connect two 2-ports with common effort at one port and common flow at the other. As one would suspect, the conductance and

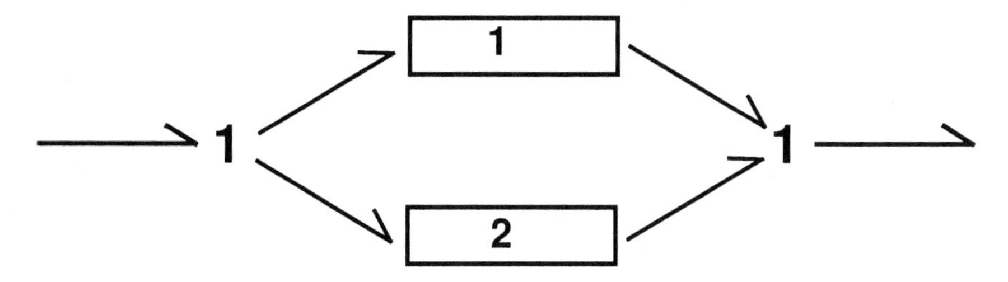

FIGURE 9.5. 2-Ports added so input and output flows are common.

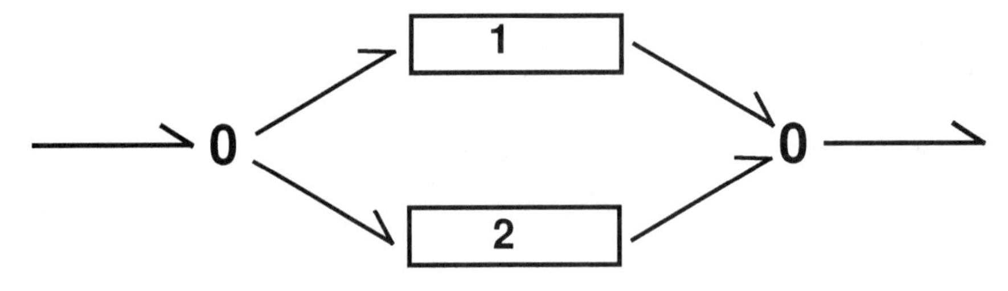

FIGURE 9.6. 2-Ports added so input and output efforts are common.

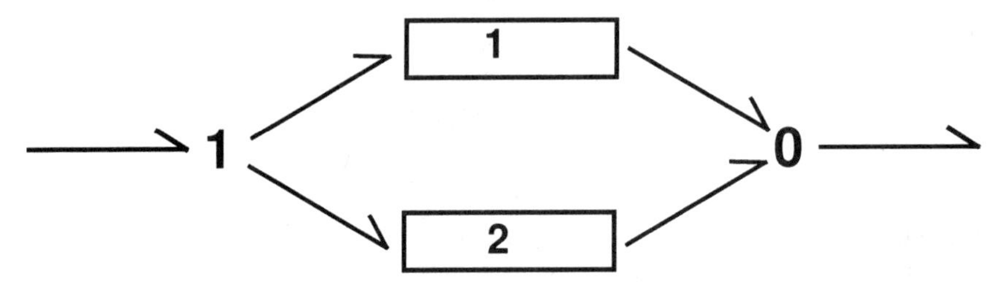

FIGURE 9.7. 2-Ports added so input flows and output efforts are common.

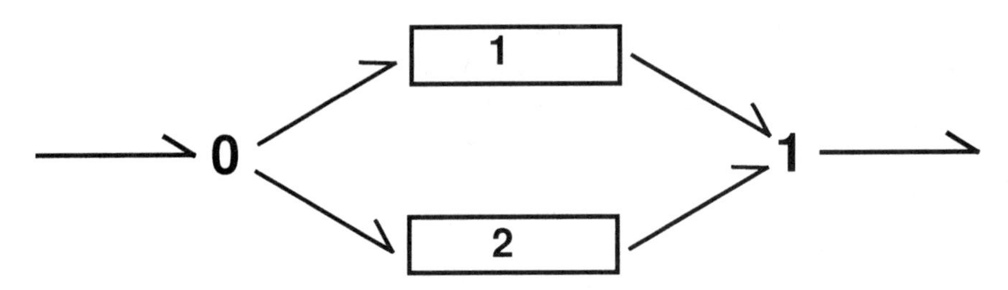

FIGURE 9.8. 2-Ports added so input efforts and output flows are common.

transconductance representations are most convenient for these sorts of connections. For the connection geometry shown in Fig. 9.7, the [**g**] add. For the geometry shown in Fig. 9.8, the [**h**] add.

One final interesting connection geometry for 2-port elements is shown in Fig. 9.9. Here the 2-ports are cascaded in a line. Using the [**a**] representation (with elements A, B, C, and D) we can write:

$$\begin{aligned} \begin{bmatrix} e_1 \\ f_1 \end{bmatrix} &= \begin{bmatrix} A^1 & B^1 \\ C^1 & D^1 \end{bmatrix} \begin{bmatrix} e_2 \\ f_2 \end{bmatrix} \\ &= \begin{bmatrix} A^1 & B^1 \\ C^1 & D^1 \end{bmatrix} \begin{bmatrix} A^2 & B^2 \\ C^2 & D^2 \end{bmatrix} \begin{bmatrix} e_3 \\ f_3 \end{bmatrix}. \end{aligned} \tag{26}$$

This allows us to put in general form the expression for cascaded elements:

$$[\mathbf{a}] = \prod_i [\mathbf{a}^i]. \tag{27}$$

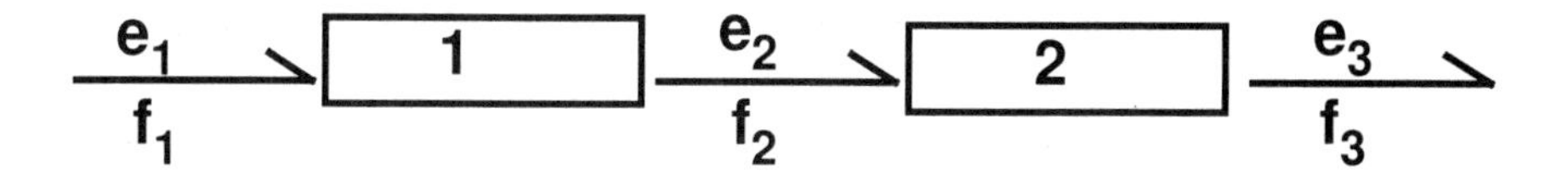

FIGURE 9.9. 2-Ports added so they are cascaded.

While a connection geometry might specify which representation for the combined 2-ports is easiest to derive, it should be kept in mind that in ideal, linear, passive systems we can easily derive any of the six representations from any other. Further, while the examples here use two connected 2-ports, the results may be extended without a change in form to any number of 2-port elements connected. For an example of the use of 2-port theory applied to a spatially continuous transducer, see Busch-Vishniac [4]. For further information on connected 2-ports, see Hilgevoord [6].

9.4 Transfer Matrix and Sensitivity

The previous section of this chapter has discussed the convenient representations for connected 2-ports. It was shown that the transfer matrix is useful for cascaded elements. In this section of the chapter we show that the transfer matrix representation is also useful for the derivation of the sensitivity of a transducer.

First, let us define transducer sensitivity as the ratio of an output variable to an input variable under specified conditions. In essence, the transducer sensitivity is nothing more than a transfer function with a particular form. For example, consider the two transfer matrices [**a**] and [**b**]. They are defined by

$$\begin{bmatrix} e_1 \\ f_1 \end{bmatrix} = [\mathbf{a}] \begin{bmatrix} e_2 \\ -f_2 \end{bmatrix}, \tag{28}$$

and

$$\begin{bmatrix} e_2 \\ f_2 \end{bmatrix} = [\mathbf{b}] \begin{bmatrix} e_1 \\ -f_1 \end{bmatrix}. \tag{29}$$

The elements of [**a**] are ratios of input variables to output variables with constraints applied to the *output* variables. Thus they are inverses of sensitivities. The elements of [**b**] are the ratios of output variables to input variables with constraints applied on the *input* variables. The [**b**] elements thus have units of transducer sensitivity.

Clearly, either the [**a**] or [**b**] matrix could be used to find the desired transducer sensitivity. The choice of which matrix to use is usually determined by the practical means available for restricting conditions at one of the ports. For sensors, we monitor the output signal and thus we typically apply conditions to the output port. This means that the [**a**] matrix is best for determining the sensitivity. For actuators, we control the transducer input and thus we are best able to restrict the input port. For actuators, then, we would determine the sensitivity from the [**b**] matrix.

As an example, consider an accelerometer. Then the input power conjugate variables are force and velocity and the output variables are voltage and current. We define two very useful transducer sensitivities: the voltage sensitivity and the current sensitivity. The voltage sensitivity is

$$\psi_v = \left.\frac{\text{voltage out}}{\text{acceleration in}}\right|_{current=0}. \tag{30}$$

It is sometimes referred to as the open circuit voltage sensitivity. We can rewrite this equation as

$$\psi_v = \left.\frac{e_2}{\partial f_1/\partial t}\right|_{f_2=0}. \tag{31}$$

Using a frequency domain view of the sensitivity, we recognize that differentiation in time produces a factor of $j\omega$. Hence we have

$$\begin{aligned}\psi_v &= \left.\frac{1}{j\omega}\frac{e_2}{f_1}\right|_{f_2=0}\\ &= \frac{1}{j\omega a_{21}}.\end{aligned} \tag{32}$$

The current sensitivity of the same accelerometer is defined as

$$\psi_c = \left.\frac{\text{current out}}{\text{acceleration in}}\right|_{\text{voltage}=0}. \tag{33}$$

It is sometimes called the short circuit current sensitivity because of the output constraints. From this definition we have

$$\begin{aligned}\psi_c &= \left.\frac{f_2}{\partial f_1/\partial t}\right|_{e_2=0}\\ &= \left.\frac{1}{j\omega}\frac{f_2}{f_1}\right|_{e_2=0}\\ &= -\frac{1}{j\omega a_{22}}.\end{aligned} \tag{34}$$

Note that in both cases, the sensitivity is defined in terms of the power conjugate variables of the transducer. Further, the sensitivity can be expressed in each case as a function of the inverse of some element of [**a**].

As a specific example, let us return to the piezoelectric compression accelerometer discussed previously in this chapter and modeled in Fig. 9.1. Using the reciprocity relations, (19), we can write the voltage sensitivity as

$$\psi_v = \frac{1}{j\omega b_{21}}, \tag{35}$$

and the current sensitivity as

$$\psi_c = -\frac{1}{j\omega b_{11}}. \tag{36}$$

We may write these sensitivities directly from (16) using the appropriate **b** matrix components. The result, after some manipulation, is as follows:

$$\begin{aligned} \psi_v &= \frac{h_{33} m}{\omega^2 m - (k + A k_p / l)} \\ \psi_c &= \frac{j\omega m \epsilon A h_{33}}{(kl - h_{33}^2 \epsilon A) + j\omega k_p A - m l \omega^2}. \end{aligned} \tag{37}$$

First consider the result given in (37) for the voltage sensitivity. It shows that the voltage sensitivity increases with the piezoelectric constant h_{33} as we would expect. Further it shows that the response is divided into two distinct regions depending on which term in the denominator of the sensitivity expression dominates. At low frequency, the sensitivity is independent of frequency, i.e., flat. At high frequencies, the sensitivity varies as $1/\omega^2$. The transition, when the two terms in the denominator are the same size, defines a resonance frequency of the transducer.

Next consider the current sensitivity of the accelerometer. This expression is more complex, but it too generally increases with increases in the piezoelectric constant, h_{33}. At low frequencies, the current sensitivity is not flat, but varies as ω. For this reason, it is common to use a *charge amplifier* rather than a *current amplifier* for such an accelerometer. Using a charge amplifier means that the charge sensitivity, which is simply the current sensitivity divided by $j\omega$, is the measure of performance. Clearly, the charge sensitivity is flat at low frequencies.

For accelerometers, the charge and voltage sensitivities might be quoted and displayed as a function of frequency. When only one sensitivity is given it is usually the charge sensitivity, $\psi_c/j\omega$, because its magnitude is greater than the voltage sensitivity at low frequencies.

9.5 Wave Matrix Representation and Efficiency

When characterizing the performance of a transducer we typically use two measures: sensitivity and efficiency. The sensitivity, as defined previously, is a measure of how well the transducer does its intended job. For accelerometers it measures the signal produced relative to the acceleration input. For pressure sensors it is a measure of the signal produced relative to the pressure input. By contrast, the efficiency of a transducer is a measure of how much energy (or power) it takes in performing the task. Formally, the efficiency is defined as the ratio of the output energy (or power) of a transducer relative to the input energy (or power). In this section of the chapter we will describe how to determine the efficiency of a transducer.

Let us begin with consideration of how we define power. So far we have defined power as the product of the power conjugate variables: $\mathcal{P} = ef$. If we use this definition to draw lines of constant power on an effort versus flow graph, as shown in Fig. 9.10, then we produce hyperboli. The farther the hyperboli from the origin, the higher the power magnitude. Given that the lines of constant power form hyperboli, this suggests an alternative form which could be used in the definition of power:

$$\mathcal{P} = c(r^2 - s^2), \tag{38}$$

where c is a constant and r and s are variables. Equation (38) is the standard definition used for hyperboli, but it requires some interpretation in the context of power. Assuming that c is a positive constant, the term cr^2 represents the positive flow of power. The term cs^2 represents the negative flow of power. The words *negative* and *positive* here refer to direction of flow. For instance, if we examine a bond between subsystems A and B, then the positive power might be defined as that moving from A to B and the negative power as that moving from B to A.

We can use the representation of power given in (38) in deriving the efficiency of a complex physical system. For this purpose we choose c to be dimensionless. Then r and s must have units of the square root of power. Using r and s as the

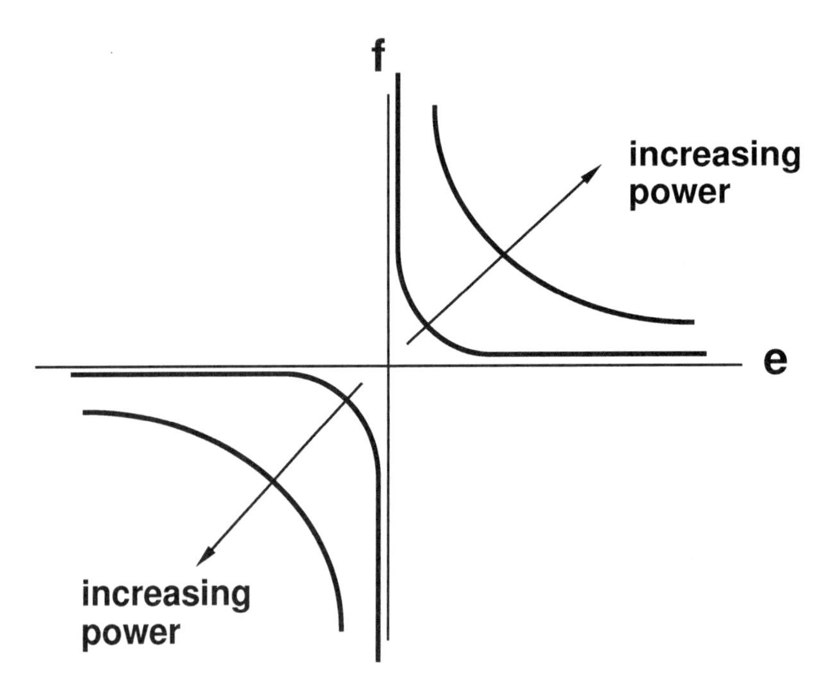

FIGURE 9.10. Graph of constant power curves.

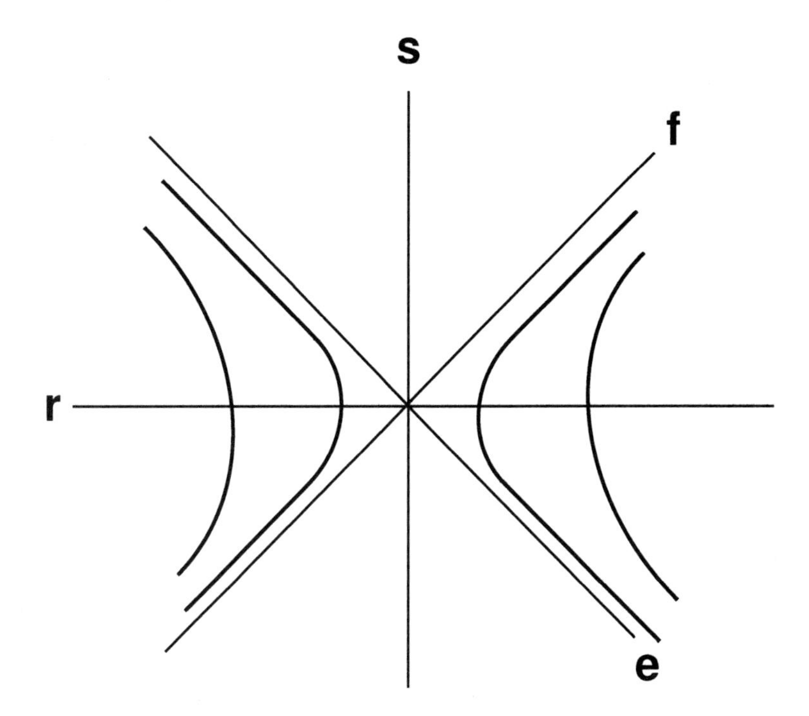

FIGURE 9.11. Constant power lines as a function of r and s.

fundamental variables, we may redraw the lines of constant power, as shown in Fig. 9.11. The fundamental change compared with the effort and flow representation is a 90° rotation.

In order to find the efficiency of a transducer, we seek a link between the r and s variables on the input and output ports (or variables like r and s). In particular we seek the following sort of representation:

$$\begin{bmatrix} r_1 \\ s_1 \end{bmatrix} = [\mathbf{M}] \begin{bmatrix} r_2 \\ s_2 \end{bmatrix}. \tag{39}$$

Note that [**M**] contains elements each of which are a ratio of an input root power to an output root power with constraints applied on the output power. These elements are identified as the inverse square roots of transducer efficiencies.

In order to find the appropriate representation, we need to do a couple of variable transformations. First, we convert from effort and flow to intrinsic effort and intrinsic flow. The intrinsic effort, ϵ, is defined as follows:

$$\epsilon = \frac{e}{\sqrt{Z_0}}, \tag{40}$$

where Z_0 is a reference impedance (e.g., 1 Ohm). The intrinsic flow is defined as

$$\phi = f\sqrt{Z_0}. \tag{41}$$

Note that with these definitions, power is still represented as the product of effort and flow (i.e., intrinsic effort and intrinsic flow). However, the units of intrinsic effort and intrinsic flow are the square root of power.

Next we transform from the intrinsic effort and flow to the wave-scatter variables $\overleftarrow{w}$ and $\overrightarrow{w}$. These variables are defined by

$$\begin{bmatrix} \overrightarrow{w} \\ \overleftarrow{w} \end{bmatrix} = \frac{1}{\sqrt{2}} \begin{bmatrix} 1 & 1 \\ 1 & -1 \end{bmatrix} \begin{bmatrix} \epsilon \\ \phi \end{bmatrix}. \tag{42}$$

The inverse transformation is

$$\begin{bmatrix} \epsilon \\ \phi \end{bmatrix} = \frac{1}{\sqrt{2}} \begin{bmatrix} 1 & 1 \\ 1 & -1 \end{bmatrix} \begin{bmatrix} \overrightarrow{w} \\ \overleftarrow{w} \end{bmatrix}, \tag{43}$$

which shows that the transformation and its inverse are identical. We refer to $\overrightarrow{w}$ as the *forewave* and to $\overleftarrow{w}$ as the *backwave*. These terms refer to the direction of power flow.

From (43) we see that

$$\epsilon\phi = \frac{1}{2}(\overrightarrow{w}^2 - \overleftarrow{w}^2). \tag{44}$$

This equation clearly defines the hyperboli of equal power. The first term is the forepower, i.e., the power flowing to the right. The second term is the backpower, or the power flowing to the left.

Now suppose that we have a transducer. We begin with a description of the transducer in terms of [**a**]:

$$\begin{bmatrix} e_{\text{in}} \\ f_{\text{in}} \end{bmatrix} = \begin{bmatrix} a_{11} & -a_{12} \\ a_{21} & -a_{22} \end{bmatrix} \begin{bmatrix} e_{\text{out}} \\ f_{\text{out}} \end{bmatrix}. \tag{45}$$

We may rewrite this expression in terms of the intrinsic variables:

$$\begin{aligned} \begin{bmatrix} \epsilon_{\text{in}}\sqrt{Z_0} \\ \phi_{\text{in}}/\sqrt{Z_0} \end{bmatrix} &= \begin{bmatrix} a_{11} & -a_{12} \\ a_{21} & -a_{22} \end{bmatrix} \begin{bmatrix} \epsilon_{\text{out}}\sqrt{Z_0} \\ \phi_{\text{out}}/\sqrt{Z_0} \end{bmatrix} \\ &= \begin{bmatrix} a_{11}\sqrt{Z_0} & -a_{12}/\sqrt{Z_0} \\ a_{21}\sqrt{Z_0} & -a_{22}/\sqrt{Z_0} \end{bmatrix} \begin{bmatrix} \epsilon_{\text{out}} \\ \phi_{\text{out}} \end{bmatrix}. \end{aligned} \tag{46}$$

Using this equation and the definitions of the forewave and backwave, (42), we may write the following:

$$\begin{bmatrix} \overrightarrow{w_{\text{in}}} \\ \overleftarrow{w_{\text{in}}} \end{bmatrix} = \frac{1}{2} \begin{bmatrix} 1 & 1 \\ 1 & -1 \end{bmatrix} \begin{bmatrix} a_{11} & -a_{12}/Z_0 \\ a_{21}Z_0 & -a_{22} \end{bmatrix} \begin{bmatrix} 1 & 1 \\ 1 & -1 \end{bmatrix} \begin{bmatrix} \overrightarrow{w_{\text{out}}} \\ \overleftarrow{w_{\text{out}}} \end{bmatrix}. \tag{47}$$

We identify the matrix produced by the multiplication of the three matrices and the factor $\frac{1}{2}$ in (47) as the wave matrix, [**W**].

Before interpreting the elements of the wave matrix, [**W**], it is useful to derive its inverse. Clearly, we may find the inverse by brute force multiplication of the three matrices in Eq. (47) and inversion of the resulting matrix. A more elegant method is to recognize that the the steps taken in deriving [**W**] could be repeated starting with the [**b**] matrix representation rather than the [**a**] matrix. This would produce

$$\begin{bmatrix} \overrightarrow{w_{\text{out}}} \\ \overleftarrow{w_{\text{out}}} \end{bmatrix} = \frac{1}{2}\begin{bmatrix} 1 & 1 \\ 1 & -1 \end{bmatrix}\begin{bmatrix} b_{11} & -b_{12}/Z_0 \\ b_{21}Z_0 & -b_{22} \end{bmatrix}\begin{bmatrix} 1 & 1 \\ 1 & -1 \end{bmatrix}\begin{bmatrix} \overrightarrow{w_{\text{in}}} \\ \overleftarrow{w_{\text{in}}} \end{bmatrix}. \tag{48}$$

From (48) we see that $[\mathbf{W}]^{-1}$ may be found by simply replacing the a_{ij} used to derive [**W**] with b_{ij}.

The elements of [**W**] and $[\mathbf{W}]^{-1}$ are directly interpretable in terms of transducer efficiency. For instance, consider W_{11}. It is defined as the ratio of the input forewave to the output forewave when the output backwave is zero. In other words, it is the square root of the ratio of the input power to the output power when no power is supplied at the output port. We recognize this ratio as the square root and inverse of the ideal efficiency of a sensor. Using (47) we can show that the sensor efficiency is given by

$$\eta_{\text{sensor}} = \left[\frac{2}{a_{11} - a_{12}/Z_0 + a_{21}Z_0 - a_{22}}\right]^2. \tag{49}$$

Since we control conditions at the input port of an actuator, we define the ideal actuator efficiency as the ratio of the output power to the input power when no power is reflected back at the input port. This may be found by squaring the 1, 1 element of $[\mathbf{W}]^{-1}$, which is the ratio of the output forewave to the input forewave when the input backwave is zero. This results in the following expression for the actuator efficiency:

$$\eta_{\text{actuator}} = \left[\frac{b_{11} - b_{12}/Z_0 + b_{21}Z_0 - b_{22}}{2}\right]^2. \tag{50}$$

As an example, let us return one last time to the piezoelectric compression accelerometer discussed previously in this chapter, and use the derived 2-port matrix to find the accelerometer efficiency using (49). First we rewrite (49) in terms of the [**b**] elements using the reciprocity relations, (19):

$$\eta = \left[\frac{2}{b_{22} - b_{12}/Z_0 + b_{21}Z_0 - b_{11}}\right]^2. \tag{51}$$

Next we recognize that the fundamental impedance in SI units is 1 Ω, so $Z_0 = 1$. Finally we use (16) to provide the b_{ij} components for substitution. The result is a very complicated expression. If we limit the efficiency expression to low frequencies, then the b_{22} term dominates, and the result is

$$\eta = \frac{4h_{33}^2}{(k + Ak_p/l)^2}. \tag{52}$$

Note, that the efficiency increases with the square of the piezoelectric constant, and decreases with the overall transducer stiffness. Because power (and energy) are proportional to the square of the power conjugate variables used to define sensitivity, we could have anticipated the dependence of the efficiency on the square of the piezoelectric constant.

As a matter of practice it is very difficult to measure efficiency because it is almost impossible to ensure that there is no back power or reflected power at the terminals. The normal method of dealing with this problem is to give efficiencies with a known load of R at the appropriate port. For this known load, the backwave is related to the forewave by

$$\overleftarrow{w} = \frac{1 - R/Z_0}{1 + R/Z_0} \overrightarrow{w}, \tag{53}$$

and the measured efficiency is

$$\eta_{\text{measured}} = \eta_{\text{theoretical}} \left[1 - \left(\frac{1 - R/Z_0}{1 + R/Z_0} \right)^2 \right]. \tag{54}$$

Equation (54) shows that the measured and theoretical efficiencies are equal if the load R is chosen equal to Z_0, i.e., if the fundamental impedance and the load match. For all other load resistances, the measured efficiency will be less than the theoretical efficiency of the transducer. If R is very large or very small compared to Z_0, then the measured efficiency approaches zero.

From the wave matrix defined above, it is a simple matter to produce the scattering matrix, [**S**]. The scattering matrix may be thought of as an alternative expression of the wave matrix in which we emphasize the causal relations.

When constructing the wave matrix we developed a fundamental relationship between the input fore- and backwave and the output fore- and backwave. From a causal perspective, the waves flowing into the transducer are the input variables and the waves flowing out of the transducer are the output variables. Hence, the input forewave and the output backwave are input variables, and the input backwave and output forewave are output variables. We designate the input variables as u_1 and u_2 and the output variables as v_1 and v_2. Hence the wave matrix representation is

$$\begin{bmatrix} u_1 \\ v_1 \end{bmatrix} = [\mathbf{W}] \begin{bmatrix} v_2 \\ u_2 \end{bmatrix}. \tag{55}$$

If we rearrange this relationship so that we express the outputs in terms of the inputs, then we have

$$\begin{bmatrix} v_1 \\ v_2 \end{bmatrix} = [\mathbf{S}] \begin{bmatrix} u_1 \\ u_2 \end{bmatrix}, \tag{56}$$

where **S** is the scattering matrix. The main advantage of the scattering representation for transducers is that it makes clear what are the port input and output variables. For further information on scattering and wave matrices, see [7]–[9].

9.6 Summary

In this chapter we have examined 2-ports from a general perspective. We have narrowed our view to those 2-ports which can be described using linear algebra. This restriction limits the transducers under consideration to those which are linear or describable in terms of linear operators (such as differentiation or integration).

Six representations for 2-ports have been presented and their relationships to one another discussed. Reciprocity relations have been presented and interpreted in terms of measurement ease. Connected 2-ports have been discussed with the aim of showing which representation is most advantageous. Also, transducer sensitivity and efficiency have been defined.

9.7 References

1. I. J. Busch-Vishniac, *A Fundamental Problem with Mobility Analysis of Vibration Isolation Systems*, J. Acoust. Soc. Am., **81**, 1801 (1987).
2. G. D. Monteah, *Applications of the Electromagnetic Reciprocity Principle* (Pergamon, Oxford, UK, 1973).
3. C. M. Harris (ed.), *Handbook of Acoustical Measurements and Noise Control*, 3rd ed. (McGraw-Hill, New York, 1991).
4. W. W. Harmon and D. W. Lytle, *Electrical and Mechanical Networks* (McGraw-Hill, New York, 1961).
5. I. J. Busch-Vishniac, *Spatially Distributed Transducers: Part I—Coupled Two-Port Models*, J. Dyn. Sys. Meas. Controls, **112**, 372 (1990).
6. J. Hilgevoord, *Dispersion Relations and Causal Description, An Introduction to Dispersion Relations in Field Theory* (North-Holland, Amsterdam, 1962).
7. G. F. Chew, *The Analytic S Matrix* (W. A. Benjamin, New York, 1966).
8. P. C. Magnusson, G. C. Alexander, and V. K. Tirupathi, *Transmission Lines and Wave Propagation*, 3rd ed. (CRC, Boca Raton, FL, 1992).
9. H. M. Paynter and I. J. Busch-Vishniac, *Wave-Scattering Approaches to Conservation and Causality*, J. Franklin Inst., **325**, 325 (1988).

10

Response Characteristics

Because transducers may take advantage of many different transduction mechanisms, we might think that it would be difficult to compare and contrast transducers in terms of performance. However, this problem is avoided by use of performance measures which can be applied generally to transducers (and, in fact, to many other sorts of devices). In this chapter we discuss the measures by which the operation of a transducer is quantified. In the first section, we define the characteristics. In the second section, we talk about calibration of transducers and errors. In the third section, we discuss time and frequency scaling.

10.1 Response Characteristics Defined

When an individual wishes to purchase a piece of equipment, he or she normally compiles a number of catalogs and compares the specifications on similar items. This information is used in assessing which item is best suited to the needs while meeting a number of constraints. In this section we define the specifications which might be used to describe transducers. As will become obvious from this discussion, care must be used in comparing equipment from different manufacturers, because the same terminology can be used to mean quite different measures of performance. Also, different phrases may be used for essentially identical performance measures.

Virtually all transducer specifications provide information on the *range* of operation. The range tells the extent of measurement ability, from the smallest to the largest amplitude signals that can be handled without the system being overloaded or saturated, and without the signal being obscured by noise. For sensors, the range usually indicates the smallest to largest input signals which can be accommodated. For actuators, the range is normally specified in terms of the output signal, although the allowable range of input variable might also be given. Further, the actuator range might be given as a function of the load on the actuator.

Range, as defined above, can be difficult to measure because we must determine at what point a signal is said to be either obscured by noise or distorted due to saturation. For this determination, as for virtually all of the response characteristic specifications, it is common to use the best possible conditions for the experimental measurement of the characteristic. Manufacturers often present an "effective," "measuring," or "working" range, which is the range of signal values for which the device produces results with an error within a specified amount. For example, a manufacturer might specify one range for 0.01% error and another, broader range for 0.1% error.

The range information provided by manufacturers is one of the most useful and reliable characteristics. Its definition is unambiguous, and manufacturers generally speak the same language when presenting range information. When less scrupulous firms try to pull the wool over the eyes of a consumer, they usually resort to describing the transducer in terms of *rangeability*. The rangeability is meant to be an indication of some range of signal values over which the device is useful. However, rangeability has no strict definition, so you have no guarantee that the range and rangeability are identical.

An example of the distinction between range and rangeability may be provided by considering the performance of a fiber optic displacement lever. In this device, half of the fibers in a probe emit light and half of them receive light. The output voltage is proportional to the intensity of the received light. A graph of the output voltage versus distance from a reflecting target is as shown in Fig. 10.1. There are two linear ranges of operation that can be defined. The range including very small displacements is referred to as the front slope, and the range of operation for larger displacements is called the back slope. The makers of fiber optic displacement levers generally provide information on the two ranges. However, some simply state that the rangeability is 10^{-6} to 5×10^{-1} inches. The rangeability as stated for such a transducer is quite misleading because there are regions within the rangeability that cannot be used due to the presence on the knee of the curve in Fig. 10.1.

Next to range, the most important response characteristic of a transducer tends to be some measure of its *accuracy*. Accuracy is a measure of the error and uncertainty in a device. In general, there are three types of error: random error, systematic error, and hysteresis error. Random error is that error which causes different output signals for apparently identical input signals. Systematic errors are errors which are correlated with some system parameter, such as alignment. Hysteresis errors are systematic errors that depend on the direction of approach (increasing or decreasing value). Accuracy includes random errors and errors due to hysteresis, but not errors due to controllable environmental quantities (such as varying temperature). However, there is no standard definition of accuracy in use commercially and, as a

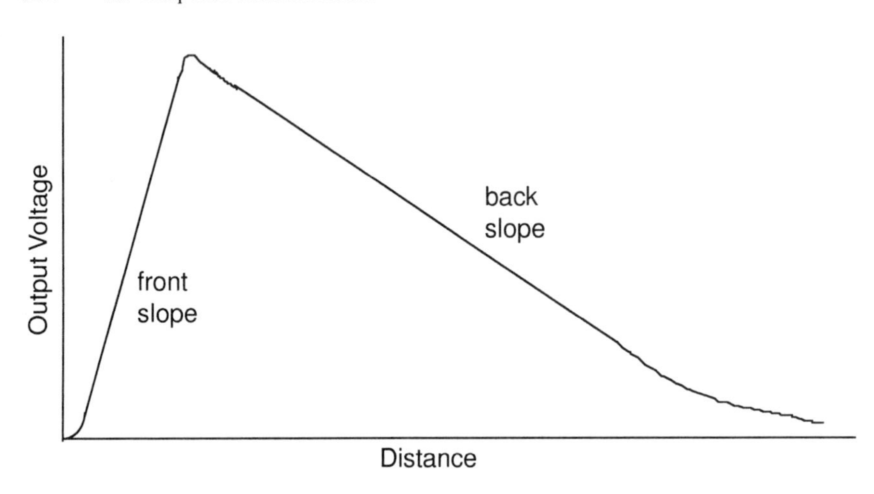

FIGURE 10.1. Response of a fiber optic lever.

result, accuracy measures are difficult to use for comparison purposes. Accuracy usually is expressed as $\pm x$ percent or $\pm y$ units, and is determined by a calibration comparison to an absolute standard. Although it is normal for a single number to be provided, it is rare that the accuracy is uniform throughout the range. Manufacturers almost never specify whether the accuracy measure given is the best, average, or worst error measured.

A measure related to accuracy, but distinctly different, is the *repeatability* of a device. The repeatability is defined as a measure of the extent to which identical input signals will produce identical output signals. The accuracy of a system can never exceed the repeatability, but the repeatability can be extremely good without the accuracy being terribly high, if a device is very consistent but in error. Repeatability is often expressed in output signal units or in a percent of full scale range. Most often, a single number is given, and it is never clear whether the repeatability cited is an average over a wide range of input signals, or the value at some specified input signal.

A response characteristic that is almost always provided on transducers is the *linearity*. In general terms the linearity is a measure of the manner in which the error varies with range in the transducer. It is generally expressed as the deviation from a straight line, either as the rms error (in percent or units) or the maximum error encountered in the transducer range. Note that this definition, ironically, means that a more linear transducer is one with a lower linearity value. There is a complication to the specification of linearity of a transducer, namely, that there are at least four straight-line references in use. These are shown in Figs. 10.2 to 10.5. Figure 10.2 shows the fit of the data to the "best straight line," i.e., the line which produces the minimum absolute error. Because the best straight line yields the lowest error, it yields the best linearity characteristic and is a very popular choice for manufacturers. There is also some justification in using this line as the standard for linearity determination, because experimental calibration tends to identify this particular linear fit. Figure 10.3 shows the same data, fit using the best straight line through the origin. This approach to determining the linearity of a transducer explicitly recognizes that there should be no output when there is no input. It generally produces a poorer linearity measure than that using the best straight-line fit. A third approach is shown in Fig. 10.4, in which the data is fit by a line that matches exactly at the terminal points. The only advantage of using this line is that it is trivially easy to draw. It cannot be justified from any other viewpoint. While this approach to finding a linear fit to data was once popular in the industry, the proliferation of computers with rapid and straightforward data processing has nearly eliminated this matching endpoints approach. Finally, Fig. 10.5 shows a fit of the measured data to a line which is based on a theoretical analysis. While this sort of determination of linearity is precisely what one uses in a scholarly article or in-house, it is rarely used in determining the linearity of a transducer because it usually produces the poorest linearity measure, as the theory used to predict performance does not take into account manufacturing, and phenomena producing small effects.

The fact that there are multiple means of generating the linearity measure of a transducer is not, in itself, a problem. However, it is extremely rare for a manufacturer to tell you how the linearity quoted in the specifications of a transducer has been determined, and this can make it quite difficult to compare devices. The current industry trend is to use either the best straight line or the best straight line through the origin.

It should also be noted that the linearity data supplied on a specification sheet is a static measure. The data which are fit to a line are determined once the system has been allowed to come to some sort of equilibrium. Further, highly precise and stable instruments are used to minimize noise and control environmental factors such as thermal drift. In the normal operation of a transducer, where the environment

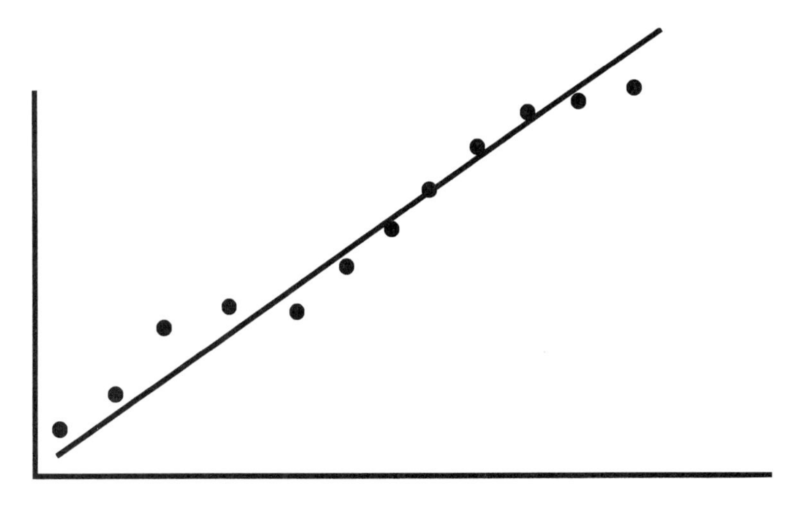

FIGURE 10.2. Best straight-line fit to data.

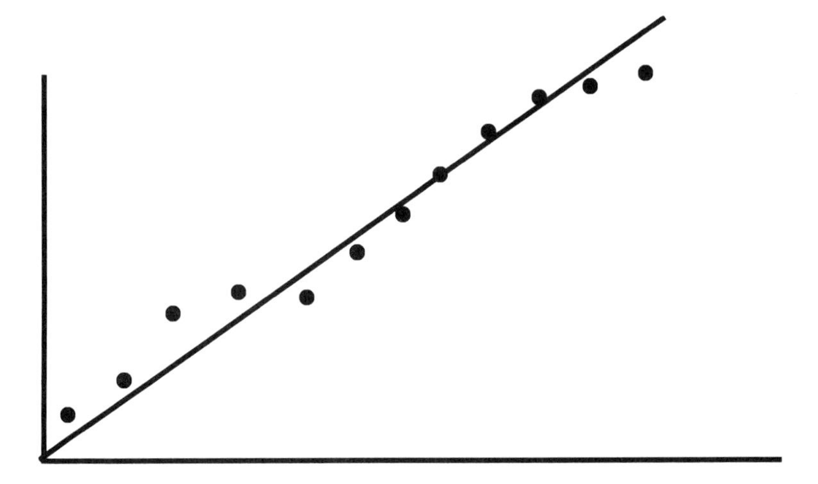

FIGURE 10.3. Fit to data using best straight-line through the origin.

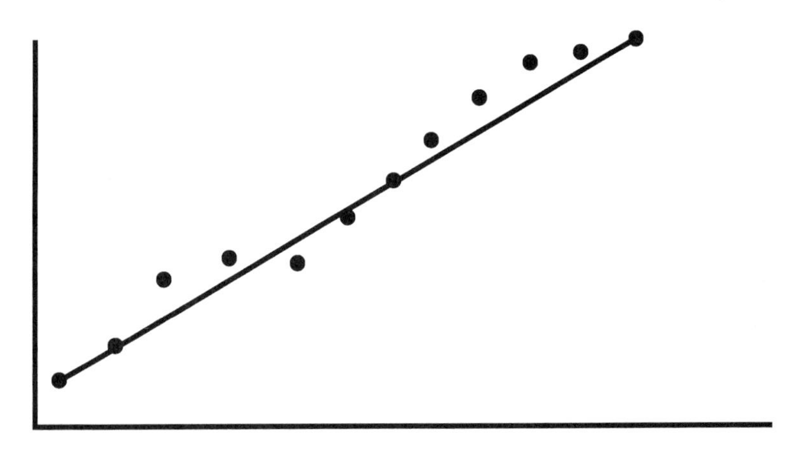

FIGURE 10.4. Fit to data using terminal based line.

and the system change dynamically, transducers tend to be less linear than their specifications would lead you to believe. Thus, we should always treat the linearity measure cited by the manufacturer as the best that we could potentially achieve using the device.

At times, manufacturers choose to present the *distortion* rather than the linearity. Distortion is a response characteristic that is nearly the complement to linearity. It is typically defined in the frequency domain (so it is a dynamic characteristic), and is a measure of the size of response at frequencies other than ω in response to excitation at ω. Thus we typically desire a low value of distortion. It should be noted, however, that just as in the measurement of linearity (time domain), the measurement of distortion typically is made under optimal conditions. This usually means using single frequency excitation (or two frequencies) and allowing

FIGURE 10.5. Fit to data using theoretical calculation.

the system to come to steady-state before determining the distortion. Thus, while it is a dynamic measure, it does not include the transient nonlinearities.

Usually, the emphasis of the distortion characteristic is on harmonic distortion in which an excitation at frequency ω causes a response at $n\omega$, n being a positive integer. Harmonic distortion is the natural result of many of the transduction mechanisms previously discussed because they are essentially quadratic. This gives rise to second harmonic distortion ($n = 2$) which is often of sufficient magnitude to measure. It is normal to express the harmonic distortion as the total harmonic distortion: the net output at all harmonic frequencies in the range, divided by the output at the excitation frequency. This ratio can be expressed either as a percentage or in decibels (20 $\log_{10}$(ratio)).

In addition to total harmonic distortion, some precision equipment specifies the intermodulation distortion. This is defined as the output at frequency ω given excitation simultaneously at two frequencies, ω_1 and ω_2, where ω is distinct from the two excitation frequencies. As such, the intermodulation distortion measures nonlinearities due to interactions between excitations. Often, but not always, the nonlinearities are harmonic in that the nonlinear response is at frequency $n_1\omega_1 \pm n_2\omega_2$, where both n_1 and n_2 are positive integers. As in the case of harmonic distortion, the intermodulation distortion is expressed by determining the ratio of the nonlinear response to the response at the excitation frequency and expressing it in decibels or as a percentage.

Note that distortion is not the only possible measure of dynamic nonlinearity. It is also possible to consider transient response characteristics (short-time dynamic measures), and to measure the amplitude linearity performance. The latter measure typically considers steady-state sinusoidal excitation, but focuses on the effect of varying the level of the input signal rather than its frequency content. By varying the input level precisely, we can monitor how well the changes in output level gain match the changes in input level gain. Any discrepancy between the two is due to nonlinearity. Typically, when this measure is present, it is expressed simply as *dynamic linearity* and the error is expressed either in decibels or as a percentage of the input signal.

Another basic measure of the performance of a transducer is its *sensitivity*. We have previously defined sensitivity as a transfer function between the output signal and the input signal of a transducer. In manufacturers's specifications it is the sensitivity magnitude that is normally given. This sensitivity typically applies over a range of operation which is quoted, or at a single input value at which it is measured.

In many cases, the manufacturers provide more information than a single value of a transfer function magnitude by providing a *frequency response* characteristic. The frequency response is a clear statement of the amplitude of the output versus frequency for a set level of the input. It is most often presented as a graph or in terms of a statement that the frequency response is "flat" to some given measure (e.g., ± 1 dB) in a range which is then stated.

For sensors the transducer *efficiency* is often not important, but for actuators the efficiency can be quite important. The efficiency is a ratio of the output power or energy to the input power or energy. Typically transducer manufacturers specify the efficiency as a single value amplitude ratio rather than as a function of frequency (as presented in Chapter 9). Further, it is not at all unusual for the efficiency value to be given with no note of whether it is the average in the range or the highest value. Nor is it common for the termination load which was used during efficiency determination to be given.

Another performance characteristic which is often supplied on transducers is the *resolution*: The resolution is the smallest signal step that can reliably be resolved (sensors) or imposed (actuators). The resolution is clearly related to the sensitivity of the device, with greater sensitivity producing larger (poorer) resolution. The resolution is also related to the amount of noise that is present, as the noise defines a threshhold value the signal must achieve to be considered a real signal. Resolution is usually quoted in terms of units of the input for sensors, or of the output for actuators. It may also be stated as a percentage of the full scale value. If the resolution is not constant within the operating range of the transducer, manufacturers might supply several values of resolution, with appropriate ranges for application of each.

As mentioned above, the resolution is a measure which combines a measure of noise and of transducer sensitivity. Just as there are separate measures of transducer sensitivity, there are also separate performance measures which describe the noise. The two accepted measures of the noisiness of a device are: its *signal-to-noise ratio* and its *noise floor*. Both indicate the amount of the output signal that is uncorrelated with the input. The noise floor is a measure of the magnitude of the time-varying output signal level which exists even in the absence of an input signal. An output signal at or below the noise floor cannot be detected without the use of sophisticated signal processing approaches. The signal-to-noise ratio (SNR) expresses the noise in terms of its comparison to a signal (and hence it is specified for a given signal level). The SNR is a decibel measure of the amplitude of the signal divided by that of the noise. An SNR of 20 dB indicates that the signal is an order of magnitude greater than the noise floor. As the name suggests, the SNR is a single number measure of noise.

It is unusual for both the SNR and the noise floor to be provided in the specification sheet on a commercial transducer and often, no noise information is provided. Even when noise data are given, it must be remembered that the noise need not be static. It is not unusual for the noise floor to change with the signal level. The noise data supplied usually is the minimum noise level, which typically corresponds to a system allowed to come to equilibrium and with no input signal whatsoever.

For some precision equipment, the manufacturer will supply a noise spectrum. This provides a measure of the noise floor as a function of frequency, and is quite useful. In general, there is more noise present at low frequencies than at high frequencies, and the noise spectrum makes it possible to know at just what threshold the signals overcome noise at any frequency. Again, the noise spectrum can be signal level dependent, although the spectrum supplied virtually always corresponds to the minimum noise level, present when there is no input signal.

The performance measures presented so far are all either static measures or measures made using steady-state sinusoidal signals. For many applications, users are interested in transient response information. There are many such performance measures which might be provided, but the two most popular are the *response time* and the *overshoot*.

The response time of any device is a measure of how quickly it responds to a step change in the input signal. It is common to present the response time in terms of time for the output signal to first be within a given fraction of the final value it will achieve, even if it subsequently moves outside the envelope of this final value. For instance, the response time for 95% might be given. Response times for transducers are assumed to be independent of the level of the input signal. It is rare for the precise testing conditions to be specified.

The overshoot of a device is a measure of how much its response goes beyond the final value before settling back down to it. This measure is usually quoted as a ratio—as in an overshoot of 10%. Overshoot is very often not provided in the manufacturers's specifications for a sensor, but is a common performance measure for actuators. Clearly, the "best" actuators are those with the fastest response times and the smallest values of overshoot. Unfortunately, these performance measures tend to work against one another and compromises must be struck.

Another characteristic of all devices is *drift*. Drift is the tendency for the output to vary monotonically and very slowly compared to times associated with the sensing or actuation. It is typically regarded as a static problem, but drift is really a dynamic problem. The causes of drift are what are known as "influence quantities" such as changing temperature and humidity in the test environment. There is no standard measure of drift. Sometimes it is treated as though it is linear and a percentage shift per time is given. At other times, drift is considered as a quadratic mechanism and a percentage shift versus the square root of time is given. Often, the measurement is presented not as drift but as *stability*, but then, ironically, the higher the stability value, the poorer the device performance.

The *offset* is the nonrandom value of the transducer output when the input is zero. In other words, some devices are designed so that they produce a dc signal when there is no input. While this might seem odd at first glance, it should be noted that a transducer can be operated most accurately when the aim is to drive the output signal to zero, i.e., null operation. Thus, some transducers permit the operator to set a dc offset which effectively sets the target value for null operation.

While there are many other pieces of information which can be supplied on the specifications for a transducer, those described above are the most likely to be provided in addition to the obvious geometric (size, weight) and economic (cost) parameters. For more information on measures of transducer performance, see [1]–[3].

10.1.1 Example of Transducer Selection Based on Specifications

The above discussion simply defines the various measures typically used in transducer performance. What is missing from this discussion is a means of putting the measures into the context of selecting a transducer. There are two important points to remember when choosing a transducer based on its specifications: the "best" choice depends upon the specific application, and seemingly small improvements in performance can be very costly. Both of these points suggest that it is vital to know precisely how you intend to use the transducer before you make a choice. For instance, if the need is for a sensor to measure displacement of the piston of an engine, then the appropriate choice may be very different from that which would result if choosing a sensor to measure the motion of a baseball thrown by a pitcher. Further, choosing better performance characteristics than needed for the specific application can dramatically increase the cost of a transducer. The decision to purchase a transducer which offers significantly more than required should be made only when it is anticipated that applications taking advantage of the additional capabilities will be coming forward soon.

In order to illustrate some of the considerations that go into choosing a transducer based on published performance characteristics, we consider a specific application and some commercially available transducers below. The aim is neither to praise nor condemn particular transducers, but to discuss their advantages and disadvantages for the chosen application.

Let us imagine that we have a new company which will revolutionize the personal computer industry by making quiet personal computers. It will do this seemingly impossible task by introducing a temperature sensor which determines when to turn on a fan to cool the electronics. Thus, instead of constantly running the computer cooling fan, the computer will be able to operate for extended periods without cooling and with relative quiet. Our task is to choose a temperature sensor suitable for this application, and our preference is to identify a commercially available product rather than to design a new sensor.

Now let us further imagine that we have narrowed the possibilities to three items, all available from one commercial vendor, Omega Engineering, Inc. [4]: an

RTD, a thermocouple, and a thermistor. Each of these are described below based on material found in the vendor catalog.

- The RTD chosen from this vendor is a thick-film platinum RTD with a temperature range of -70 to $+600°C$. The sensor is 3.2 mm wide $\times$ 25.4 mm long $\times$ 1.3 mm thick, and attached to nominally 8.0 mm long leads. For this RTD, $\alpha = 0.00385$, and the nominal resistance is 100 Ω at 0°C. The RTD time constant is less than 0.1 s, and the surface insulation is 1000 MΩ at room temperature and 250 V. The self-heating effect is given as 0.006°C/mW in stirred water at ice point. The maximum deviation from the calibration given is ± 0.12 Ω (correponding to ± 0.3°C) at 0°C, and ± 0.30 Ω (± 0.8°C) at 100°C. The stability cited is less than 0.05% after 10 cycles from -70°C to 600°C. The device cost is $22.
- The best thermocouple available from the same vendor is a J-type, unsheathed thermocouple with a recommended temperature range of 0 to 760°C. The thermocouple uses 305 mm long $\times$ 0.127 mm diameter wire and the bead diameter is roughly 0.38 mm. For J-type thermocouples, the sensitivity is roughly 54 μV/°C. The nominal resistance is 14.20 Ω at room temperature, and the time constant is less than 1.0 s in still air. The minimum insulation resistance wire-to-wire is 1.5 MΩ at 500 V dc. The unit meets ANSI standards for the limits of the error of thermocouples. The thermocouple is available in a package of five units for $17.
- A number of thermistors are available from the same vendor, with very similar specifications. For instance, one particular thermistor, which could be quite useful, has a range of -30 to $+100$°C. The thermistor end of the device has maximum dimensions of 2.03 mm wide $\times$ 2.03 mm thick $\times$ 3.81 mm, and is attached to wires with a nominal length of 152.4 mm. For this device, operated in the resistance mode, the sensitivity is given by $R = -17.115T + 2768.23$, where T is the temperature in centigrade. In the voltage mode, the thermistor sensitivity is $V_{out} = 0.00680 V_{in} T + 0.3489 V_{in}$. The maximum input voltage is 2.0 V and the maximum current is 625 μA. The time constant for this thermistor is less than 10 s. The power needed to raise the thermistor 1°C above the surrounding environment (known as the dissipation constant) is 1 mW in still air. The accuracy and interchangeability over the full range are given as ± 0.15°C and the linearity deviation is ± 0.216°C. This thermistor is available at a cost of $19.

Now we must use this information on the three sensors to determine which device to use. We do this by consideration of the specific requirements of the application, in this case, using a temperature sensor in a personal computer in order to switch on and off a cooling fan. Given this application, all three sensors have a temperature range adequately broad. Further, all three have fast enough response times, even though these response times differ by over two orders of magnitude. Thus, it is the remaining information on sensitivity, size, cost, and accuracy that will determine which device to choose.

First consider the sensitivity of the three temperature sensors. For the RTD, the sensitivity is determined by α. From the data reported above, this yields a sensitivity of 0.385 Ω/°C. For the thermocouple, the sensitivity cited is 54 μV/°C. For the thermistor, the sensitivity is either 0.00680V_{in} V/°C, or 17.115 Ω/°C. From these numbers, we see that the thermistor is far more sensitive than the RTD (by roughly a factor of 50 in the thermistor resistance mode). Using the maximum allowable input voltage, the sensitivity of the thermistor in the voltage mode is higher than that of the thermocouple by over 250 times. Hence, from the view of best sensitivity, the thermistor is the best choice.

Size is of vital importance in this example application, as real estate interior to a computer is very expensive. In terms of the size of the transducer element (excluding the leads) the thermocouple has the smallest footprint, then the thermistor, then the RTD. Both the thermocouple and thermistor are likely to be acceptably small, but the RTD is quite probably too long. Hence, at this point, we eliminate the RTD from further consideration and focus on the thermistor and thermocouple comparison.

The remaining important issues are cost and accuracy. From a cost perspective, the thermocouple is the clear winner. It is about one-fifth the cost of the thermistor. From an accuracy viewpoint, although the vendor simply states that the thermocouples comply with a standard and the novice would need to find this standard, the thermocouple is more accurate than the thermistor.

Putting all of this information together, the thermocouple is the best choice in every way except sensitivity. If the thermocouple sensitivity is adequate (which is almost certainly the case) then this is the device to purchase. If the thermocouple is deemed to be inadequately sensitive, then the thermistor should be used. The determination of a specific sensitivity need would be set by knowledge of the temperature at which the electronics must be cooled. Finding this specific temperature is beyond the scope of the current example.

10.2 Calibration

Calibration is the process by which a sensor or actuator is provided with a factual statement about the errors the instrument will have. Historically, calibration dates back to Babylonian times. The Chinese used jade templates for calibration of shadow lengths cast by sundials. We have progressed significantly since these ancient times and we now have many standards labatories and organizations. These include NIST (formerly NBS), ASTM, ANSI, ISO and many others. For a history of calibration, see Sydenham [5].

In general there are three types of calibration methods that can be used to determine the response parameters defined in the previous section of this chapter: comparison method, absolute direct method, and absolute reciprocal method. In the comparison method we compare the transducer performance to that of a similar transducer with known performance. In absolute direct calibration a special calibrating machine is used to calibrate directly the transducer. In absolute reciprocal calibration of a transducer, the transducer is calibrated using other transducers but, through the use of special methods, the calibration of the other transducers is rendered irrelevant. In each of these methods we seek to use a calibrator which is five to ten times more accurate than the minimum to which we seek to calibrate the new transducer. Below we discuss each of the calibration methods.

First consider the comparison method of calibration. This approach has the advantage of being rapid and generally straightforward. However, it is substantially less accurate than the absolute calibration methods. For example, consider the calibration of a new accelerometer using the comparison method. If a calibrated accelerometer and a shaker are available, then we would attach the calibrated accelerometer and the new accelerometer to the shaker, usually just stacking the two one on top of the other. A signal would be supplied to the shaker and the output signals of the calibrated accelerometer, $O_c(\omega)$, and the new accelerometer, $O_n(\omega)$, measured. These responses can be represented using the transfer functions

$$\begin{aligned} O_c(\omega) &= ET_c, \\ O_n(\omega) &= ET_n, \end{aligned} \tag{1}$$

where E is the input signal and T_c and T_n are the transfer functions for the calibrated and new accelerometer, respectively. Using (1) we can determine the unknown transfer function T_n:

$$T_n = \frac{O_n}{O_c} T_c. \tag{2}$$

Note that the right-hand side is entirely known.

While the comparison method of calibration is fast and eliminates the need for specialized equipment, its major disadvantage is that it is more prone to errors than the other calibration methods. There are three types of errors that often become important: errors in measurement, errors due to changing excitation, and errors due to slipping out of range. We may consider each of these separately.

Suppose that the transfer function representing the calibrated transducer is given by

$$T_c = \hat{T}_c + \epsilon_1, \tag{3}$$

where $\hat{T}_c$ is the actual transfer function and ϵ_1 is the error that was made in determining the transfer function. Similarly there is a measurement error in the calibration process for the new transducer:

$$\frac{O_n}{O_c} = \hat{R} + \epsilon_2, \tag{4}$$

where $\hat{R}$ is the actual value of the ratio and ϵ_2 is the error. Combining these two equations we see that T_n is

$$\begin{aligned} T_n &= (\hat{R} + \epsilon_2)(\hat{T}_c + \epsilon_1) \\ &= \hat{R}\hat{T}_c + \epsilon_2\hat{T}_c + \epsilon_1\hat{R} + \epsilon_1\epsilon_2. \end{aligned} \tag{5}$$

We recognize the first term as the actual transfer function for the new transducer. The remaining three terms are errors due to measurement.

A second source of error in the comparison method is due to changing excitation. This error is only significant when it is not possible to test simultaneously both the calibrated and new transducer. For cases where the output signals of the calibrated and new transducer are obtained sequentially, rather, than in parallel, there is the risk that the load seen by the excitor is different and hence the excitation supplied is not identical. Suppose that the calibrated transducer is subjected to an excitation of E and that the new transducer receives $E + \alpha$. Then the error in measurement is

$$\frac{O_n}{O_c} = \frac{T_n}{T_c} + \frac{T_n}{T_c}\frac{\alpha}{E}. \tag{6}$$

The second term in this equation is the error. Note that it depends on the relative difference in the excitations provided, as expected. It also depends on the relative size of the transducer transfer functions. If the new transducer is much more sensitive than the calibrated transducer, then the error is more significant.

A third source of error in the comparison method is due to the slipping out of range of one of the transducers. For example, suppose that the excitation provided is flat across a broad range of frequencies and that the transfer function of the two transducers is as shown in Fig. 10.6. Then the theoretical ratio of the output from the new transducer to that of the calibrated transducer is as shown in Fig. 10.7. However, in the measurement there is probably a noise floor which will be important at high and low frequencies for both transducers. This will cause serious distortion in the resulting ratio in those frequency regimes. For the example given, the distortion due to noise means that the results are least accurate where the transfer function ratio shown in Fig. 10.7 has peaks. Ideally, we can avoid this problem by always using a more accurate transducer as the comparison calibrator for less accurate new transducers. However, this approach is not always practical, as improvements in technology result in more accurate transducers.

Next consider the absolute direct method of calibration. In this approach a special piece of calibrating equipment produces a well-controlled input signal which is used to calibrate a particular type of transducer. For example, an accelerometer calibrator formerly available from GenRad [6] used a pair of matched pistons that was moved sinusoidally with high precision. Weights were provided so that the load on each piston could be kept the same during the calibration process. Another example is calibration of a microphone using a pistonphone. A pistonphone uses a high precision motor to move a piston at a single frequency. The piston motion causes a volumetric change of a cavity which produces a known sound pressure.

The advantage of absolute direct calibration is that it is fast and quite precise. However, it tends to be an expensive approach to calibration since specialized equipment is needed, and absolute calibration is usually performed at only one frequency or a few frequencies. We still need the comparison method to complete the calibration at all frequencies. However, now the comparison is to the response of the transducer at the frequency at which the absolute direct calibration was used.

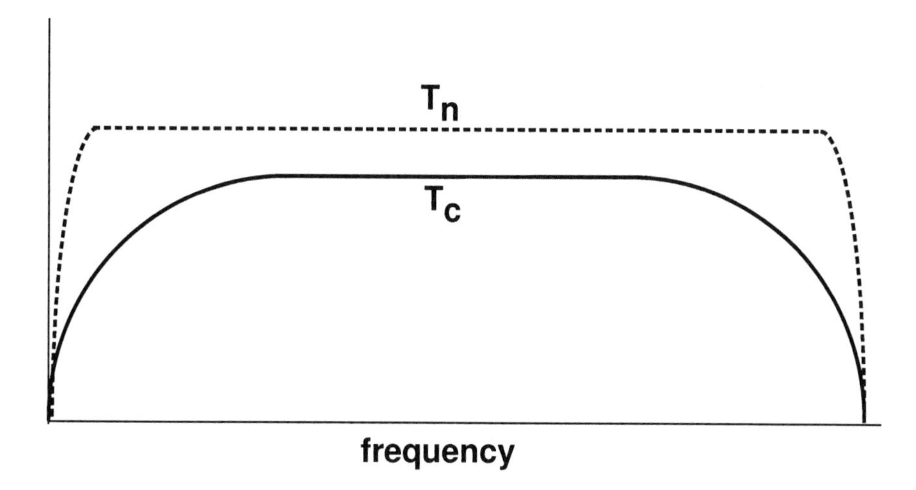

FIGURE 10.6. Theoretical transfer functions of the calibrated and new transducer.

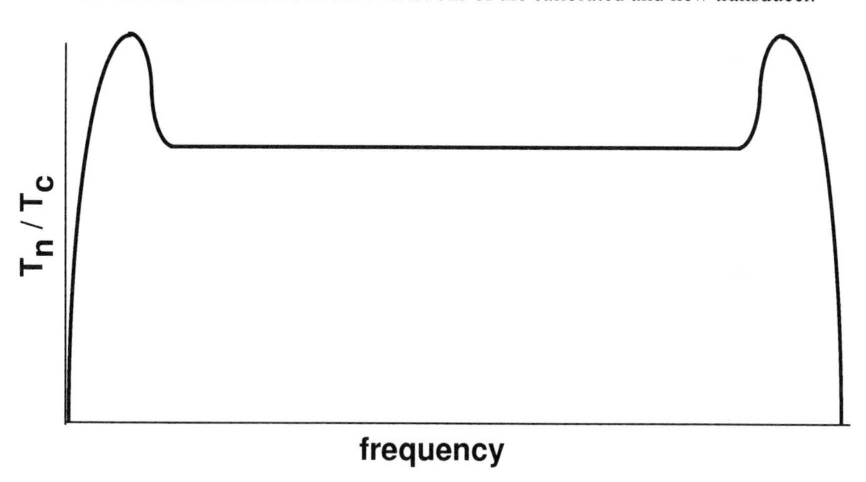

FIGURE 10.7. Theoretical ratio of output responses of the calibrated and new transducer.

Finally, consider absolute reciprocal calibration. This method uses at least one reciprocal transducer and relies on reciprocity relationships in determining the calibration. Historically it was developed for microphone calibration, although it is broadly applicable (See MacLean [7]). Let us consider the microphone example. Reciprocity calibration of a microphone requires three steps: first a reversible transducer, B, is placed a distance d from loudspeaker C. The open circuit voltage response of B is measured. This is shown graphically in Fig. 10.8. This produces the relation

$$e_B = T_B P_C, \tag{7}$$

where P_C is the sound pressure at B produced by C. Next transducer B is replaced by the microphone we wish to calibrate, A, and the measurement repeated. This is shown in Fig. 10.9 and produces

$$e_A = T_A P_C. \tag{8}$$

Assuming that the two transducers disturb the sound field an equal amount, we can assume that P_C is identical in the two steps. Hence, combining (7) and (8)

$$T_A = \frac{e_A}{e_B} T_B. \tag{9}$$

Finally loudspeaker C is replaced by B, used as a sound source, and again, the voltage response of transducer A is measured. This is shown in Fig. 10.10 and yields

$$e'_A = T_A P_B \tag{10}$$

$$P_B = \gamma_B i_B. \tag{11}$$

Here P_B is the pressure produced at A by B, i_B is the input current to B, and γ_B is the speaker sensitivity of B. Combining the two equations above, we get

$$T_A = \frac{e'_A}{\gamma_B i_B}. \tag{12}$$

What remains is analysis based on the measurements. First, note that (9) and (12) both provide expressions for T_A, which is the transfer function we desire to find. We multiply the two equations to get

$$T_A^2 = \frac{e_A e'_A}{e_B i_B} \frac{T_B}{\gamma_B}. \tag{13}$$

Note that the first factor in this equation is entirely comprised of measured quantities. The second is made of quantities that relate to B. It is this second factor that must be reformulated so that the nature of transducer B is inconsequential. This

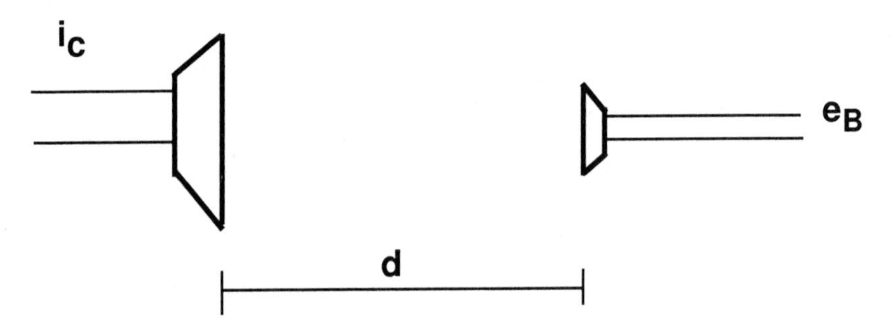

FIGURE 10.8. First measurement in reciprocity calibration of a microphone.

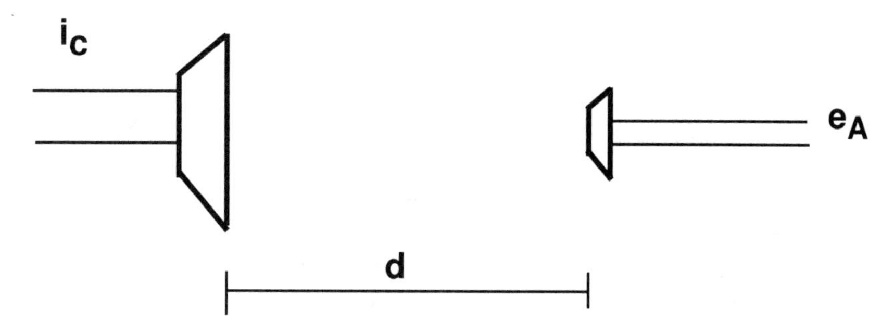

FIGURE 10.9. Second measurement in reciprocity calibration of a microphone.

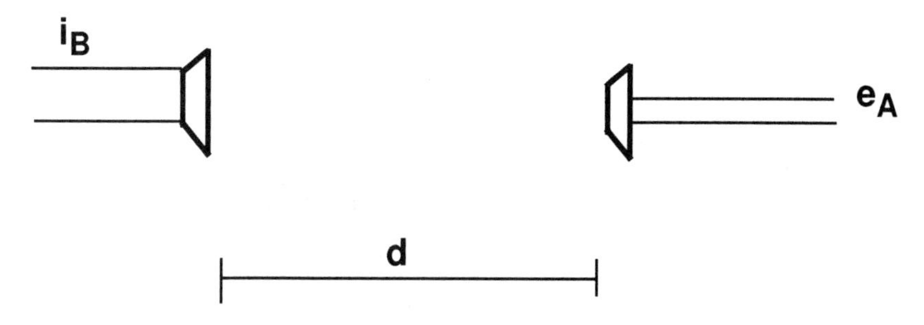

FIGURE 10.10. Third measurement in reciprocity calibration of a microphone.

reformulation begins with the use of the following reciprocity relation:

$$\left.\frac{e_1}{e_2}\right|_{f_2=0} = -\left.\frac{f_2}{f_1}\right|_{e_1=0}. \tag{14}$$

In the first step of the reciprocity calibration we related P_C and e_B. Using the reciprocity relation we have

$$\begin{aligned}\left.\frac{P_C}{e_B}\right|_{i_B=0} &= \frac{1}{T_B} \\ &= -\left.\frac{i_B}{Q_C}\right|_{P_C=0}. \end{aligned} \tag{15}$$

Here Q_C is the volume velocity produced at B by C. It is related to P_B by

$$Q_C = \frac{2d\lambda}{\rho_0 c} P_B, \tag{16}$$

where λ is the sound wavelength, and $\rho_0 c$ is the specific acoustic impedance of air (415 rayls). Using (15), (11), and (16) we have

$$\frac{T_B}{\gamma_B} = \frac{2d\lambda}{\rho_0 c}, \tag{17}$$

which can be used in (13) to express the transducer B parameters in terms of known quantities. The final result is an equation for T_A in terms of measured and known quantities.

The advantage of reciprocity calibration is that it is quite accurate and does not require specialized equipment. However, as the above analysis suggests, it is tedious and the correct reciprocity calibration method must be found for each transducer. Reciprocity calibration methods have been developed for accelerometers and other vibration sensors, but absolute direct and comparison calibration methods are still the preferred choice of most vendors.

10.3 Frequency and Time Scaling

Once a new transducer has been designed, it is common to build a prototype that is tested in the laboratory. This prototype may be larger or smaller than the intended new transducer, in order to make it possible to study the performance of the transducer more easily. The prototype is calibrated and its response characteristics determined. These results are then used to predict the response of the full-scale version of the transducer. In addition to using a scaled-up or scaled-down model, we may also build a purely electrical version of the transducer (when possible). This process is referred to as analog simulation and can be quite useful in identifying problems in transducer design. Of course, numerical simulation is also used to predict system performance. Both analog and digital (numerical) simulation tend to be more useful in design than in prototype testing, because they are only very crude representations of the real transducer system. In this section we discuss scaling in the frequency and time domains with the aim of providing a means of going from model results to predictions for behavior of a full-scale device.

First let us consider the analog simulation of a mechanical system using an electrical circuit. Clearly, based on our system of analogies, forces translate to voltages, velocities to currents. We are free to pick two scaling factors, one for each of these relations:

$$\begin{aligned} F &= K_e e, \\ V &= K_i i, \end{aligned} \tag{18}$$

where K_e and K_i are the scaling factors for conversion to voltage and current, respectively. Having chosen these two factors, the impedance scaling is set:

$$\begin{aligned} Z_m &= \frac{K_e}{K_i} Z_{el} \\ &= K_Z Z_{el}. \end{aligned} \tag{19}$$

With these definitions we are able to examine the scaling for mechanical to electrical ideal elements. For example, consider a mechanical resistance, R_m. With the scaling factors chosen, the suitable electrical resistance must relate to the mechanical resistance by

$$R_m = K_Z R_{el}. \tag{20}$$

Similarly, a mass has impedance ms and the analogous inductor has impedance Ls. Given the impedance scaling of (19), we must have

$$m = K_Z L. \tag{21}$$

A mechanical compliance has impedance $1/C_m s$ and the electrical capacitor which is analogous has impedance $1/C_{el}s$. Again using (19) we must have

$$\frac{1}{C_m} = K_Z \frac{1}{C_{el}}. \tag{22}$$

We may examine the scaled electrical model of the mechanical system and ask whether we have fundamentally changed its behavior from that of the original system. In particular, has the resonant frequency been affected? While not all systems have a resonance frequency exactly equal to $\sqrt{1/IC}$, most systems have a resonance proportional to that product. Hence let us consider $\sqrt{1/IC}$. For the original mechanical system this yields $\sqrt{1/mC_m}$ which is, from (21) and (22), identical to $\sqrt{1/LC_{el}}$. Hence, regardless of the choice of scaling factors for the force and velocity, the fundamental behavior of the electrical analog of a mechanical system is unchanged.

Now suppose that we wish to change the frequency or time behavior of a prototype transducer system. We might opt to increase the resonance frequency or speed up the temporal response in order to make measurements more rapidly. Then we proceed by introducing a frequency scaling factor defined by

$$\omega_{\text{test}} = K_\omega \omega_{\text{actual}}, \tag{23}$$

where ω_{test} is the frequency at which we choose to simulate or test and ω_{actual} is the frequency of the real device. This frequency scaling will affect the choice of inductance and capacitance elements. For example, we wish to scale the inductive impedance from $Lj\omega_{\text{actual}}$ to $Lj\omega_{\text{test}}$. This means that we have

$$L_{\text{test}} = \frac{K_Z}{K_\omega} L_{\text{actual}}. \tag{24}$$

Capacitive elements scale as well:

$$C_{\text{test}} = \frac{1}{K_Z K_\omega} C_{\text{actual}}. \tag{25}$$

Resistive elements, since their impedance does not depend on frequency, are unaffected by the frequency scaling. Using these scalings, we can determine how to build a prototype so as to scale the frequency response as suits us. Alternatively, if we choose to build a prototype at a scale different from the actual scale of the new transducer, we can determine how to scale the frequency response measurements from the scaling relations. Note that the equations above show that if we change the inductance elements and the capacitance elements arbitrarily in a prototype, then the test results have no simple frequency scaling that can be applied to predict the response of the full scale transducer.

Temporal scaling is not independent of frequency, since they are, in some senses, the inverses of each other. Hence frequency scaling as above results in time scaling

$$t_{\text{test}} = \frac{t_{\text{actual}}}{K_{\omega}}, \tag{26}$$

where t_{test} is the time scale at which we choose to test or simulate, and t_{actual} is the time scale of the real device. Differentiation then scales as

$$\frac{dy}{dt_{\text{test}}} = K_{\omega} \frac{dy}{dt_{\text{actual}}}, \tag{27}$$

and integration as

$$\int y \, dt_{new} = \frac{1}{K_{\omega}} \int y \, dt_{\text{actual}}. \tag{28}$$

The net result of using the techniques described here is that we can choose the scaling from mechanical forces to voltages, the scaling from mechanical velocities to currents, and the temporal or frequency scaling, all independently. Once the scaling factors are chosen, we could develop an analog electrical model of the transducer, or build a prototype transducer whose frequency or temporal characteristics are chosen to be convenient for testing.

10.4 Summary

In this chapter we have discussed the response characteristics generally used for transducers. We have also discussed the calibration of transducers, focusing on three approaches: comparison calibration, absolute direct calibration, and absolute reciprocity calibration. Finally, we discussed analog simulation and frequency and time scaling. The use of frequency or time scaling is in determining the characteristics of a prototype transducer which may be easily tested.

10.5 References

1. E. O. Doebelin, *Measurement Systems, Application and Design*, 3rd ed. (McGraw-Hill, New York, 1983).
2. J. B. Bentley, *Principles of Measurement Systems*, 2nd ed. (Longman Scientific and Technical with Wiley, New York, 1988).
3. C. L. Nachtigal (ed.), *Instrumentation and Control* (Wiley, New York, 1990).
4. Omega Engineering Inc., PO Box 4047, Stamford, CT 06907-0047, USA.
5. P. H. Sydenham (ed.), *Handbook of Measurement Science* (Wiley, New York, 1982).
6. GenRad Inc., 300 Baker Ave., MS No. 11, Concord, MA 01742-2174, USA.

7. W. R. MacLean, *Absolute Measurement of Sound Without a Primary Standard*, J. Acoust. Soc. Am., **12**, (1940), 143. Reprinted in H. B. Miller (ed.), *Acoustical Measurements* (Hutchinson Ross, Stroudsburg, PA, 1982).

11
Practical Considerations

The previous chapters in this book considered electromechanical transducers in detail: the fundamental energy conversion mechanisms, performance characteristics, and models which permit performance prediction. While this myopic view can be justified from the perspective of the design of electromechanical sensors and actuators, it does not lay the foundation for the practical use of such devices. This chapter is intended to be an introduction to the practical considerations of electromechanical transducers, particularly as they are used as components in larger systems.

The transducers described so far in this book all have one thing in common—they are analog devices. Indeed, virtually all of the fundamental mechanisms associated with the conversion of energy and information from electrical to mechanical form or vice versa are analog in nature. However, we usually choose to measure and control system performance using computers which are suited to handling digital information. Thus there is a basic incompatibility between sensors and actuators and the hardware to which they are typically attached. In the first section of this chapter we address some of the repercussions of this incompatibility by discussing electrical and mechanical means of digitizing the signals out of sensors. In the second section we discuss signal processing issues that arise from digitization as well as other issues of signal conditioning, including amplification and filtering. The chapter continues with a discussion of novel sensing mechanisms, in which the detection scheme involves indirect measures of the signal produced by a sensor. For instance, we might choose to detect frequency shift due to a change in the capacitance of a sensor rather than to sense the capacitance change directly. Finally, the chapter ends with a discussion of spatially distributed transducers—arrays and large individual elements.

11.1 Digitization of Analog Signals

All of the transduction mechanisms discussed in Part II of this book are fundamentally analog in nature. They are able to accept input energy that varies continuously over the range, and to respond with an output that reflects infinitesimal differences in the input to the fullest extent that the presense of noise permits. This is distinctly different from digital devices, in which typically both the output and input information are quantized so that the signals may assume only certain discrete values.

While the distinction between analog and digital hardware is mostly transparent to a user, there are a number of issues that can arise due to the seemingly minor incompatibilities. Typically, the first place the incompatibility between analog and digital information arises is at the interface between a transducer and a computer. Here *computer* is used in a very broad way to indicate processing hardware that is digital in nature. It may take the form of a recognizable personal computer, or it might be a digital oscilloscope or controller. As processing equipment has become progressively dominated by digital hardware in the last two decades, the incompatibility of processing equipment and the transducers attached has become more common.

Since digital signals effectively lose some information compared to analog signals (due to the discrete values that can be assigned), it is reasonable to think of solutions to the digital/analog incompatibility in terms of converting the analog signal into a digital signal. There are two basic means of achieving this conversion for an electromechanical transducer: electrical digitization of the signal, and mechanical digitization. Of these two options, electrical digitization is by far the more common. This technique is usually incorporated into the digital hardware in the form of an analog-to-digital converter (ADC). Mechanical digitization, by contrast, is usually incorporated into the transducer itself so that the electrical signal produced (or accepted) is in digital form.

When electrical conversion of an analog signal is employed, the typical scheme is as shown in Fig. 11.1, which also shows the production of an analog signal for an actuator from a digital signal. Note that the key items are analog amplifiers and filters, and converters of signals from analog to digital (ADC) or from digital to analog (DAC). The analog amplifiers and filters will be discussed in the next section of this chapter. Here we focus on the ADC and DAC, describing them in their simplest form.

Analog to digital signal converters consist of three main components: a sample and hold circuit, a comparator, and encoding logic. The sample and hold circuit is intended to do much as its name implies: it tracks an analog input for a short period of time, and then holds the last value tracked for the remainder of the duty cycle. The sample and hold circuit has a control input which determines how quickly the analog signal is sampled. Its output signal is still in analog form. A comparator is a special purpose operational amplifier which is usually used without any feedback. Its purpose is to compare two input signals and decide which is larger. For analog to digital conversion, one of the input signals to a comparator is a reference signal

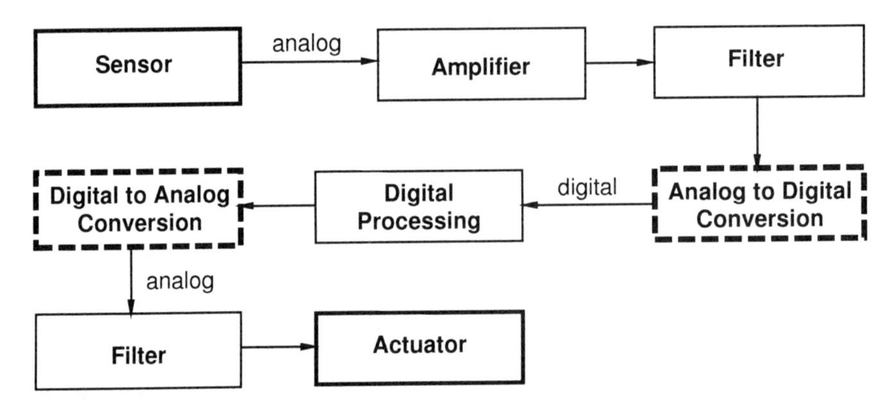

FIGURE 11.1. Electromechanical transducers with a digital processing scheme.

which sets the threshold for the choice of a high (1) or low (0) value. In essense then, the comparator is a single bit ADC. The encoding logic in an ADC permits the output signals of the comparators to be converted to the appropriate ones and zeros of a digital sequence representing the numerical value of the signal as represented by the sample.

Figure 11.2 shows one design for an ADC, including the comparator circuit and the encoding logic in the schematic. This figure, taken from Scheingold [1], is for a parallel converter sometimes referred to as a flash converter because of its rapid response. In this conceptually simple converter, there are as many comparators as there are bits in the data representation. Each of the comparators receives the input signal from the sample and hold circuit simultaneously—hence the name parallel converter. The comparators are said to be latched, as a resistive voltage divider controls the reference input signals to the comparators. Thus a change in the circuit reference voltage, V_{ref}, affects all of the reference voltages at the comparators.

Digital-to-analog converters are simpler devices than their ADC sisters because there is no need for a sample and hold circuit or for a comparator. Instead DACs are usually built from resistive networks, with one resistor and one switch per bit in the data representation. Additionally, the DAC requires an operational amplifier to sum the currents received from the network. For instance, Fig. 11.3, taken from Sheingold [1], shows an example of a DAC. In this example, a reference voltage is supplied to a set of resistors and switches. The switch positions are controlled by the values of each bit: 0 or 1. If the value is 1, then the switch for that bit is closed and current flows through the resistor to the op-amp input which is maintained at zero volts. By choosing the resistors as shown in Fig. 11.3, the appropriate weighting is given to each bit in the data sequence. Using the digital processor clock to change the data stream at a set speed, the analog output signal is created as a series of discrete values each of which lasts until the next sequence is given to the DAC. Thus, the analog output signal is stepwise continuous. If the step size is small and the refresh rate of the data sequence high, then the resulting output signal looks nearly identical to an analog signal.

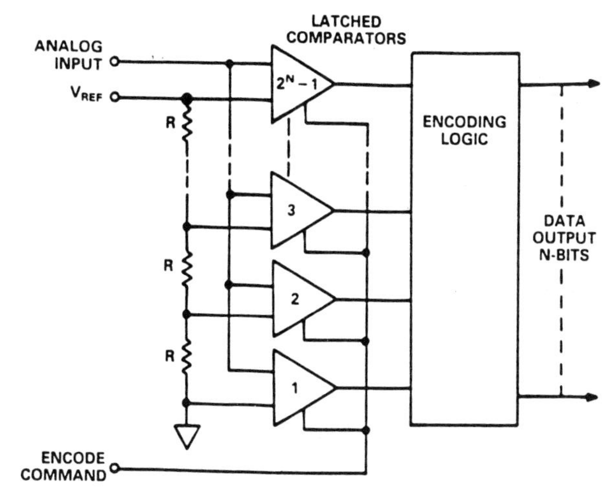

FIGURE 11.2. Parallel ADC schematic. Figure reproduced with permission from D. H. Sheingold (ed), **Analog-Digital Conversion Handbook**, Prentice-Hall, Englewood Cliffs, NJ, 1986.

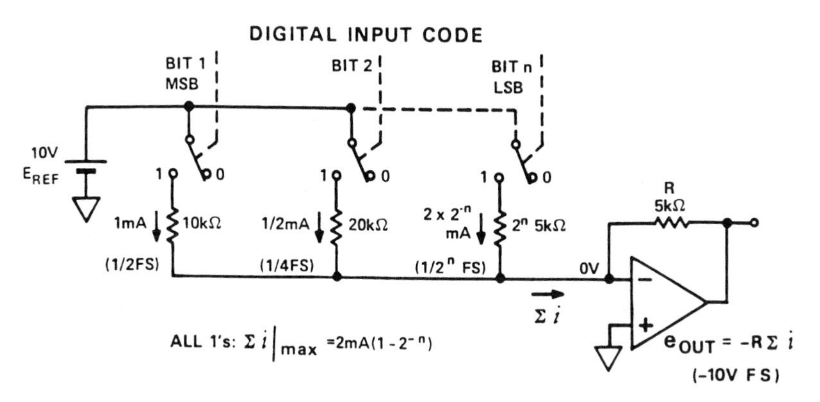

FIGURE 11.3. Example of a DAC. Figure reproduced with permission from D. H. Sheingold (ed), **Analog-Digital Conversion Handbook**, (Prentice-Hall, Englewood Cliffs, NJ, 1986).

Clearly, the topic of analog-to-digital electronic conversion is well studied and documented. The purpose here has been to present a very brief introduction to what is involved in the hardware. For further information, the reader is referred to the vast literature on digital hardware, including Sheingold [1], Lin [2], and Bouwens [3].

While the vast majority of systems using transducers use electrical digitization approaches, there are some interesting alternatives that opt to incorporate a mechanical digitization approach into the transducer itself. In general, such devices add a component to an analog transducer that may be thought of as a mechanical modulator whose sole purpose is discretization of an otherwise continuously variable signal. It is simplest to imagine using a modulator when there is a physical separation between essential components of the electromechanical transducer.

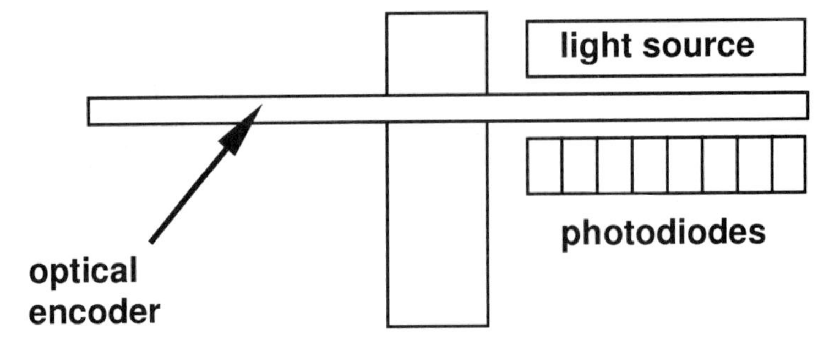

FIGURE 11.4. Operation of an optical encoder.

The classic example of this structure in an electromechanical transducer is found in optomechanical sensors where there is often a physical separation between the light source and the sensor.

An example of a mechanical modulator used to digitize an otherwise analog signal is demonstrated by the optical encoder shown in Fig. 11.4. Here a light source is separated from a set of photodiodes by a path which is broken by an optical encoder disk. This disk can take many forms, but a particularly interesting one is for the disk to be coded so as to provide an output in which the light detectors directly provide digital numerical sequences that indicate rotational position. For example, Fig. 11.5, taken from Slocum [4], shows a three-bit encoder in which the most significant bit is represented by the inside ring and the least by the outside ring. Here white portions of the disk permit light to be transmitted and correspond to a low value for that bit. Dark portions produce a high value.

In the optical encoder, the output of each photodiode is still an *analog* signal. However, because the distance between the source and the photodiodes is fixed, and the material properties of the encoder disk are also constant, the output of each photodiode has been effectively limited to two values only. Thus, the distinction

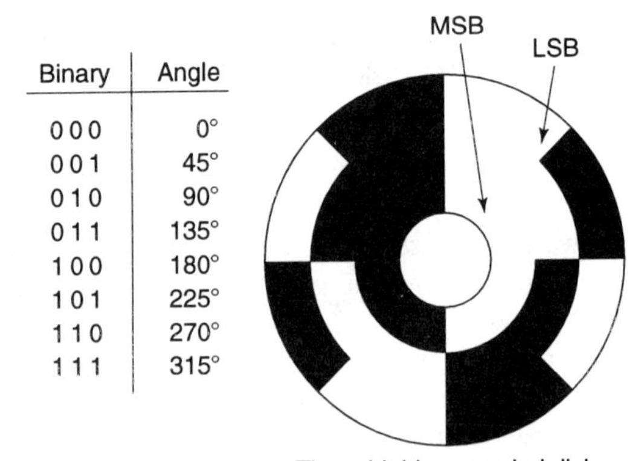

Binary	Angle
0 0 0	0°
0 0 1	45°
0 1 0	90°
0 1 1	135°
1 0 0	180°
1 0 1	225°
1 1 0	270°
1 1 1	315°

FIGURE 11.5. Example of a three-bit binary optical encoder disk. Figure taken from Slocum [4].

between analog and digital signals becomes moot at this point, provided that we can ignore those times when a photodiode is exposed to an edge between the transparent and reflecting fields. In practice, we can prevent the edge transitions from having much impact by incorporating a threshold-type of circuit in the electronic processing.

While optical encoders are the most common sorts of electromechanical transducers with digitizing modulators incorporated, they are by no means the only mechanically digitized transducers conceivable. For instance, we could imagine a vibration sensor which consists of a series of vibrating reeds, each a cantilever beam with metal deposited onto it. As the sensor is subjected to vibration excitation, each reed vibrates. By way of digitization of this sensor, we could simply apply a voltage between the tip of each beam and a pad at a fixed distance. Thus, if the reed vibrates enough to touch the pad (close the switch) current flows. Otherwise, no current flows. By varying the size or material of the reeds, we could ensure that the reeds have different thresholds for switching from high- to low-current flow. While this device would not produce a digital sequence that gives the numerical value of the measurand, it would be a mechanical replacement for the comparators shown in Fig. 11.2, and would require only encoding for the conversion of a binary sequence to the correct binary representation of the number.

11.2 Signal Conditioning

The general schematic of a closed-loop system, using a digital processor and both sensors and actuators shown in Fig. 11.1, shows two key elements in addition to the transducers and the converters: filters and an amplifier. Filters and amplifiers are two of the most basic signal conditioning elements available, and they tend to be present in virtually all transducer processing loops. In a very real sense, both amplifiers and filters are rendered necessary as a result of the conversion of analog to digital signals and vice versa. Filters are necessary as a result of sampling the signal at discrete times, while amplifiers at the input stage are needed in order to achieve the maximum signal-to-noise ratio. These concepts are explained more fully below.

11.2.1 Filtering

First let us consider the effect of representing a continuous signal by a set of discrete values. Our initial, continuous signal is described by $s(t)$, but that signal is assumed to be sampled at intervals of τ so that we save the values of s at times 0, $\tau, 2\tau, ..., (N-1)\tau$. The result is that we have N samples of s which we write as s_m, where m is the integer indicating sample number. In other words, s_i is $s(t = i\tau)$. Because the time domain has been effectively discretized, the frequency domain must also make a transition from continuous to discrete. This is clear from the

Fourier transform of the discrete sequence, which can be written as,

$$S_k = \frac{1}{N}\sum_{n=0}^{n=N-1} s_n e^{-j2\pi kn/N} = \frac{1}{N}\sum_{n=0}^{n=N-1} s_n e^{-j(2\pi k/N\tau)(n\tau)}. \quad (1)$$

The last expression in (1) is intentionally written to facilitate interpretation. The last term in the exponent is time, and thus the summation looks like the discrete version of integration over time. The other term in parentheses in the exponent must then be frequency. Thus the sampling in time means that the frequency components which are present are those which correspond to the various values of k in Eq. (1). These are

$$\omega = 0, \frac{2\pi}{N\tau}, \frac{4\pi}{N\tau}, \ldots, \frac{2(N-1)\pi}{N\tau} \approx \frac{2\pi}{\tau}.$$

The discrete frequencies are spaced uniformly since the time samples are uniformly spaced, and the highest frequency present in the signal is determined by the sampling rate.

Now consider the effect of this conversion from a continuous signal (in time) to a discrete sequence of numbers. For instance, Fig. 11.6(a) shows a continuous sinusoidal signal and Fig 11.6(b) shows what is would look like if its sampled values were plotted. We can see that even connecting the sampled values by straight lines would produce a plot closely resembling the continuous signal. Thus, we would expect the sampled signal to be a good approximation to the continuous one. However, suppose the signal were very high in frequency, so that exactly one-half period of the sinusoid elapsed between each sample. This is shown in Fig. 11.7. Now the continuous signal and discrete signal bear little resemblance to one another.

What Fig. 11.7 shows is that there is a clear problem with representing a continuous signal by its discrete samples if those samples are spaced too far apart in time. Indeed, one can show (as is done in virtually all texts on digital signal processing) that we must sample faster than half the period of a sinusoidal signal in order to see all of the zero crossings clearly, i.e., in order to define the frequency clearly. It is this concept that forms the basis for the sampling theorem and the definition of the Nyquist frequency as the top frequency which can be sensed unequivocably in a periodically sampled signal. The Nyquist frequency can be written as $f_{\mathrm{Ny}} = 1/2\tau$.

If a continuous signal contains frequencies higher than the Nyquist frequency, then those frequency components will be mapped erroneously into lower frequencies in much the same way that the continuous signal in Fig. 11.7 was mapped into a signal at zero frequency. This distortion of a signal due to a low sampling rate is referred to as aliasing. It can be avoided only by using a signal to remove those high-frequency components from the signal before sampling. It is for this reason that filters (and more specifically *low-pass filters*) are placed between every analog signal and the ADC. To be sure, the filter also distorts the time continuous signal, but it does so in a known way and therefore is preferred compared to the unknown distortion produced by aliasing. The sampling theorem also explains why ADCs with higher (faster) sampling rates are more desirable than those with low sam-

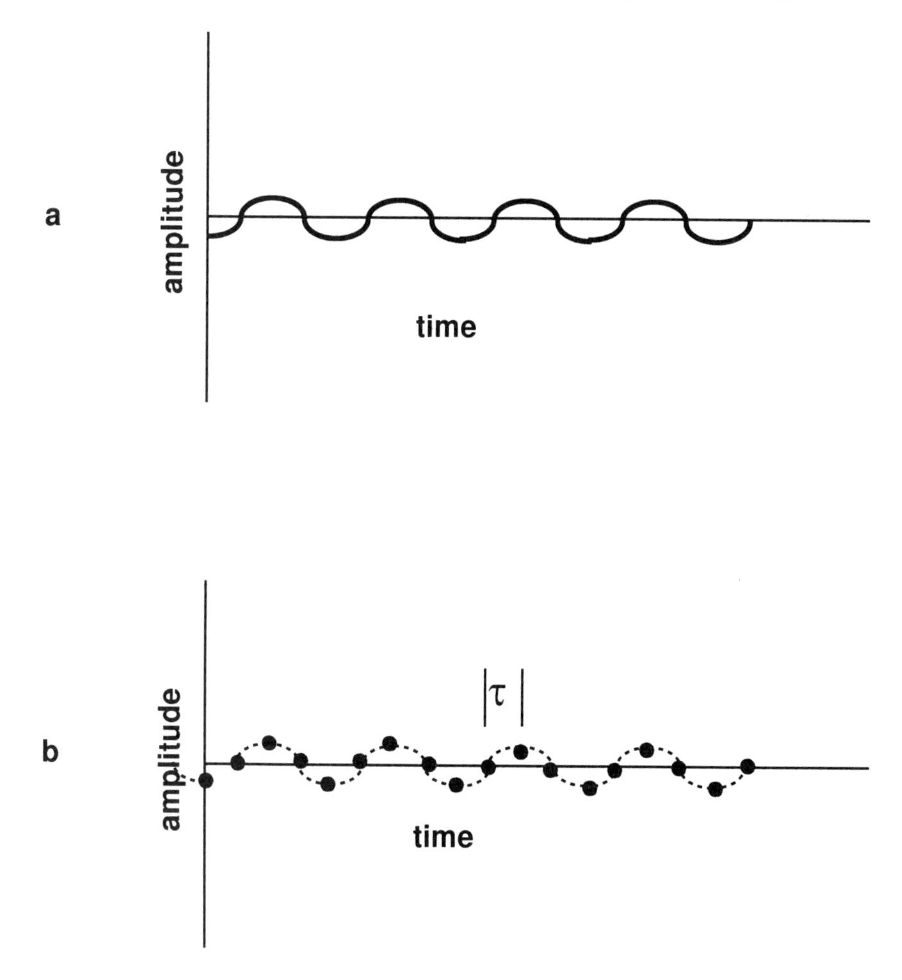

FIGURE 11.6. (a) Continuous signal as a function of time, and (b) its discrete representation.

pling rates. It is because the bandwidth of the high sampling rate ADCs is higher. For a more complete explanation of aliasing, the reader is referred to any text on digital signal processing, certainly including Oppenheim and Schafer [5], Rabiner and Gold [6], Morgera and Krishna [7], and McGillem and Cooper [8].

The above discussion shows that there is a key relation between time and frequency that must be considered when analog signals are being digitized. However, aliasing is not the only time/frequency problem that arises in signal processing. Another issue relates to the finite duration of the signal. This introduces what is known as a windowing problem. It can be understood by consideration of a sinusoidal signal that starts and stops at a given time, as shown in Fig. 11.6(a). This signal could be represented as the product of two signals: a sinusoidal signal that goes on for all time, and a rectangular window that turns on (goes from 0 to 1) at time $t = 0$ and turns off at time $t = (N - 1)\tau$. The Fourier transform of a product

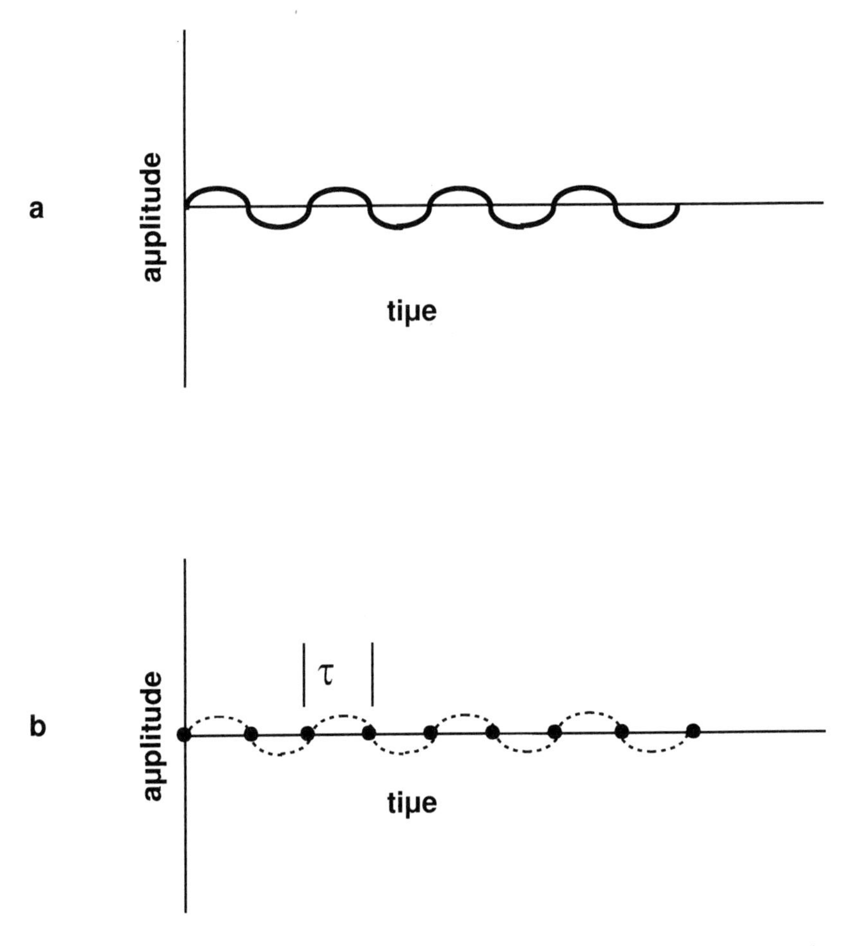

FIGURE 11.7. (a) High-frequency continuous signal as a function of time, and (b) its discrete representation using too low a sampling rate.

is the convolution of the Fourier transforms of each function. Since the Fourier transform of a rectangular function is rich, consisting of a main peak at $\omega = 0$ and several sidelobes, the effect of the convolution is to convert the spectrum of the infinite-time sinusoid from a single peak at the sinusoidal frequency Ω to a peak at that frequency with lots of sidelobes for the finite-duration signal. Thus, finite duration distorts the signal in the frequency domain, while preserving it in the time domain.

Another way to view this problem is to think about the process that is undergone in constructing a Fourier transform using a signal of finite duration, such as in performing an FFT or DFT on a computer. Suppose that we take some arbitrary signal such as is shown in Fig. 11.8(a) and we desire its Fourier transform. Then the process assumes that the signal from $t = 0$ to $t = T$ is representative of a continuous signal that can be made by just repeating the signal over and over again, as shown in Fig. 11.8(b). Note that in building up a continuous signal this

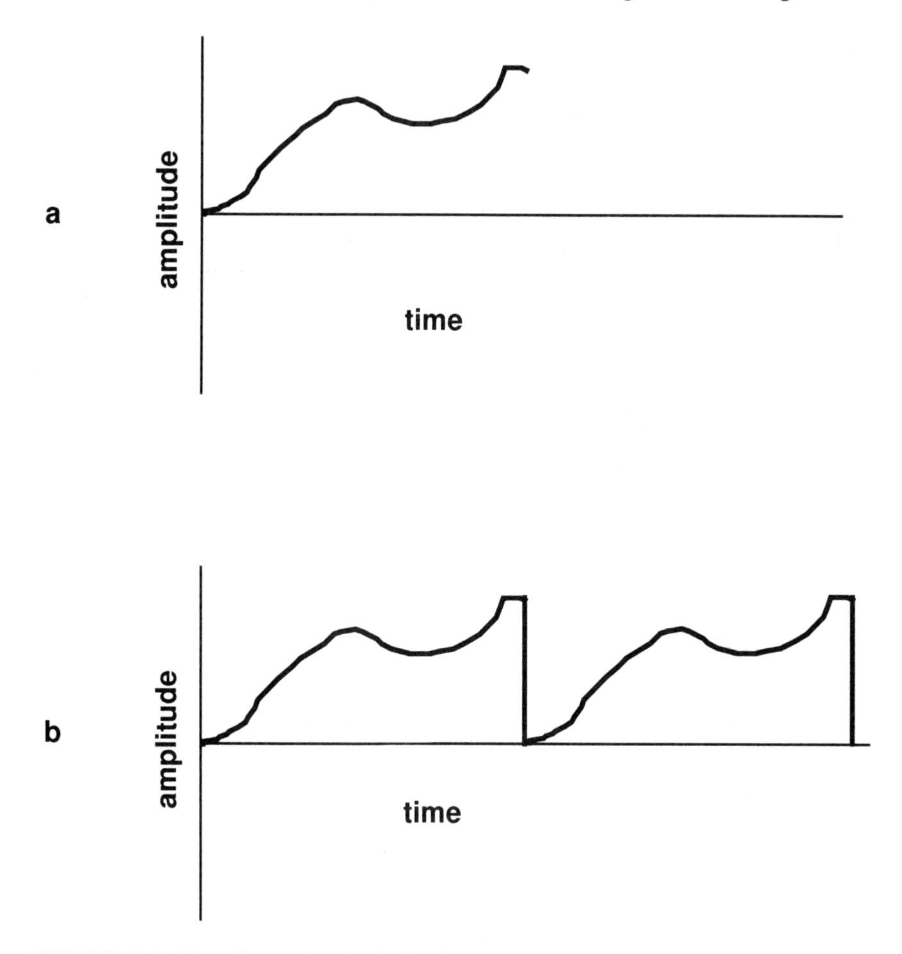

FIGURE 11.8. (a) Arbitrary signal of finite duration, and (b) the continuous time signal the FFT process imagines it comes from.

way, we introduce a problem at the junction between one signal and the next repetition. This is because the signal started at an amplitude of zero but does not end at zero. Thus as the signal is repeated there is a very rapid transition (like the edge of a step function), complete with lots of erroneous high-frequency content.

The solution to the windowing problem is hinted at by the discussion above. The key is to prevent rapid transitions due to repetition of the sampled signal in forming the presumed continuous signal. This may be done by changing the rectangular window into one which starts at zero and ends at zero, but has a more gradual rise to 1 and fall to 0 than does a rectangular function. Options include a squared cosine function (one-half period) or other smooth functions. A number of alternatives have been developed over the years, each window bearing the name of the person most responsible for its definition (Hanning, Hamming, Tchebyshev, etc.), and each having its own set of advantages and disadvantages. All of these windows have been implemented in hardware and/or software. All except the rectangular window distort the signal in the time domain in order to produce less distortion of the frequency domain representation. Thus, the choice to use a window other than the rectangular window should only be made if frequency domain signal processing is going to be used.

A final note about filtering hardware is in order before leaving the subject. Filters come in a large variety of types and implementations. The measures which are typically used to represent these devices are their bandwidth, skirt slope, and ripple. The bandwidth is a measure of what frequencies are passed essentially unscathed. In most filters, the bandwidth is adjustable. The skirt slope is a measure of how much attenuation of signals outside the bandwidth is available. For instance, a low pass filter might have skirts of −40 dB/decade, meaning that at the frequency 10 times above the "knee" in the filter, the signal is attenuated by 40 dB. The ripple is a measure of smoothness in the bandpass region of the filter. The less ripple, the better the filter in general. In addition to these measures, there are times when the phase distortion of the signal is important and this metric is sometimes provided by manufacturers.

11.2.2 Amplifying

The section above on filtering began with a discussion of the effects of discretization in the time domain. In addition to discretizing an analog signal in the time domain, the process of converting a signal from analog to digital represents its amplitude at any sample time using a discrete set of values. Thus the signal is discretized in both time and amplitude. The need for amplification for analog signals going into an ADC is a consequence (in large part) of the discretization in the amplitude representation.

Each bit of a digital representation can have only two possible values: 0 or 1. We can imagine we are representing a signal confined between 0 and 1 without loss of generality because we can certainly scale a signal. If the analog signal has a value S at the moment it is sampled, where $0 \leq S < \frac{1}{2}$, then the digital representation is 0. Otherwise S is represented by 1. Consideration of this assignment of discrete value to the continuous range of 0–1 leads us to conclude that the largest possible error is $\frac{1}{2}$.

In general we could continue using the same logic as more bits are added to the numerical representation. For a two-bit representation, the maximum error in the last bit would be $1/2^2$. For an m-bit representation, the maximum error is $1/2^m$. This error, called the quantization error, defines the unavoidable noise which is introduced in the process of discretizing the amplitude of an analog signal.

From the quantization noise we may estimate the signal-to-noise ratio of the conversion system. For a single bit, the maximum value of the signal is 1, while the maximum noise is $\frac{1}{2}$. Hence the signal-to-noise ratio is 2 or 6 dB ($20 \log_{10} 2$). Each additional bit improves this ratio by the same amount, i.e., by 6 dB. Thus for an 8-bit converter, the best signal-to-noise ratio we can achieve is 48 dB, for a 12-bit converter, the best signal-to-noise ratio is 72 dB, and for a 16-bit converter the best signal-to-noise ratio is 96 dB.

Given that the range of a numerical representation is set by the number of bits, and that the quantization noise and signal-to-noise ratio are also set by the number of bits in the conversion process, the remaining question is how to achieve the cleanest signal. The answer is to make sure that the analog signal going into an ADC is making full use of the dynamic range of the converter. It is for this reason that an amplification stage is normally put between the sensor creating the analog signal and the ADC.

There are many forms of signal amplifiers varying from single computer chips to large instrument amplifiers. The appropriate amplifier generally is dependent upon the nature of the signal to be amplified. For electromechanical sensors and actuators, the main question is whether the signal is a voltage signal, a current signal, or a charge signal. Voltage amplifiers are by far the most common type of signal amplifier and they are far too numerous in kind to describe here. Instead the reader is referred to books on electronic instrumentation such as Diefenderfer [9], Jones and Chin [10], and Coombs [11].

The standard approach to amplification of a current or charge signal is to convert it to a voltage signal for which lots of amplifier types are available. Conversion of current to voltage is accomplished by means of a transimpedance amplifier, which can be as simple as an operational amplifier with a feedback resistor. Conversion of a charge signal to a voltage is very useful for capacitive sensors such as piezoelectric devices. Here again there are many amplifiers available, including operational amplifiers with feedback capacitors and a capacitor to ground in parallel with the charge source.

The key measures of the quality of an amplifier are its linearity, gain factor, frequency response, stability, and accuracy. The most expensive amplifiers are those with adjustable gain over a wide range, wide bandwidth over which there is little distortion, and high stability.

11.3 Novel Sensing/Actuation Techniques

Part II of this book assumed that actuation or sensing is accomplished by a direct means. For instance, if a sensor produces a change in capacitance and an associated change in voltage, it was tacitly assumed that the voltage signal was directly monitored. In this section we briefly examine indirect methods of sensing or actuation. In general these indirect methods are often chosen because they provide enhancements in performance, in particular, improving accuracy, linearity, or cost. The two indirect approaches presented in some detail are frequency detection, and time of flight methods.

11.3.1 Frequency Detection Schemes

Of all of the parameters that we can detect, the two we measure most accurately and economically are frequency and time. For this reason, electromechanical sensors are often used in measurement schemes that relate signals to frequency shifts or to lapsed time rather than sensing voltage, current, or charge.

The most common technique for frequency detection uses an inductor and capacitor to establish a resonant oscillator. One of these devices, either the inductor or the capacitor is the transducer while the other component is fixed. As the transducer operates its value of capacitance or inductance changes dynamically. This causes a shift in the oscillator frequency which is sensed using a frequency counter circuit.

It is interesting to note that the resonant frequency of a simple oscillator circuit varies as

$$\omega_n = \sqrt{\frac{1}{IC}}, \tag{2}$$

for low values of resistance in the oscillator. Hence, there is clearly a nonlinear relation between the natural frequency and the transducer property—capacitance or inductance. However, one can use the standard means of expressing the transducer property as a nominal value and a dynamically varying portion so that the resonant frequency becomes

$$\omega_n = \sqrt{\frac{1}{I_0 C_0 (1+\alpha)}} \approx \sqrt{\frac{1}{I_0 C_0}} \left[1 - \frac{\alpha}{2 I_0 C_0}\right], \tag{3}$$

where α is the fractional variation in the transducer property, and C_0 and I_0 are the nominal circuit values of capacitance and inductance. Hence, it is possible to establish a reasonably linear relationship between changes in the resonant frequency of the oscillator circuit and the change in the transducer property. The linear approximation gets progressively better as the $I_0 C_0$ product increases.

While virtually any of the devices described in Chapters 3 to 6 could be used in a frequency detection scheme, one slightly unusual example is presented here to make the frequency detection concept clearer. This example is an odor sensor described by Matsuno et al. [12] and shown in Fig. 11.9. In this sensor, a quartz crystal is loaded by a membrane with a selective ability to adsorb odorants. As odorant molecules are adsorbed, the membrane mass increases, thus decreasing the resonant frequency of the system. It is this frequency shift that is sensed, and linearly related to the odorant mass per unit area.

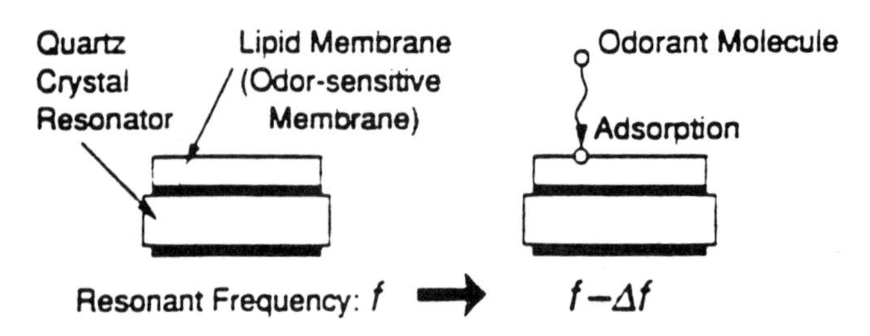

FIGURE 11.9. Odor sensor described by Matsuno et al. [12].

In addition to the frequency shift produced in a resonant circuit or structure, another common frequency detection approach is to make use of a Doppler shift. This is particularly useful for determination of a linear velocity and is the basis for laser Doppler velocimeters, some ultrasonic flowmeters, and even the dreaded radar guns used by police in determining the speed of a vehicle. In each of these cases, a wave striking a moving object causes the wavefronts to be either spread out (object receding) or compressed (object approaching), thus producing a change in the apparent frequency. This drop is directly related to the speed of the object, so it is possible to detect the object's velocity from the frequency shift. For waves in a real medium (e.g., sound) the apparent frequency, f', is given by

$$f' = \frac{f}{\sqrt{1 \pm v/c}}, \tag{4}$$

where v is the speed of the moving object, f is the driving frequency, and c is the wave speed in the medium. For electromagnetic waves (including light) the apparent frequency shift is given by

$$f' = f\sqrt{\frac{1 \mp v/c}{1 \pm v/c}}, \tag{5}$$

where the top signs apply for the receding object.

Car speed detectors make fairly direct use of the relations above, as do a few ultrasonic velocity detectors. Laser Doppler velocimeters are somewhat more complex, as they aim to monitor the velocity of liquids or gases. In these devices, an example of which is shown in Fig. 11.10, taken from Doebelin [13], the fluid is typically seeded with particles that are used to reflect light. One (or more) laser(s) is used to produce multiple light beams which define a detection region where they cross and produce interference fringes. It is the motion of the fringes which is related to the velocity of the particles.

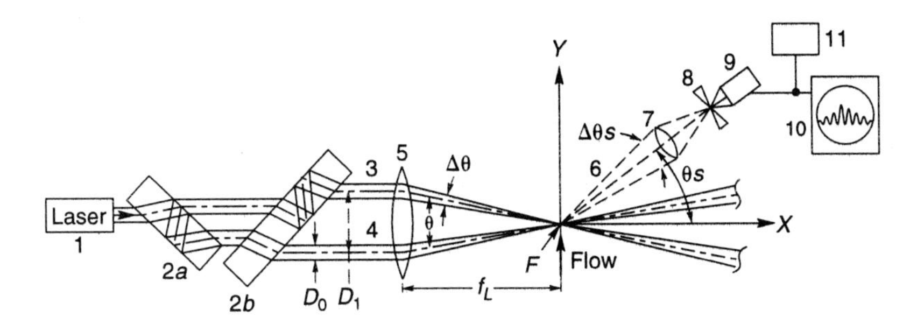

FIGURE 11.10. Example of a laser Doppler velocimeter. Figure taken from E. O. Doebelin, **Measurement Systems: Application and Design**, 3rd ed., (McGraw-Hill, NY, 1983). Reproduced with permission by The McGraw-Hill Companies.

11.3.2 Lapsed Time Detection

In the time domain, the rough equivalent of sensing a frequency shift is sensing lapsed time. This approach to sensing is often referred to as "time-of-flight" measurement, and has been used in a variety of applications, from thickness gauging to flow measurement. In lapsed time sensing, it is imperative that the measurand affect the time required for something to happen. As a simple example, consider the Polaroid® transducer used in automatic focusing of their cameras. This device generates a sound and simply monitors the time lapsed for a return echo. Given a known speed of sound propagation, the distance to the object to be captured on film is known once the lapsed time is sensed. A similar approach is used in ultrasonic gauging of materials. Sound sources placed on two sides of an object determines the object's thickness by monitoring the time for return signals on both sides.

More sophisticated lapsed-time sensors use a medium which creates a time bias, or a situation in which there is relative motion between the source and the receiver. An example of the former approach is used in some ultrasonic flowmeters, as pictured schematically in Fig. 11.11, taken from Doebelin [13]. Here there are pairs (at least one) of sound sources and sound sensors spaced a known distance apart. The time for propagation of the sound wave between the source and sensor is given by

$$t = \frac{L}{c + v}, \tag{6}$$

where L is the distance traveled, v is the fluid velocity, and c is the speed of sound propagation. The fluid velocity is said to be positive if the flow is in the direction from the source to the sensor. The velocity is negative if the fluid motion is counter to the source to flow direction. The advantage of using two pairs of sensors and sources aligned as shown in Fig. 11.11 is that we can detect the difference in the lapsed times, producing a twofold increase in the linear signal.

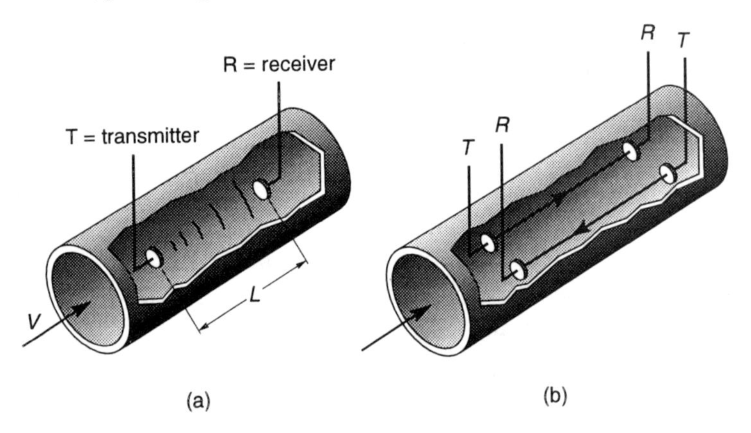

FIGURE 11.11. Ultrasonic flowmeter based on lapsed time measurement. Figure taken from E. O. Doebelin, **Measurement Systems: Application and Design**, 3rd ed., (McGraw-Hill, NY, 1983). Reproduced with permission by The McGraw-Hill Companies.

An example of a lapsed time sensing approach which depends on relative motion between the source and the receiver is found in Russell [14], which describes a magnetostrictive position sensor. In this sensor, a short current pulse is run through a magnetostrictive waveguide which is near some external magnet. The current pulse establishes a magnetic field which interacts with the external magnet to induce stresses in the waveguide. This launches a sound wave which is detected at a sensor at one end of the waveguide. Since the speed of electromagnetic propagation is many orders of magnitude greater than sound speed propagation, the current pulse can be thought to generate a magnetic field in the waveguide instantaneously. Thus, the time from the current pulse to the sound wave detection at the waveguide end corresponds to the position of the external magnet.

Lapsed time and frequency sensing approaches can be used with virtually any of the transducers described in this book. These techniques offer an alternative to direct sensing of currents, voltages, or charges, and are useful in many circumstances.

11.4 Spatially Distributed Transducers

The material presented previously in this book has tacitly focused on a single transducer. It is common to assume that a transducer exists at a point in space. This means that the spatial extent of the device is generally taken to be insignificant in the conversion of energy and information. There are two circumstances when this assumption is intentionally violated: when using arrays of transducers to achieve a performance in the ensemble which the individual element is not capable of producing, and when a single element is made large compared to dynamic scale lengths. We investigate each of these circumstances briefly below.

First consider arrays of transducers. Arrays are multiple transducer elements arranged in lines, rings, planes, and even three-dimensional shapes such as spheres. They are not uncommon for sensors or actuators. There are a number of reasons why we might choose to build an array of transducers, but the most common are to gather (or produce) more than one signal simultaneously, and to introduce an ability to sense (or generate) signals with nonuniform directional distribution. The optical encoder shown in Fig. 11.4 is an example of an array designed to permit multiple signals to be gathered simultaneously. Here there are multiple optical sensors, each corresponding to one bit of information. The linear array of microphones shown in Fig. 11.12 is a classic example of a transducer array created for the purpose of providing directional information. The response of this array, assuming equally sensitive elements and a simple summing of the output electrical signals, is a maximum for sound which approaches the array at angle $\theta = 0$ (broadside), and gradually decreases as the angle increases.

For arrays of transducers which produce a single signal output, for instance the linear array of microphones, the output is very much a function of the manner in which the multiple signals of the transducer elements are combined to form a

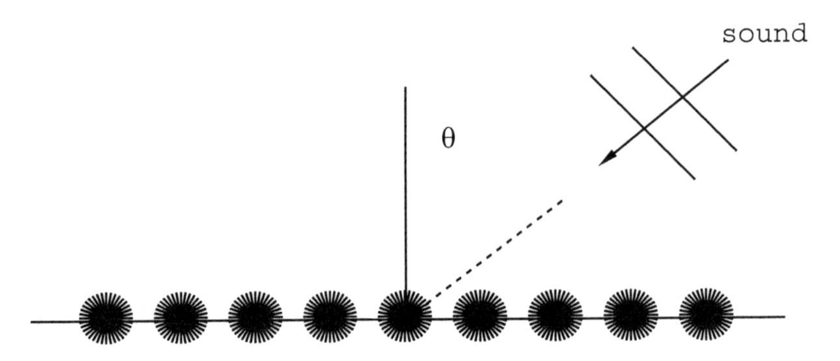

FIGURE 11.12. Linear array of microphones.

single array output. For instance, the signals from the elements of the linear array in Fig. 11.12 can be amplified with differing amounts of gain or can be delayed by differing amounts of time. Further, the spacing between the elements, which is shown as constant in Fig. 11.12, can be intentionally made nonuniform. The results of these changes are manifest through a directional sensitivity which can be greatly affected. For instance, the peak in response can be made to vary from broadside to endfire (sound propagating parallel to the line) simply by introducing these signal processing modifications. For further information on arrays and signal processing, readers are referred to Pillai [15], Haykin et al. [16], and Johnson and Dudgeon [17].

For arrays of transducers in which multiple signals are combined to form one, it is sometimes possible to replace the array of small transducers with a single transducer which has significant spatial extent. For instance, Busch-Vishniac et al. [18] describe a microphone built for teleconferencing in which a linear array of 28 microphones is replaced by a single microphone with the same total spatial extent as the line array. The amplification differences for elements in the array are transformed into a capacitor electrode whose width is a function of position. This is shown in Fig. 11.13. By varying the electrode width, the area of the capacitive microphone and thus the sensitivity is effectively varied as a function of position. Similar approaches to spatially distributed sensors and actuators are described by Bailey and Hubbard [19], who have used the sensors and actuators for the active damping of vibrating structures.

Finally, some comments on the modeling of transducer arrays and single transducer elements with significant spatial extent are in order. Transducer arrays generally may be thought of as independent elements which are connected only at the input and output ports to the individual elements. A reasonable model of this situation uses the connected 2-port models presented in Chapter 9. This is described in detail in Busch-Vishniac [20]. For spatially distributed transducers, there is usually a mechanical coupling between positions in the transducer. One set of models for this situation has been presented by Busch-Vishniac [21] and extends transmission line theory to include interactions with the environment all along the line. This produces a model capable of including input and output sig-

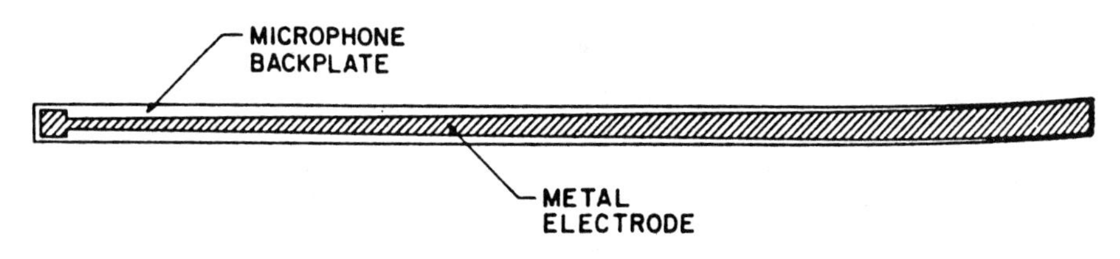

FIGURE 11.13. A single microphone with large spatial extent can replace a line array of microphones. This picture shows the electrode pattern for a parallel plate capacitor microphone. The sensitivity at any position is a function of the electrode width. Figure reprinted with permisssion from I. J. Busch-Vishniac, J. E. West and R. L. Wallace, Jr., *A New Approach to Transducer Design Applied to a Foil Electret Acoustic Antenna,* J. Acoust. Soc. Am., **76**, 1609 (1984). Copyright 1984 American Institute of Physics.

nal contributions, as well as mechanical and electrical coupling throughout the device.

11.5 Summary

This chapter has addressed a few of the important issues in the practical usage of electromechanical sensors and actuators. In particular, this chapter has focused on the digitization of signals from analog transducers and the affect of filtering and amplification in this conversion process. In addition, indirect methods of sensing and actuating using resonant frequency shifts and lapsed time monitoring were presented. Finally, spatially distributed transducer systems were discussed, in terms both of arrays of transducers and of single transducers with large spatial extent.

11.6 References

1. D. H. Sheingold (ed.), *Analog-Digital Conversion Handbook* (Prentice Hall, Englewood Cliffs, NJ, 1986).
2. W. C. Lin, *Handbook of Digital System Design for Scientists and Engineers: Design with Analog, Digital, and LSI* (CRC, Boca Raton, FL, 1981).
3. A. J. Bouwens, *Digital Instrumentation* (McGraw-Hill, New York, 1984).
4. A. H. Slocum, *Precision Machine Design* (Prentice Hall, Englewood Cliffs, NJ, 1992).
5. A. V. Oppenheim and R. W. Schafer, *Digital Signal Processing* (Prentice Hall, Englewood Cliffs, NJ, 1975).
6. L. R. Rabiner and B. Gold, *Theory and Application of Digital Signal Processing* (Prentice Hall, Englewood Cliffs, NJ, 1975).

7. S. D. Morgera and H. Krishna, *Digital Signal Processing: Applications to Communications and Algebraic Coding Theories* (Academic Press, Boston, MA, 1989).

8. C. D. McGillem and G. R. Cooper, *Continuous and Discrete Signal and System Analysis*, 2nd ed. (Holt, Rinehart and Winston, New York, 1984).

9. A. J. Diefenderfer, *Principles of Electronic Instrumentation* (Saunders, Philadelphia, PA, 1972).

10. L. D. Jones and A. F. Chin, *Electronic Instrumentation and Measurements* (Prentice Hall, Englewood Cliffs, NJ, 1991).

11. C. F. Coombs, Jr. (ed.), *Electronic Instrument Handbook*, 2nd ed. (McGraw-Hill, New York, 1995).

12. G. Matsuno, D. Yamazaki, E. Ogita, K. Mikuriya, and T. Ueda, *A Quartz Crystal Microbalance-Type Odor Sensor Using PVC-Blended Lipid Membrane*, IEEE Trans. Instrum. and Meas., **44**, 739 (1995).

13. E. O. Doebelin, *Measurement Systems: Application and Design* (McGraw-Hill, New York, 1983).

14. J. Russell, *New Developments in Magnetostrictive Position Sensors*, Sensors, 44, June 1997.

15. S. U. Pillai, *Array Signal Processing* (Springer-Verlag, New York, 1989).

16. S. Haykin, J. H. Justice, N. L. Owsley, J. L. Yen, and A. C. Kak, *Array Signal Processing* (Prentice Hall, Englewood Cliffs, NJ, 1985).

17. D. H. Johnson and D. E. Dudgeon, *Array Signal Processing: Concepts and Techniques* (Prentice Hall, Englewood Cliffs, NJ, 1993).

18. I. J. Busch-Vishniac, J. E. West, and R. L. Wallace, Jr., *A New Approach to Transducer Design Applied to a Foil Electret Acoustic Antenna,* J. Acoust. Soc. Am., **76**, 1609 (1984).

19. T. Bailey and J. E. Hubbard, *Distributed Piezoelectric-Polymer Active Vibration Control of a Cantilever Beam*, AIAA J. Guid. Cont. Dyn., **16**, 605 (1985).

20. I. J. Busch-Vishniac, *Spatially Distributed Transducers I—Coupled Two-Port Models,* J. Dyn. Sys. Meas. Control, **112**, 372 (1990).

21. I. J. Busch-Vishniac, *Spatially Distributed Transducers II—Augmented Transmission Line Models,* J. Dyn. Sys. Meas. Control, **112**, 381 (1990).

Index

Accelerometer 8–9, 48–49, 164–167, 224–225, 263–268
accuracy 301–302
actuator definition 9
airbag sensors 9
Ampere's law 103–104
amplifiers 330–331
analog to digital converters 321–323
anemometer 9, 243
angular position sensors 81, 115–116, 202, 214–215
attitude sensor 214–215
arrays 335–337
automobile sensors 7, 9

Biot–Savart law 101
bond graph models 34–49, 64, 71–74, 106–109, 112–115, 124, 159–163, 165, 190–191, 195–196, 211–213, 280

Calibration 310–316
capacitance definition 63
causality 35, 38–45, 73–74, 287–288
cermets 221
charged particle interactions 188–190
circuit models 29–34
copier 179–180
crank engine 246
Curies, Jacque and Pierre 141
cylindrical capacitor 89–92

Digital to analog converters 321–322
digitization 321–325
disk drive 192–193
disk heads 242–243
displacement sensors 68–69
displacement variables 21–22
distortion 304–305
distributed transducers 335–337
drift 307

Earphone 7
eccentricity sensor 128–129
eddy current sensors 134–137
efficiency 293–299, 306
effort source 22–23
elastoresistance 217–220
electret transducers 68, 74–74, 89–90
electric displacement 62
electric fields 59–62
electric polarization field 62
electrostatic motors 83–85
electrostriction 96–97

energy domains 1–5, 17–19
engine igniter 173–174

Faraday's law 103–104
fiber optic sensors 256–273
 intensity modulated 262–273
 phase modulated 271–273
filters 325–330
Firestone 19–20
flaw detector 137–138
flextensional sound source 133–134, 171
flow meter 194–198, 203, 340–341
flow sensors 74–75, 122, 191–194, 204 235, 237–238, 243, 268–269
flow source 22
force heads 9
force sensors 68–70, 77, 274–275
frequency detection 331–333
frequency response 51–53, 306

Gauss's law 60–62
geartooth sensors 204
geophone 272
gradient transduction 92–95, 129–130
 normal 93–94, 129–130
 tangential 94–95, 129–130
gyrators 43–45, 191

Hall effect sensors 5, 195–204, 247
Hamiltonian variables 21–22
harmonic distortion 67–68, 76, 85, 304–306
heat flux sensor 239–240
hot wire anemometer 231–232
humidity sensor 14, 88, 212–213
hydrophones 71–72, 167–168, 173–174

Ideal capacitance 23–27
ideal inertance 23–27
ideal resistance 24–27
impedance analogy 20

Kirchhoff's laws 30–31

Laser Doppler velocimeter 333
laser profile sensor 179–180
lead magnesium niobate 97
linear velocity transducer 194–196
linearity 302–304
 dynamic 305–306
liquid level sensors 120–121, 209–210, 233–234, 270–271
liquid velocity sensor 193–194, 204
load cells 127–128, 222, 224
loudspeakers 8, 190–191
LVDTs 125–127
LVT 194–196

Magnetic circuits 106–107
magnetic fields 62, 100–103
magnetic induction 61–62, 100–103
magnetic materials 105–106
magnetic systems 100–106
magnetoresistivity 242–243
magnetostriction 131–134
Maxwell 20
measurement 11–13
mechanical relay 112–115, 273–274
mechatronics 8
mercury switch 210–211
microbend sensors 268–270
microphones 71–72, 74–75, 170, 175–176, 211–212, 335–337
microvalves 74–75, 116–117, 176–177, 246–247
mobility analogy 19–20
momentum variables 21–22
motors 9, 83–85, 88–90
 electret 89–90
 electrostatic 83–85, 88–89

Nail gun 28–29
noise floor 306–307

Odor sensor 332
offset 308
one-port elements 22–27, 36
optics 250–252
overshoot 307

Parallel plate capacitor 63–89
 varying area 79–85
 varying gap 63–79
 varying permittivity 85–89
Peltier effect 240–242
phonograph cartridge 173
photoconductive sensors 253–256
photodiodes 258
photostriction 273–274

piezoelectric materials 147–155
piezoelectric structures 154–159
piezoelectric transducers 163–177
piezoelectricity 140–147, 162–165
piezomagnetism 185–187
piezoresistivity 15, 216–228
position sensor 14, 86–87, 216, 255–256, 265–268, 324
positioners 77–78, 132–133, 156–158, 174–175
potentiometric transducers 213–216
power conjugate variables 18–21
pressure sensors 114–115, 167–168, 205, 208–210, 223–224, 266–267
printer heads 70–71, 127–128, 169–170
profiling actuator 77–78
proximity sensor 87–88, 115–116, 136
pumps 9
pyroelectricity 177–178
pyromagnetism 187

Quantization noise 330–331
quantum optical sensors 252–256

Range 301
reciprocity 285–288
reed switches 119–121
relays 112–115, 273–274
repeatability 302
resolution 306
response time 307
robotic actuator 246
rotary position sensor 81–82, 115–116
RTDs 228–233, 308–309
RVDTs 126–127

Sampling 325–329
scaling 316–318
scattering matrix 298
Seebeck effect 234–237
sensitivity 291–293, 306
sensor definition 8–9
shape memory alloys 243–247
signal-to-noise ratio 306
signal conditioning 325–331
slip sensor 81–82
solenoids 118–120
solid-state transducers 10–11, 69–72, 75–78, 83–84, 116–117, 175–177, 223–224, 230–233, 238, 241, 253–254, 256–257
sound source 193, 133–134, 171–172
speed detector 175–176
staging problem 9–10
state equations 50–53
state variables 51
strain gauges 219–225

Tactile sensor 255, 257
telephone 7
temperature sensors 228–242
Terfenol 132–133
thermal systems 46–48
thermistors 229–237, 308–309
thermocouples 237–240, 308–309
thermoelectric actuation 10
thermoelectricity 234–243
thermoresistivity 228–234
through and across analogy 19 " 20
tilt sensors 203
time-of-flight measurement 334–335
torque sensors 223–225, 267–270
transducer definition 1–6
transfer functions 52
transfer matrix 291–293
transformers 44–45
Trent 19–20
tunnel current transducer 211–212
two-port elements 43–49
two-port representations 279–285, 288–291

Ultrasonic cleaner 170–171
ultrasonic sound source 171–172

Vacuum sensor 208
variable conductance transducers 208–213
variable reluctance transducers 107–129
  with varying area 121–123
  with varying gap 107–121
  with varying permeability 123–129
voice coil actuator 191–192

Water level gauge 87–88
wave matrix 293–299
Wheatstone bridges 225–228
windowing 328–330

Thermal Contact Conductance
C.V. Madhusudana

Transport Phenomena with Drops and Bubbles
S.S. Sadhal, P.S. Ayyaswamy, and J.N. Chung

Fundamentals of Robotic Mechanical Systems:
Theory, Methods, and Algorithms
J. Angeles

Electromagnetics and Calculations of Fields
J. Ida and J.P.A. Bastos

Mechanics and Control of Robots
K.C. Gupta

Wave Propagation in Structures:
Spectral Analysis Using Fast Discrete Fourier Transforms, 2nd ed.
J.F. Doyle

Fracture Mechanics
D.P. Miannay

Principles of Analytical System Dynamics
R.A. Layton

Composite Materials:
Mechanical Behavior and Structural Analysis
J.M. Berthelot

Modern Inertial Technology:
Navigation, Guidance, and Control, 2nd ed.
A. Lawrence

Dynamics and Control of Structures:
A Modal Approach
W.K. Gawronski

Electromechanical Sensors and Actuators
I.J. Busch-Vishniac

Nonlinear Computational Structural Mechanics:
New Approaches and Non-Incremental Methods of Calculation
P. Ladevèze